Implementing Microsoft Azure Architect Technologies: AZ-303 Exam Prep and Beyond

Second Edition

A guide to preparing for the AZ-303 Microsoft Azure Architect Technologies certification exam

Brett Hargreaves
Sjoukje Zaal

BIRMINGHAM—MUMBAI

Implementing Microsoft Azure Architect Technologies: AZ-303 Exam Prep and Beyond
Second Edition

Commissioning Editor: Rahul Nair
Senior Editor: Shazeen Iqbal
Content Development Editor: Ronn Kurien
Technical Editor: Aurobindo Kar
Copy Editor: Safis Editing
Project Coordinator: Neil Dmello
Proofreader: Safis Editing
Indexer: Tejal Daruwale Soni
Production Designer: Joshua Misquitta

First published: January 2020
Second edition: December 2020
Production reference: 1171220

Published by Packt Publishing Ltd.
Livery Place
35 Livery Street
Birmingham
B3 2PB, UK.

ISBN 978-1-80056-857-0

www.packt.com

Contributors

About the authors

Based in the UK, **Brett Hargreaves** is a lead Azure consultant who has worked for some of the world's biggest companies for over 25 years, helping them design and build cutting-edge solutions. With a career spanning infrastructure, development, consulting, and architecture, he has been involved in projects covering the entire solution stack, including hardware, virtualization, databases, storage, software development, and the cloud. He loves passing on his knowledge to others through books, blogging, and his online training courses, which have over 20,000 students (and counting!).

Sjoukje Zaal is a CTO, Microsoft Regional Director, and Microsoft Azure MVP with over 20 years' experience in architecture-, development-, consultancy-, and design-related roles. She currently works at Capgemini, a global leader in consultancy, technology services, and digital transformation.

She loves to share her knowledge and is active in the Microsoft community as a co-founder of the user groups Tech Daily Chronicle, Global XR Community, and the Mixed Reality User Group. She is also a board member of Azure Thursdays and Global Azure. Sjoukje is an international speaker and is involved in organizing many events. She has written several books and writes blogs.

About the reviewers

Ricardo Cabral is a licensed computer engineer with several Microsoft certifications, and is also a **Microsoft Certified Trainer (MCT)**. Having worked in both administration and development roles, with several years' experience in IT management, development, and projects, he now works as an IT consultant and trainer. In his spare time, he actively participates in, and volunteers and/or speaks at, technical community meetings.

> *I would like to thank all my friends and a special individual (you know who you are) who helped guide me in my decisions. I would also like to thank Packt Publishing for the opportunity to review this wonderful book.*

Aprizon has been working with Microsoft Infrastructure technologies for more than 20 years, starting with Windows Server through to the cloud platform, including Office 365, Microsoft Azure, and Microsoft Enterprise Mobility Security, and is passionate about business process transformation. He has worked in an IT consulting company as an expert by leveraging Office 365, enterprise mobility security, and Microsoft Azure to increase efficiency and implement changes from the ground up. He has also worked as a Microsoft Certified Trainer and delivers Microsoft Official Curriculum training.

Above all else, he is a father, husband, son, brother, and friend.

Derek Campbell works as a senior solution architect in the advisory team at Octopus Deploy. He has worked all over the globe, including in London, Melbourne, and Singapore, and now from his home in Glasgow, Scotland.

Derek originally started in operations 15 years ago, becoming a system architect. Over time, he moved into DevOps. He has been automating infrastructure configuration, working with Azure and CI/CD pipelines, for about 8 years and has helped lead and implement CI/CD across multiple companies during his time in DevOps and automation consultancy.

Packt is searching for authors like you

If you're interested in becoming an author for Packt, please visit authors. packtpub.com and apply today. We have worked with thousands of developers and tech professionals, just like you, to help them share their insight with the global tech community. You can make a general application, apply for a specific hot topic that we are recruiting an author for, or submit your own idea.

Table of Contents

3

Implementing and Managing Virtual Machines

4

Implementing and Managing Virtual Networking

5
Creating Connectivity between Virtual Networks

6
Managing Azure Active Directory (Azure AD)

7
Implementing Multi-Factor Authentication (MFA)

8
Implementing and Managing Hybrid Identities

Section 2: Implement Management and Security Solutions

9
Managing Workloads in Azure

10

Implementing Load Balancing and Networking Security

11

Implementing Azure Governance Solutions

14

Implementing Authentication

Section 4: Implement and Manage Data Platforms

15

Developing Solutions that Use Cosmos DB Storage

16

Developing Solutions that Use a Relational Database

Preface

This book is the successor of *Microsoft Azure Architect Technologies – Exam Guide AZ-300*. The new exam, AZ-303, is mostly the same; however, Microsoft has shifted the focus away from much of the theory AZ-300 covered and is now more practical. An example of this is the removal of the messaging architecture requirement; this has instead been moved to the AZ-304 exam, which is more focused on the choice of technologies.

As Azure is an ever-developing platform, with new services being continually introduced and enhanced, the AZ-303 update also includes the requirement to understand services such as Azure Bastion, Azure Blueprints, and Azure Front Door. These were relatively new, or in preview, when AZ-300 was first released and are now generally available.

This book will therefore prepare you for the updated AZ-303 exam, which is the most practical exam of the Azure Architect Expert series. By reading this book, you will get updated on all the new functionalities, features, and resources. This book will cover all the exam objectives, giving you a complete overview of the objectives that are covered in the exam.

This book will start with implementing and monitoring infrastructure in Azure. You will learn how to analyze resource utilization and consumption. You will learn about storage accounts, Azure Virtual Network, Azure **Active Directory** (**AD**), and integrating on-premise directories. Next, you will learn about implementing management and security in Azure and how to implement governance. The focus of this book will then switch to implementing web-based solutions with Azure native technologies such as Web Apps, Functions, and Logic Apps. Finally, we look at how to develop data solutions for the cloud using SQL and NoSQL technologies.

Each chapter concludes with a *Further reading* section, which is an integral part of the book because it will give you extra, and sometimes crucial, information for passing the AZ-303 exam. As the exam questions will change slightly over time, and this book will eventually become outdated, the *Further reading* section will be the place that provides access to all the updates.

Who this book is for

This book targets Azure solution architects who advise stakeholders and translate business requirements into secure, scalable, and reliable solutions. They should have advanced experience and knowledge of various aspects of IT operations, including networking, virtualization, identity, security, business continuity, disaster recovery, data management, budgeting, and governance. This role requires managing how decisions in each area affect an overall solution.

What this book covers

Chapter 1, Implementing Cloud Infrastructure Monitoring, covers how to use Azure Monitor, how to create and analyze metrics and alerts, how to create a baseline for resources, how to configure diagnostic settings on resources, how to view alerts in Log Analytics, and how to utilize Log Search Query functions.

Chapter 2, Creating and Configuring Storage Accounts, covers Azure storage accounts, creating and configuring a storage account, installing and using Azure Storage Explorer, configuring network access to the storage account, generating and managing SAS, and how to implement Azure storage replication.

Chapter 3, Implementing and Managing Virtual Machines, covers virtual machines, availability sets, provisioning VMs, VM scale sets, modifying and deploying ARM templates, deployment using Azure DevOps, Dedicated Host, and how to configure Azure Disk Encryption for VMs.

Chapter 4, Implementing and Managing Virtual Networking, covers Azure VNet, IP addresses, how to configure subnets and VNets, configuring private and public IP addresses, and user-defined routes.

Chapter 5, Creating Connectivity between Virtual Networks, covers VNet peering, how to create and configure VNet peering, VNet-to-VNet, how to create and configure VNet-to-VNet, verifying virtual network connectivity, and compares VNet peering with VNet-to-VNet.

Chapter 6, Managing Azure Active Directory (Azure AD), covers how to create and manage users and groups, adding and managing guest accounts, performing bulk user updates, configuring self-service password reset, working with Azure AD join, and how to add custom domains.

Chapter 7, Implementing Multi-Factor Authentication (MFA), covers Azure MFA, how to configure user accounts for MFA, how to configure verification methods, how to configure fraud alerts, configuring bypass options, and how to configure trusted IPs.

Chapter 8, Implementing and Managing Hybrid Identities, covers Azure AD Connect, how to install Azure AD Connect, managing Azure AD Connect, and how to manage password sync, password writeback, and Azure AD Connect Health.

Chapter 9, Managing Workloads in Azure, covers Azure Migrate, the different Azure Migrate tools, migrating on-premises machines to Azure, VM Update Management, and Azure Backup.

Chapter 10, Implementing Load Balancing and Network Security, covers Azure Load Balancer and Application Manager, multi-region load balancing with Traffic Manager and Azure Front Door, Azure Firewall, Azure Bastion, and Network Security Groups.

Chapter 11, Implementing Azure Governance Solutions, covers how to manage access to Azure resources using management groups, **role-based access control** (**RBAC**), Azure Policy, and Azure Blueprints.

Chapter 12, Creating Web Apps Using PaaS and Serverless, covers App Service, App Service plans, WebJobs, how to enable diagnostics logging, Azure Functions, and Azure Logic Apps.

Chapter 13, Designing and Developing Apps for Containers, covers Azure Container Instances, how to implement an application that runs on an Azure Container Instances, creating a container image by using a Docker file, publishing an image to Azure Container Registry, web apps for containers, Azure Kubernetes Service, and how to create an Azure Kubernetes service.

Chapter 14, Implementing Authentication, covers App Service authentication, how to implement Windows-integrated authentication, implementing authentication by using certificates, OAuth2 authentication in Azure AD, how to implement OAuth2 authentication, implementing tokens, managed identities, and how to implement managed identities for Azure resources' Service Principal authentication.

Chapter 15, Developing Solutions that Use Cosmos DB Storage, covers how to create, read, update, and delete data by using the appropriate APIs, partitioning schemes, and how to set the appropriate consistency level for operations.

Chapter 16, Developing Solutions that Use a Relational Database, covers Azure SQL Database and how to provision and configure an Azure SQL database; how to create, read, update, and delete data tables by using code; how to configure elastic pools for Azure SQL Database; how to set up failover groups; and Azure SQL Database Managed Instance.

Chapter 17, Mock Exam Questions, contains sample exam questions.

Chapter 18, Mock Exam Answers, contains answers to the sample exam questions.

To get the most out of this book

An Azure subscription is required to get through this book, along with the following software/tools:

Software/Hardware covered in the book	OS Requirements
PowerShell	Windows, macOS, and Linux (any)
Azure CLI	Windows, macOS, and Linux (any)
Visual Studio	Windows (although there is a version that runs on macOS, the examples in the book will not work as per the instructions)
Visual Studio Code	Windows, macOS, and Linux (any)

If you are using the digital version of this book, we advise you to type the code yourself or access the code via the GitHub repository (link available in the next section). Doing so will help you avoid any potential errors related to the copying and pasting of code.

Ideally, you should have a basic understanding of Azure, either through hands-on experience or by completing the AZ900 courses and books.

Download the example code files

You can download the example code files for this book from GitHub at `https://github.com/PacktPublishing/Microsoft-Azure-Architect-Technologies-Exam-Guide-AZ-303`. In case there's an update to the code, it will be updated on the existing GitHub repository.

We also have other code bundles from our rich catalog of books and videos available at `https://github.com/PacktPublishing/`. Check them out!

Download the color images

We also provide a PDF file that has color images of the screenshots/diagrams used in this book. You can download it here: `http://www.packtpub.com/sites/default/files/downloads/9781800568570_ColorImages.pdf`.

Conventions used

There are a number of text conventions used throughout this book.

`Code in text`: Indicates code words in text, database table names, folder names, filenames, file extensions, pathnames, dummy URLs, user input, and Twitter handles. Here is an example: "Mount the downloaded `WebStorm-10*.dmg` disk image file as another disk in your system."

A block of code is set as follows:

```
@using System.Security.Claims
@using System.Threading

<div class="jumbotron">
        @{
            var claimsPrincipal = Thread.CurrentPrincipal as
ClaimsPrincipal;
            if (claimsPrincipal != null && claimsPrincipal.
Identity.IsAuthenticated)
            {
```

When we wish to draw your attention to a particular part of a code block, the relevant lines or items are set in bold:

```
Set-AzResource `
-PropertyObject $PropertiesObject `
-ResourceGroupName PacktAppResourceGroup `
-ResourceType Microsoft.Web/sites/sourcecontrols `
-ResourceName $webappname/web `
-ApiVersion 2015-08-01 `
-Force
```

Any command-line input or output is written as follows:

```
$ mkdir css
$ cd css
```

Bold: Indicates a new term, an important word, or words that you see onscreen. For example, words in menus or dialog boxes appear in the text like this. Here is an example: "From the left menu, select **Azure Active Directory**."

> **Tips or important notes**
> Appear like this.

Get in touch

Feedback from our readers is always welcome.

General feedback: If you have questions about any aspect of this book, mention the book title in the subject of your message and email us at `customercare@packtpub.com`.

Errata: Although we have taken every care to ensure the accuracy of our content, mistakes do happen. If you have found a mistake in this book, we would be grateful if you would report this to us. Please visit `www.packtpub.com/support/errata`, selecting your book, clicking on the Errata Submission Form link, and entering the details.

Piracy: If you come across any illegal copies of our works in any form on the Internet, we would be grateful if you would provide us with the location address or website name. Please contact us at `copyright@packt.com` with a link to the material.

If you are interested in becoming an author: If there is a topic that you have expertise in and you are interested in either writing or contributing to a book, please visit `authors.packtpub.com`.

Reviews

Please leave a review. Once you have read and used this book, why not leave a review on the site that you purchased it from? Potential readers can then see and use your unbiased opinion to make purchase decisions, we at Packt can understand what you think about our products, and our authors can see your feedback on their book. Thank you!

For more information about Packt, please visit `packt.com`.

Section 1: Implement and Monitor Azure Infrastructure

From the implementation to the monitoring of your services, this section covers the core aspects of the Azure platform and how to ensure it runs at optimal health.

This section contains the following chapters:

1
Implementing Cloud Infrastructure Monitoring

This book will cover all of the exam objectives for the **AZ-303 exam**. When relevant, you will be provided with extra information and further reading guidance about the different topics of this book.

This chapter introduces the first objective, which is going to cover **Implement Cloud Infrastructure Monitoring**. It will cover the various aspects of **Azure Monitor**. You will learn how to create and analyze *metrics* and *alerts* and how to create a baseline for resources. We are going to look at how to create action groups and how to configure diagnostic settings on resources. We are going to cover **Azure Log Analytics** and how to utilize log search query functions; finally, we will look at monitoring security events, networking, and cost management.

Being able to monitor all aspects of your solution is important for service health, security, reliability, and costs. With so much data available, it's important to know how to set up alerts and query logs effectively.

The following topics will be covered in this chapter:

- Understanding Azure Monitor
- Creating and analyzing metrics and alerts
- Creating a baseline for resources
- Configuring diagnostic settings on resources
- Viewing alerts in Log Analytics
- Utilizing log search query functions
- Using Network Watcher
- Monitoring security
- Managing costs

Technical requirements

The demos in this chapter use an Azure Windows VM. To create a Windows VM in Azure, refer to the following walk-through: `https://docs.Microsoft.com/en-us/azure/virtual-machines/windows/quick-create-PowerShell`.

Understanding Azure Monitor

Azure Monitor is a monitoring solution in the Azure portal that delivers a comprehensive solution for collecting, analyzing, and acting on telemetry from the cloud and on-premises environments. It can be used to monitor various aspects (for instance, the performance of applications) and identify issues affecting those applications and other resources that depend on them.

The data that is collected by Azure Monitor fits into two fundamental types: metrics and logs. Metrics describe an aspect of a system at a particular point in time and are displayed in numerical values. They are capable of supporting near real-time scenarios. Logs are different from metrics. They contain data that is organized into records, with different sets of properties for each type. Data such as events, traces, and performance data are stored as logs. They can then be combined for analysis purposes.

Azure Monitor supports data collection from a variety of Azure resources, which are all displayed on the overview page in the Azure portal. Azure Monitor provides the following metrics and logs:

- **Application monitoring data**: This consists of data about the functionality and performance of the application and the code that is written, regardless of its platform.

- **Guest OS monitoring data**: This consists of data about the OS on which your application is running. This could be running in any cloud or on-premises environment.

- **Azure resource monitoring data**: This consists of data about the operation of an Azure resource.

- **Azure subscription monitoring data**: This consists of data about the operation and management of an Azure subscription, as well as data about the health and operation of Azure itself.

- **Azure tenant monitoring data**: This consists of data about the operation of tenant-level Azure services, such as Azure Active Directory.

> Important note
> Azure Monitor now integrates the capabilities of Log Analytics and Application Insights. You can also keep using Log Analytics and Application Insights on their own.

The following diagram gives a high-level view of Azure Monitor. On the left, there are the sources of monitoring data, in the center are the data stores, and on the right are the different functions that Azure Monitor performs with this collected data, such as analysis, alerting, and streaming to external systems:

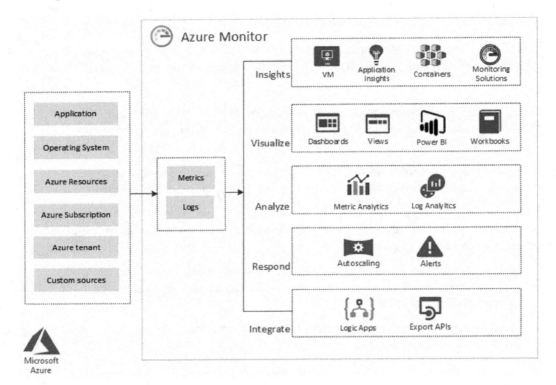

Figure 1.1 – Overview of Azure Monitor capabilities

Now that we have some basic knowledge about Azure Monitor, we are going to look at how to analyze alerts and metrics across subscriptions.

Creating and analyzing metrics and alerts

To analyze alerts and metrics across Azure Monitor, we need to go to the monitoring resource inside the Azure portal. In the upcoming sections, we will set up metrics and alerts and show you how to analyze them.

Metrics

Metrics describe an aspect of a system at a particular point in time and are displayed in numerical values. They are collected at regular intervals and are identified with a timestamp, a name, a value, and one or more defining labels. They are capable of supporting near real-time scenarios and are useful for alerting. Alerts can be fired quickly with relatively simple logic.

Metrics in Azure Monitor are stored in a time-series database that is optimized for analyzing timestamped data. This makes metrics suited for the fast detection of issues. They can help to detect how your service or system is performing, but to get the overall picture, they typically need to be combined with logs to identify the root cause of issues.

You can use metrics for the following scenarios:

- **Analyzing**: Collected metrics can be analyzed using a chart in Metric Explorer. Metrics from various resources can be compared as well.

- **Visualizing**: You can create an Azure Monitor workbook to combine multiple datasets into an interactive report. Azure Monitor workbooks can combine text, Azure metrics, analytics queries, and parameters into rich interactive reports.

- **Alerting**: Metric alert rules can be configured to send out notifications to the user. They can also take automatic action when the metric value crosses a threshold.

- **Automating**: To increase and decrease resources based on metric values that cross a threshold, autoscaling can be used.

- **Exporting**: Metrics can be streamed to an Event Hub to route them to external systems. Metrics can also be routed to logs in the Log Analytics workspace in order to be analyzed together with the Azure Monitor logs and to store the metric values for more than 93 days.

- **Retrieving**: Metrics values can be retrieved from the command line using PowerShell cmdlets and the CLI, and from custom applications using the Azure Monitoring REST API.

- **Archiving**: Metrics data can be archived in Azure Storage. It can store the performance or health history of your resource for compliance, auditing, or offline reporting purposes.

There are four main sources of metrics that are collected by Azure Monitor. Once they are collected and stored in the Azure Monitor Metric database, they can be evaluated together regardless of their source:

- **Platform metrics**: These metrics give you visibility of the health and performance of your Azure resources. Without any configuration required, a distinct set of metrics is created for each type of Azure resource. By default, they are collected at one-minute intervals. However, you can configure them to run at different intervals as well.

- **Guest OS metrics**: These metrics are collected from the guest OS of a virtual machine. To enable guest OS metrics for Windows machines, the **Windows Diagnostic Extension (WAD)** agent needs to be installed. For Linux machines, the **InfluxData Telegraf** agent needs to be installed.

- **Application metrics**: These metrics are created by Application Insights. They can help to detect performance issues for your custom applications and track trends in how the application is being used.

- **Custom metrics**: These are metrics that you define manually. You can define them in your custom applications that are monitored by Application Insights or you can define custom metrics for an Azure service using the custom metrics API.

> **Tip**
>
> For more information about the **InfluxData Telegraf** agent, go to the *InfluxData* website, `https://www.influxdata.com/time-series-platform/telegraf/`.

Multi-dimensional metrics

Metrics data often has limited information to provide context for collected values. This challenge is addressed by Azure Monitor using multi-dimensional metrics. The dimensions of the metrics are name-value pairs that store additional data that describe the metric value. For example, a metric called *available disk space* could have a dimension called *Drive* with the values *C:* and *D:* stored inside. This value would allow the viewing of available disk space across all drives, or each drive individually.

In the next section, we are going to create a metric in the Azure portal.

Creating a metric

To display the metrics for a Windows VM (if you followed the creating a VM walk-through detailed in the *Technical requirements* section) in Azure Monitor, follow these steps:

1. Navigate to the Azure portal by opening `https://portal.azure.com`.

2. In the left-hand menu, select **Monitor** to open the **Azure Monitor** overview blade.

3. First, we're going to look at metrics. Therefore, in the left-hand menu, select **Metrics** or select the **Explore Metrics** button from the overview blade.

4. In the **Metrics** overview blade, the **Select a Scope** may be automatically displayed; if not, click on the **+ Select a scope** button. A new blade will open up where you can select the subscription, the resource group, and the resource type. Select the subscription that is used for the Windows VM, select the resource group, and then select the VM. You can filter by other resource types, as well:

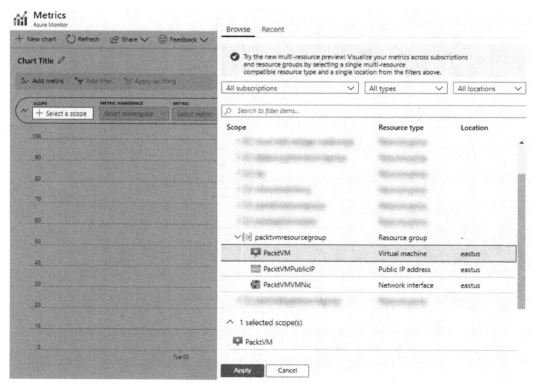

Figure 1.2 – Selecting the resources

5. Click on **Apply**.

6. Then you can select the metric type. Select **CPU Credits Consumed**, for instance:

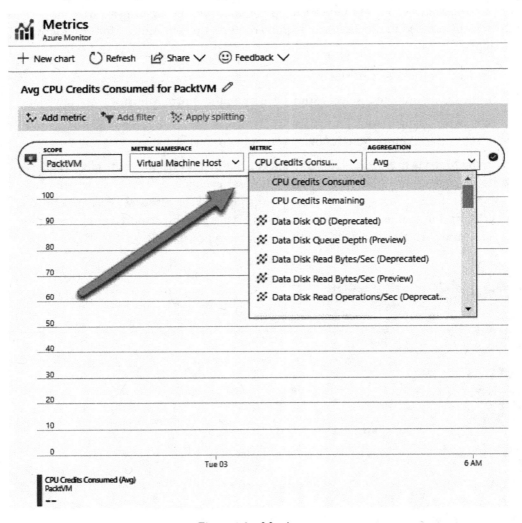

Figure 1.3 – Metric type

> **Tip**
> Take some time to look at the different metrics that you can choose from. This may be a part of the exam questions.

7. You can select a different type of aggregation as well, such as the count, average, and more, in the filter box. At the top-right of the blade, you can select a different time range for your metric as well:

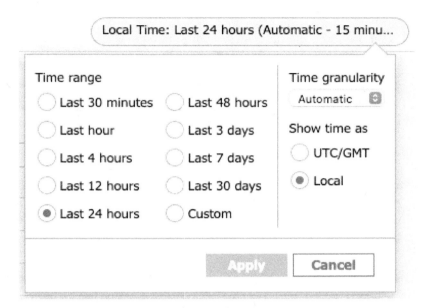

Figure 1.4 – Time ranges

8.   You can also pin this metric to the overview dashboard in the Azure portal. Therefore, click on the **Pin to dashboard** button, and then choose to pin it to the current dashboard or create a new dashboard for it. For now, select **Pin to current dashboard**:

Figure 1.5 – Pinning a metric to a dashboard

9.   If you now select **Dashboard** from the left-hand menu, you'll see that this metric is added to it. This way, you can easily analyze this metric without needing open Azure Monitor.

> **Important note**
> Metrics are also available directly from the Azure resource blades. So, for instance, if you have a VM, go to the VM resource by selecting it. Then, in the left-hand menu, under **Monitoring**, you can select **Metrics**.

In the next section, we're going to look at how to set up and analyze alerts in Azure Monitor.

Alerts

With alerts, Azure can proactively notify you when critical conditions occur in the Azure or on-premises environment. Alerts can also attempt to take corrective actions automatically. Alert rules that are based on metrics will provide near real-time alerting, based on the metric. Alerts that are created based on logs can merge data from different resources.

The alerts in Azure Monitor use action groups, which are unique sets of recipients and actions that can be shared across multiple rules. These action groups can use webhooks to start external actions, based on the requirements that are set up for this alert. These external actions can then be picked up by different Azure resources, such as Runbooks, Functions, or Logic Apps. Webhooks can also be used to add these alerts to external **IT Service Management** (**ITSM**) tools.

You can also set alerts for all of the different Azure resources. In the following sections, we are going to create an alert.

Creating an alert and an action group

To create an alert, follow these steps:

1. From the **Azure Monitor** overview blade, in the left-hand menu, select **Alerts**. You can also go to the alerts settings by clicking on **Create alert** to create an alert directly.

2. In the **Alerts** blade, click on **+ New alert rule** in the top menu:

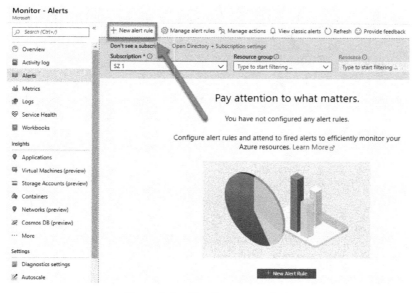

Figure 1.6 – Creating a new alert

3. The **Create rule** blade is displayed. Here, you can create the rule and action groups. To create a new rule, you need to first select the resource. Click on the **Select** button in the **RESOURCE** section:

Figure 1.7 – Creating a new rule

4. In the next blade, you can filter by the subscription and resource type. Select **Virtual machines**:

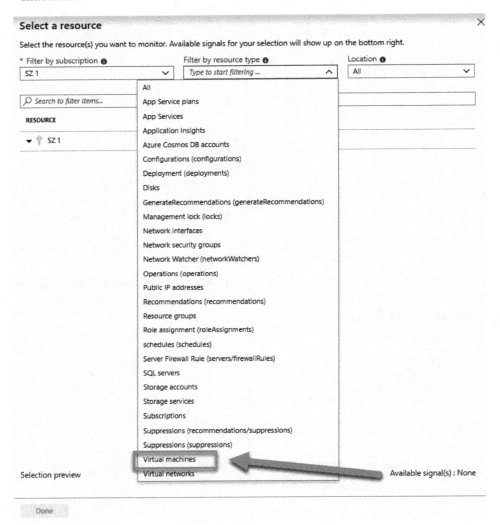

Figure 1.8 – Filtering by subscription and resource type

5. Select the VM from the list and click **Done**.

6. Now that we have selected a resource, we're going to set up the condition. Click on **Add condition**.

7. The condition blade is open, and so we can filter by a certain signal. Select **Percentage CPU** and click **Done**:

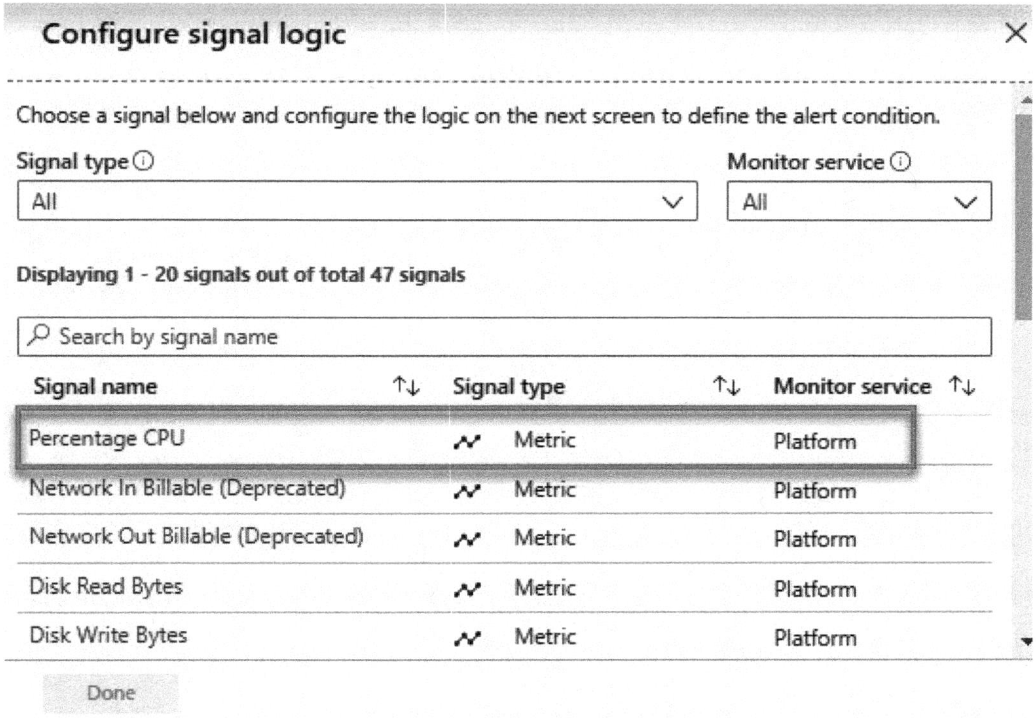

Configure signal logic ✕

Choose a signal below and configure the logic on the next screen to define the alert condition.

Signal type ⓘ	Monitor service ⓘ
All ∨	All ∨

Displaying 1 - 20 signals out of total 47 signals

🔍 Search by signal name

Signal name ↑↓		Signal type ↑↓	Monitor service ↑↓
Percentage CPU	∿	Metric	Platform
Network In Billable (Deprecated)	∿	Metric	Platform
Network Out Billable (Deprecated)	∿	Metric	Platform
Disk Read Bytes	∿	Metric	Platform
Disk Write Bytes	∿	Metric	Platform

Done

Figure 1.9 – Filtering on a signal

8. Next, you can set the alert logic for this alert. You can choose multiple operators, set the aggregation type, and set the threshold value for this alert. Set the following:

  a) **Threshold: Static** (in the next section, we are going to cover the difference between static and dynamic thresholds)

  b) **Operator: Greater than**

  c) **Aggregation type: Average**

  d) **Threshold Value: 90%**

9. Leave **Evaluated based on** with its default settings.

10. This alert will notify you when the CPU usage of the VMs is greater than 90% over a 5-minute period. Azure Monitor will check this every minute:

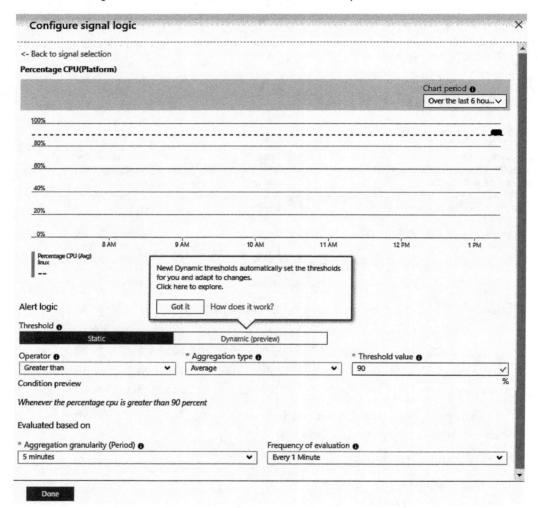

Figure 1.10 – Setting condition values

11. Click on **Done** to create this condition.

12. Now, we have to create an action group to send the alert to. This is then responsible for handling the alert and taking further action on it. The action group that you create here can be reused across other alerts as well. So, in our case, we will create an email action group that will send out an email to a certain email address. After it has been created, you can add this action group to other alerts. Under **Action group**, select the **Create new** button.

13. In the **Action Group** blade, add the following settings:

    a) **Action group name**: Type `Send email`.

    b) **Short name**: Type `email`.

    c) **Subscription**: Select the subscription where the VM is created.

    d) **Resource group**: Select **Default-ActivityLogAlerts** (to be created).

14. Then, we have to provide the actual action. Add the following values:

    a) **Action name**: `email`

    b) **Action type**: **Email/SMS/Push/Voice**

15. Then, select **Edit details** and select the **Email** checkbox. Provide an email address and click on the **OK** button:

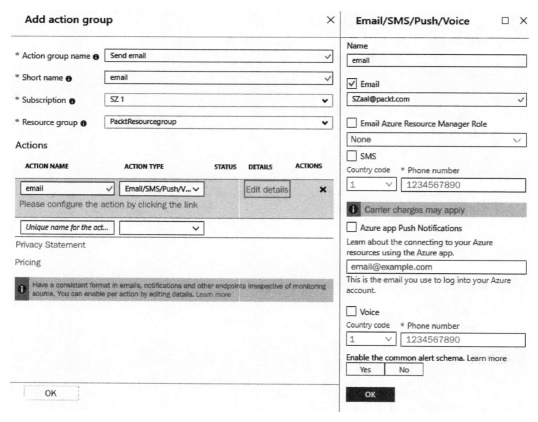

Figure 1.11 – Creating an action group

16. Click on **OK** again.

17. Finally, you have to specify an alert name, set the severity level of the alert, and click on **Create alert rule**:

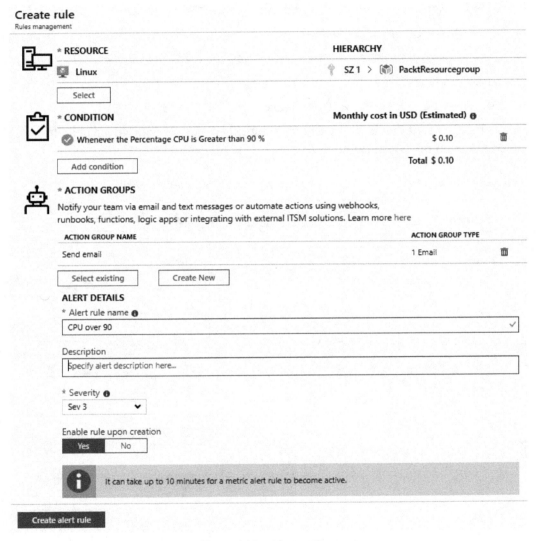

Figure 1.12 – Alert settings

We have now created an alert and an action group that will alert a user via email when the CPU goes over 90%. In the next section, we're going to create a baseline for resources.

Creating a baseline for resources

To create a baseline for your resources, Azure offers Metric Alerts with Dynamic Thresholds. Using Dynamic Thresholds, you don't have to manually identify and set thresholds for alerts, which is an enhancement to Azure Monitor Metric Alerts. Advanced machine learning capabilities are used by the alert rule to learn the historical behavior of the metrics while identifying patterns and anomalies that indicate possible service issues. With Dynamic Thresholds, you can create an alert rule once and apply it automatically to different Azure resources during the creation of the resources.

In the following overview, you will find some scenarios when Dynamic Thresholds to metrics alerts are recommended:

- **Scalable alerting**: Dynamic Thresholds are capable of creating tailored thresholds for hundreds of metric series at a time. However, this is as easy as creating an alert rule for one single metric. They can be created using the Azure portal or **Azure Resource Manager** (**ARM**) templates and the ARM API. This scalable approach is useful when applying multiple resources or dealing with metric dimensions. This will translate to a significant time-saving on the creation of alert rules and management.

- **Intuitive Configuration**: You can set up metric alerts using high-level concepts with Dynamic Thresholds, so you don't need to have extensive domain knowledge about the metric.

- **Smart Metric Pattern Recognition**: By using a unique machine learning technology, Azure can automatically detect metric patterns and adapt to metric changes over time. The algorithm used in Dynamic Thresholds is designed to prevent wide (low recall) or noisy (low precision) thresholds that don't have an expected pattern.

In the next section, we're going to configure diagnostic settings on resources.

Configuring diagnostic settings on resources

You can also configure diagnostic settings on different Azure resources. There are two types of diagnostic logs available in Azure Monitor:

- **Tenant logs**: These logs consist of all of the tenant-level services that exist outside of an Azure subscription. An example of this is the Azure Active Directory logs.

- **Resource logs**: These logs consist of all of the data from the resources that are deployed inside an Azure subscription, for example, VMs, storage accounts, and network security groups.

The contents of the resource logs are different for every Azure resource. These logs differ from guest OS-level diagnostic logs. To collect OS-level logs, an agent needs to be installed on the VM. The diagnostic logs don't require an agent to be installed; they can be accessed directly from the Azure portal.

The logs that can be accessed are stored inside a storage account and can be used for auditing or manual inspection purposes. You can specify the retention time in days by using the resource diagnostic settings. You can also stream the logs to event hubs to analyze them in Power BI or insert them into a third-party service. These logs can also be analyzed with Azure Monitor. Then, there will be no need to store them in a storage account first.

Enabling diagnostic settings

To enable the diagnostic settings for resources, follow these steps:

1. Navigate to the Azure portal by opening `https://portal.azure.com`.

2. Go to the VM again. Make sure that the VM is running, and in the left-hand menu, under **Monitoring**, select **Diagnostic settings**.

3. The **Diagnostic settings** blade will open up. You will need to select a storage account where the metrics can be stored.

4. Click on the **Enable guest-level monitoring** button to update the diagnostic settings for the VM:

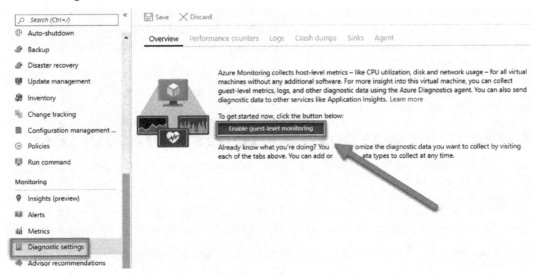

Figure 1.13 – Enabling diagnostic settings for a VM

5. When the settings are updated, you can go to **Metrics** in the top menu to set the metrics that are collected.

6. New metrics will be available from the metrics blade after enabling diagnostic logging in Azure Monitor. You can analyze them in the same way that we did earlier in this chapter, in the **Metrics** section.

In the next section, we're going to look at the Azure Log Analytics service, which is now a part of Azure Monitor as well.

Viewing alerts in Log Analytics

Azure Log Analytics is a service that collects telemetry data from various Azure resources and on-premises resources. All of that data is stored inside a Log Analytics workspace, which is based on Azure Data Explorer. It uses the Kusto Query Language, which is also used by Azure Data Explorer to retrieve and analyze the data.

Analyzing this data can be done from Azure Monitor. All of the analysis functionalities are integrated there. The term Log Analytics now primarily applies to the blade in the Azure portal where you can analyze metric data.

Before we can display, monitor, and query the logs from Azure Monitor, we need to create a Log Analytics workspace. For that, we have to follow these steps:

1. Navigate to the Azure portal by opening `https://portal.azure.com`.

2. Click on **Create a resource**.

3. Type `Log Analytics` in the search box and create a new workspace.

4. Add the following values:

    a) **Log Analytics workspace**: Type `PacktWorkspace` (the name for this Log Analytics workspace needs to be unique; if the name is already taken, specify another name).

    b) **Subscription**: Select a subscription.

    c) **Resource group**: Create a new one and call it `PacktWorkspace`.

    d) **Location**: Select **West US**.

    e) **Pricing tier**: Keep the default one, which is **per GB**.

5. Click on the **OK** button to create the workspace.

> **Important note**
> You can also create this workspace from Azure Monitor. Go to the **Azure Monitor** blade, and under **Insights** in the left-hand menu, select **More**. When no workspace has been created, Azure will ask to create one.

Now that we have created a Log Analytics workspace, we can use it inside Azure Monitor to create some queries to retrieve data. We will do this in the next section.

Utilizing log search query functions

Azure Monitor is now integrated with the features and capabilities that Log Analytics was offering. This also includes creating search queries across the different logs and metrics by using the Kusto Query Language.

To retrieve any type of data from Azure Monitor, a query is required. Whether you are configuring an alert rule, analyzing data in the Azure portal, retrieving data using the Azure Monitor Logs API, or being notified of a particular condition, a query is used.

The following list provides an overview of all of the different ways queries are used by Azure Monitor:

- **Portal**: From the Azure portal, interactive analysis of log data can be performed. There, you can create and edit queries and analyze the results in a variety of formats and visualizations.

- **Dashboards**: The results of a query can be pinned to a dashboard. This way, results can be visualized and shared with other users.

- **Views**: By using the View Designer in Azure Monitor, you can create custom views of your data. This data is provided by queries as well.

- **Alert rules**: Alert rules are also made up of queries.

- **Export**: Exports of data to Excel or Power BI are created with queries. The query defines the data to export.

- **Azure Monitor Logs API**: The Azure Monitor Logs API allows any REST API client to retrieve log data from the workspace. The API request includes a query to retrieve the data.

- **PowerShell**: You can run a PowerShell script from command line or an Azure Automation runbook that uses `Get-AzOperationalInsightsSearchResults` to retrieve log data from Azure Monitor. You need to create a query for this cmdlet to retrieve the data.

In the following section, we are going to create some queries to retrieve data from the logs in Azure Monitor.

Querying logs in Azure Monitor

To query logs in Azure Monitor, perform the following steps:

1. Navigate to the Azure portal by opening `https://portal.azure.com`.

2. In the left-hand menu, select **Monitor** to open the **Azure Monitor** overview blade. Under **Insights**, select **More**. This will open the Log Analytics workspace that we created in the previous step.

3. On the overview page, click on **Logs** in the top menu. This will open the Azure Monitor query editor:

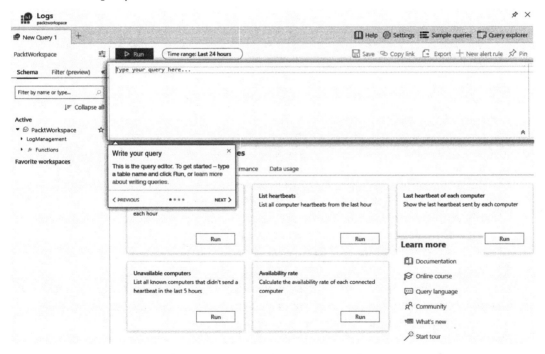

Figure 1.14 – Azure Monitor query editor

4. Here, you can select some default queries. They are displayed at the bottom part of the screen. There are queries for retrieving unavailable computers, the last heartbeat of a computer, and much more. Add the following queries to the query editor window to retrieve data:

The following query will retrieve the top 10 computers with the most error events over the last day:

```
Event | where (EventLevelName == "Error") | where
(TimeGenerated > ago(1days)) | summarize ErrorCount =
count() by Computer | top 10 by ErrorCount desc
```

The following query will create a line chart with the processor utilization for each computer from the last week:

```
Perf | where ObjectName == "Processor" and CounterName
== "% Processor Time" | where TimeGenerated between
(startofweek(ago(7d)) .. endofweek(ago(7d)) ) | summarize
avg(CounterValue) by Computer, bin(TimeGenerated, 5min) |
render timechart
```

> **Tip**
> Be careful, Kusto is case sensitive!

> **Important note**
> A detailed overview and tutorial on how to get started with the Kusto Query Language are beyond the scope of this book. If you want to find out more about this query language, you can refer to `https://docs.microsoft.com/en-us/azure/azure-monitor/log-query/get-started-queries`.

Log Analytics provides a powerful tool to explain what is happening within your Azure Infrastructure. Next, we will look at how we can use the built-in networking tools to help identify and resolve communication issues between components.

Using Network Watcher

Azure provides the **Network Watcher** tool for monitoring and investigating problems between devices on a **Virtual Network (VNET)**, including the following:

- Connection Monitoring
- Performance Monitoring
- Diagnostics
- Network Security Group flow logs

> **Important note**
>
> A VNET is a private network you can create in your Azure subscription. VNETs are defined with set IP ranges, which in turn can be sub-divided into subnets. Some Azure services, such as VMs, must be connected to a VNET. Other services, such as App Services and Azure SQL, can optionally use VNETs to ensure traffic between them is direct and secure.

You can also see a topology map of devices to understand better the various components involved in the communication flow.

The first step in setting up the Network Watcher capabilities is to ensure it has been enabled for the region(s) you are using by following these steps:

1. Navigate to the Azure portal by opening `https://portal.azure.com`.

2. In the left-hand menu, select or search for `Network Monitor`:

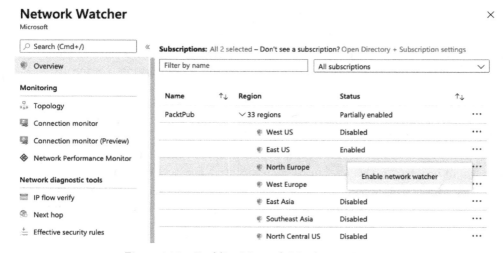

Figure 1.15 – Enabling Network Watcher per Region

3. If the region that contains the resources you wish to monitor is set to **Disabled**, click the ellipses at the right and select **Enable network watcher**.

4. On the left-hand menu, select the **Topology**, then select a resource group that contains resources you wish to view. In the following screenshot, I am choosing a group that includes a simple VM:

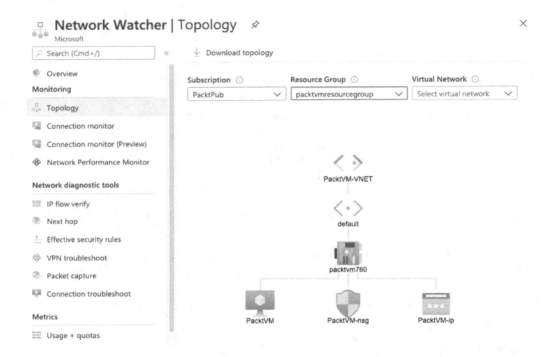

Figure 1.16 – Network Monitor Topology view

Once Network Watcher has been enabled for your region, we can now start to use the different tools, such as Connection Monitor, to troubleshoot and fix common communication problems.

Connection Monitor

When services are running, you may want to be alerted to issues with connectivity. An example might be a web server that needs to maintain a connection to a backend database server. However, the destination can be another VM, a URI, or an IP address. The URI or IP address can be either an internal resource in your Azure subscription or an external resource.

Connection Monitor allows us to set up continual monitors that can trigger alerts when communications are interrupted:

1. Still in Network Watcher, on the left-hand menu, select **Connection Monitor**.

2. Click **Add** to create a Connection Monitor.

3. Complete the details to define a **source**, **target**, and **port**. For this example, instead of monitoring connection to another server, we will monitor connections to the internet, specifically to the *Packt Publishing* website:

    a) **Name**: InternetConnection

    b) **Virtual Machine**: Source VM you wish to monitor

    c) **Destination**:

       --**Specify Manually**

       --**URI**: www.packtpub.com

    d) **Port**: 443

4. Click **Add**:

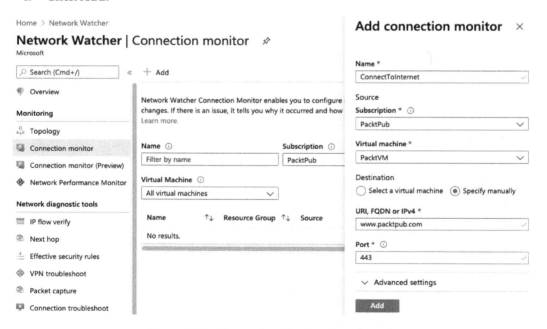

Figure 1.17 – Connection Monitor setup example

Once set up, you can select the Connection Monitor you have just created, and it will show basic details of the status and flow of traffic from source to destination. You can also set the time period to see data from the past hour up to the past 30 days as shown in the following screenshot:

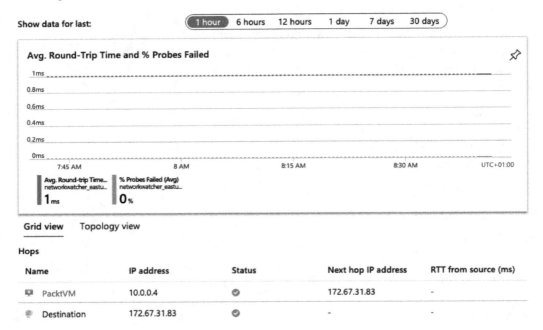

Figure 1.18 – Connection Monitor example

Connection Monitor is great for initial investigations and for setting up alerts; for more complex issues we use advanced options such as the Diagnostics tooling.

Diagnostics

When you encounter connectivity issues, Network Watcher diagnostics offers a range of tools to help pinpoint the problem.

The first step in troubleshooting connectivity issues is to confirm that traffic is flowing.

IP Flow Verify

IP Flow Verify allows you to confirm the flow of traffic from a source to a destination is working. Set up a typical test by performing the following steps:

1.   From the Network Watcher blade, select **IP flow verify** from the left-hand menu.

2.   Select your VM and network interface you wish to test.

3. Select the desired protocol (**TCP** or **UDP**).
4. Select the direction of traffic you want to check.
5. Confirm the local (source) IP address and port your traffic flows on.
6. Enter the remote (destination) IP address and port.

The following figure shows an example request. When the **Check** button is clicked, we can see a status response returned. In the example, we can see the request has failed, but importantly we see it failed because of the **DenyAllOutbound** Network Security Group rule:

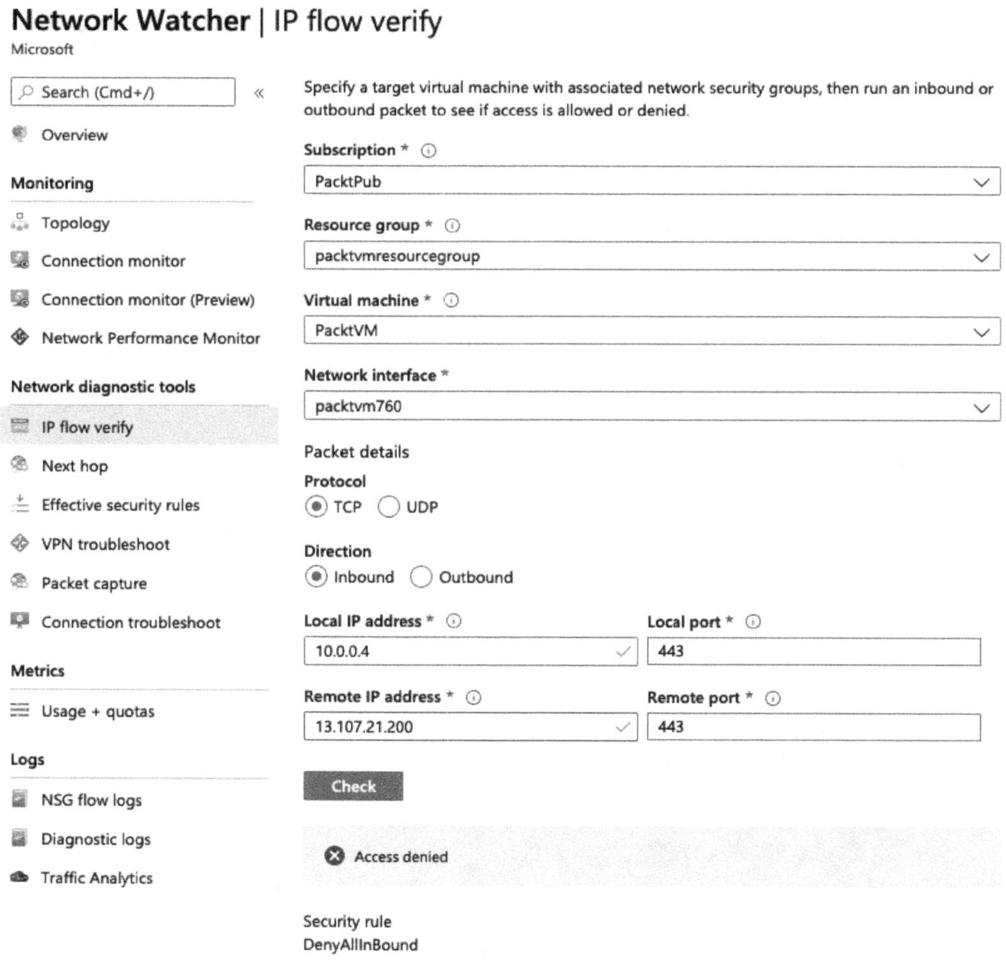
Network Watcher | IP flow verify
Microsoft
Search (Cmd+/)
Overview
Monitoring
Topology
Connection monitor
Connection monitor (Preview)
Network Performance Monitor
Network diagnostic tools
IP flow verify
Next hop
Effective security rules
VPN troubleshoot
Packet capture
Connection troubleshoot
Metrics
Usage + quotas
Logs
NSG flow logs
Diagnostic logs
Traffic Analytics
Specify a target virtual machine with associated network security groups, then run an inbound or outbound packet to see if access is allowed or denied.
Subscription *
PacktPub
Resource group *
packtvmresourcegroup
Virtual machine *
PacktVM
Network interface *
packtvm760
Packet details
Protocol
TCP
UDP
Direction
Inbound
Outbound
Local IP address *
10.0.0.4
Local port *
443
Remote IP address *
13.107.21.200
Remote port *
443
Check
Access denied
Security rule
DenyAllInBound

Figure 1.19 – IP flow verify example

IP flow verify helps to confirm that end-to-end communication is functioning, but if you do find problems you can use other Network Watcher tools to continue your investigations.

Next Hop

The subsequent step in identifying communications issues could be to understand the route traffic takes from point *a* to point *b*, and the Next Hop service helps with this:

1.   Still in Network Watcher, in the left-hand menu, click **Next Hop**.

2.   Define the source VM you wish to check connectivity *from*.

3.   Enter the IP address of the service you are attempting to reach and click the **Next Hop** button.

The example in the following screenshot shows the next hop to the IP address (one of the Bing.com addresses) is the Azure Internet egress appliance, and the route to it has been defined in the system route table (route tables will be covered in *Chapter 4, Implementing and Managing Virtual Networking*):

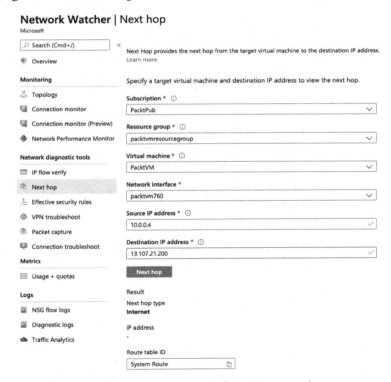

Figure 1.20 – Next Hop configuration example

Depending on the results from the IP flow verify and Next Hop tools, the next step in your troubleshooting process may be to look at access issues.

Viewing effective security rules

A common cause of issues is a misconfiguration of **Network Security Groups (NSG)** between devices. We cover NSGs in *Chapter 10, Implementing Load Balancing and Network Security*. In brief, they allow you to define firewall rules on VNETs or devices.

Restrictions on allowed IP addresses and ports can be set in multiple layers, and as such, can become complex and challenging to manage. For this reason, you can use the effective security rules option:

1.   Still in Network Watcher, in the left-hand menu, click **Effective security rules**.

2.   Select the **Subscription**, **Resource group**, and **Virtual machine** you wish to check. The following screenshot shows an example:

Figure 1.21 – Configuring the effective security rules option

3.   Once your VM has been selected, the effective rules will be listed, separated by NSG, **Inbound rules**, and **Outbound rules**. The following screenshot shows a typical result:

PacktVM-nsg

Inbound rules

Name ↑↓	Priority ↑↓	Source	Source Ports ↑↓	Destination	Destination Ports ↑↓	Protocol ↑↓	Access ↑↓
SSH	300	0.0.0.0/0,0.0.0.0/0	0-65535	0.0.0.0/0,0.0.0.0/0	22-22	TCP	⊘ Allow
AllowVnetInBound	65000	Virtual network (2 prefixes)	0-65535	Virtual network (2 prefixes)	0-65535	All	⊘ Allow
AllowAzureLoadBalancerInBound	65001	Azure load balancer (2 prefixes)	0-65535	0.0.0.0/0,0.0.0.0/0	0-65535	All	⊘ Allow
DenyAllInBound	65500	0.0.0.0/0,0.0.0.0/0	0-65535	0.0.0.0/0,0.0.0.0/0	0-65535	All	⊘ Deny

Outbound rules

Name ↑↓	Priority ↑↓	Source	Source Ports ↑↓	Destination	Destination Ports ↑↓	Protocol ↑↓	Access ↑↓
AllowVnetOutBound	65000	Virtual network (2 prefixes)	0-65535	Virtual network (2 prefixes)	0-65535	All	⊘ Allow
AllowInternetOutBound	65001	0.0.0.0/0,0.0.0.0/0	0-65535	Internet (236 prefixes)	0-65535	All	⊘ Allow
DenyAllOutBound	65500	0.0.0.0/0,0.0.0.0/0	0-65535	0.0.0.0/0,0.0.0.0/0	0-65535	All	⊘ Deny

Figure 1.22 – Example of effective NSG rules in action

We will now have a look at using Packet Capture to examine the data.

Packet Capture

When everything looks OK but you are still experiencing issues, you may need to look in detail at the actual traffic being sent and received. Specialist tools are available for analyzing packet information, and through the Network Watcher, you can set up Packet Capture to collect data for a specific amount of time and then examine that traffic:

1.   Still in Network Watcher, in the left-hand menu, click **Packet Capture**.

2.   Select your VM.

3.   Choose whether to store the Packet Capture data in a storage account (we cover storage accounts and how to create them in *Chapter 2, Creating and Configuring Storage Accounts*), in the VM itself, or both.

4. Optionally set the maximum and minimum bytes per capture or a time limit.
    The following screenshot shows an example of what this looks like:

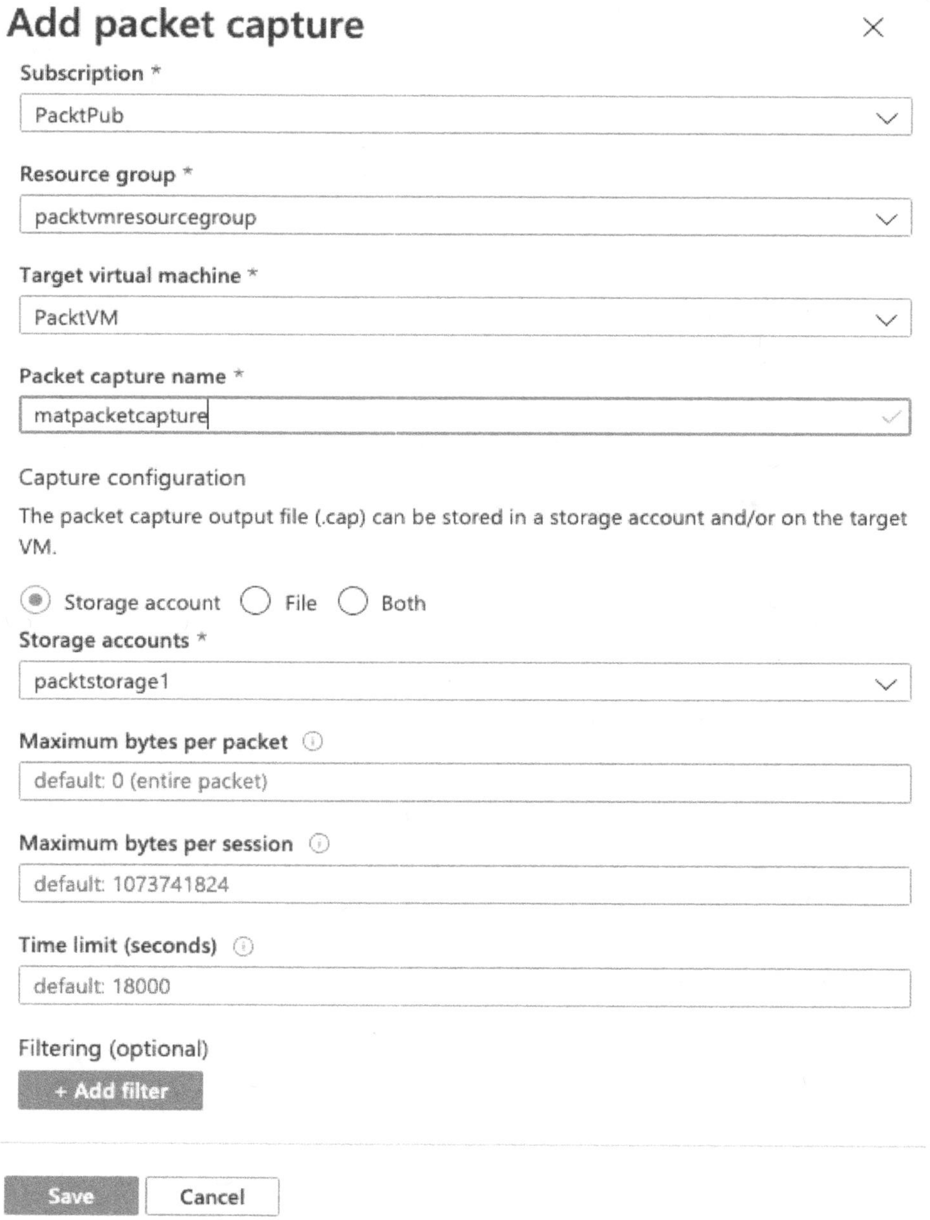

Add packet capture   ✕

Subscription *

    PacktPub   ⌄

Resource group *

    packtvmresourcegroup   ⌄

Target virtual machine *

    PacktVM   ⌄

Packet capture name *

    matpacketcapture   ⌄

Capture configuration

The packet capture output file (.cap) can be stored in a storage account and/or on the target VM.

⦿ Storage account   ◯ File   ◯ Both

Storage accounts *

    packtstorage1   ⌄

Maximum bytes per packet ⓘ

    default: 0 (entire packet)

Maximum bytes per session ⓘ

    default: 1073741824

Time limit (seconds) ⓘ

    default: 18000

Filtering (optional)

    + Add filter

    Save   Cancel

Figure 1.23 – Example packet capture setup

5. Optionally click **+Add Filter** to enter more precise details of the source and destination for which you want to capture data, as in the following screenshot:

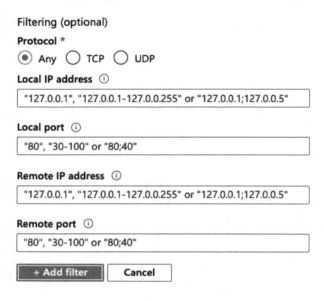

Figure 1.24 – Packet capture filters

6. Click **Save**.

7. The capture session will automatically start; let it run for a few minutes.

8. Stop the capture by clicking the ellipsis at the right of the session you created and click **Stop**.

9. Click on the session, and the file link will be presented in the lower pane.

10. Click on the capture link to download it.

    The following screenshot shows an example of how this might look:

Figure 1.25 – Example Packet Capture

The Packet Capture can then be opened in a viewing tool for a detailed examination of the traffic. The following screenshot shows an example of how this might look:

No.	Time	Source	Destination	Protocol	Length	Info
18	1.147696	216.58.213.99	192.168.1.4	TCP	66	443 → 51094 [ACK] Seq
19	1.203112	192.168.1.4	168.63.129.16	TCP	354	49719 → 32526 [PSH,
20	1.203920	168.63.129.16	192.168.1.4	HTTP	79	HTTP/1.1 100 Continue
21	1.204002	192.168.1.4	168.63.129.16	HTTP	425	PUT /status HTTP/1.1
22	1.209293	168.63.129.16	192.168.1.4	HTTP	157	HTTP/1.1 200 OK
23	1.209530	192.168.1.4	168.63.129.16	TCP	355	49719 → 32526 [PSH, A
24	1.210218	168.63.129.16	192.168.1.4	HTTP	79	HTTP/1.1 100 Continue
25	1.210296	192.168.1.4	168.63.129.16	HTTP	3173	PUT /status HTTP/1.1
26	1.210454	168.63.129.16	192.168.1.4	TCP	54	32526 → 49719 [ACK] S
27	1.214303	168.63.129.16	192.168.1.4	HTTP	157	HTTP/1.1 200 OK
28	1.255726	192.168.1.4	168.63.129.16	TCP	54	49719 → 32526 [ACK] S
29	1.256672	192.168.1.4	168.63.129.16	HTTP	145	GET /machine/plugins?
30	1.267312	168.63.129.16	192.168.1.4	HTTP	207	HTTP/1.1 200 OK (tex
31	1.287004	192.168.1.4	216.58.213.1	TCP	55	51096 → 443 [ACK] Seq
32	1.288500	216.58.213.1	192.168.1.4	TCP	66	443 → 51096 [ACK] Seq
33	1.318293	192.168.1.4	168.63.129.16	TCP	54	49724 → 80 [ACK] Seq
34	1.425276	170.194.32.60	192.168.1.4	TCP	60	10400 → 3389 [ACK] Se

Figure 1.26 – Example traffic details from a Packet Capture

As we have seen, Network Watcher is a robust set of tools to help identify issues with connectivity and to provide a detailed analysis of the flow of traffic. Also, it is important for monitoring traffic and events for security purposes.

Monitoring security

Azure manages and protects many aspects of your solutions for you; however, it is still crucial that you monitor for intrusion events either at the platform level or in your hosted applications.

To help you monitor and protect your environment, you can use the Azure **Activity log**.

Activity log

Every action you perform in Azure, either directly in the portal, via PowerShell, the Azure CLI, using *DevOps pipelines*, or even as a result of an automated task, is logged.

These logs can be viewed at the resource level, resource group level, or subscription level. The process is the same for them all, but the following is an example of how to view subscription events:

1. Navigate to the Azure portal by opening `https://portal.azure.com`.

2. In the left-hand menu, select or search for `Subscriptions`.

3. Select the subscription you wish to view.

4. In the left-hand menu, click **Activity log**.

As the following screenshot shows, you are presented with a list of events showing what happened, when, and who or what initiated it. Events are grouped by the operation name, and clicking on the operation will provide more granular details of the events:

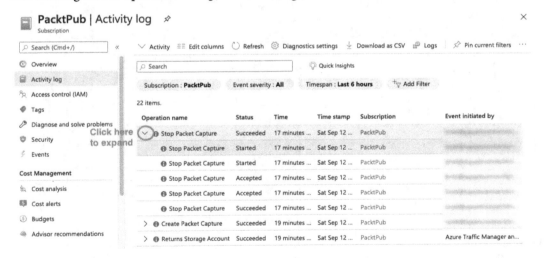

Figure 1.27 – Azure Activity Logs

Above the events are a series of filters to set the level you wish to view, over what time span, and a severity setting.

The severity can be filtered by *Critical, Warning, Error,* and *Informational.*

You can also add additional filters by clicking the **Add Filter** button, which then allows you to filter by the following properties:

- **Resource Group**
- **Resource**
- **Resource Type**
- **Operation**
- **Event Initiated By**
- **Event Category**

To see more detail of a particular event, follow these steps:

1. From the list of events, expand the **Operation Name** group by clicking on the arrow, as shown in the previous screenshot.

2. Now click on the event to show the summary.

3. Click **JSON** to see more details of the event. The following screenshot shows an example:

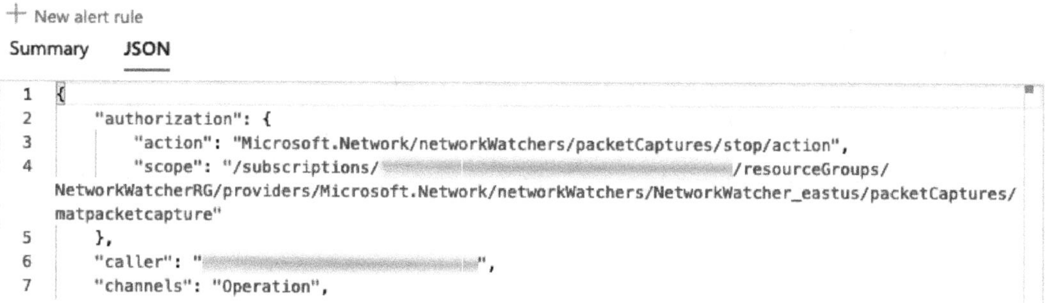

Figure 1.28 – Example event details in JSON

4. If you want to be alerted whenever this event occurs, click **New Alert Rule** and then create the alert as before.

Using the Event viewer and creating relevant alerts will help identify inappropriate activities within Azure, either via the console or other methods.

Monitoring security is an important and critical activity to ensure the safety of your systems and data. In the following section, we look at another equally important task—keeping control of your costs.

Managing costs

Because Azure is a *Pay As You Go* service, and due to the range and power of the available components, it can be easy to lose sight of costs.

Fortunately, Azure provides several tools to help monitor and alert you on your ongoing and forecast spend:

1. Navigate to the Azure portal by opening `https://portal.azure.com`.

2. In the left-hand menu, select or search for **Subscriptions**.

3. Select the subscription you wish to view.

The overview page of a subscription shows you a high-level view of your current spend, broken down by your most costly resources, and a forecast based on currently deployed infrastructure, as shown here:

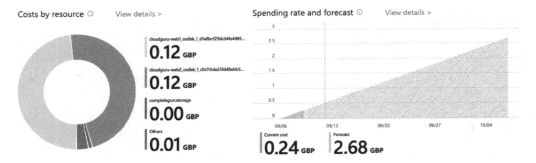

Figure 1.29 – Example costs dashboard

For a more detailed analysis, you can use the **Cost Analysis** tools.

Cost Analysis

In the **Subscription** blade, under **Cost Management**, click the **Cost Analysis** left-hand menu option. This view shows a more detailed and customizable set of charts. As shown in the following screenshot, the default view provides a breakdown by **Service name**, **Location**, and **Resource group name**:

Figure 1.30 – Cost analysis details

This view can be configured as per your requirements using a mixture of the filtering and selection options in the top menu. You can view costs as charts or text and set timespans or grouping options.

As an example of what can be achieved through this view, we shall create a monthly cost per resource table that can then be exported to CSV:

1. From the current **Costs Analysis View**, click the **View** menu, which by default will be **Accumulated costs**. The following screenshot shows the available options:

Figure 1.31 – Example subscription costs

2.   Change the view to **Cost by Resource**.

3.   Now select the **Granularity** option and change it from **None** to **Monthly**:

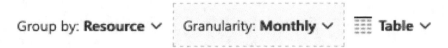

Figure 1.32 – Changing the Granularity to Monthly

4.   From the top menu, click **Download**, choose your export format, such as CSV, then click **Download data**.

Now that we have created a monthly cost per resource table, let's take a look at budgets.

Budgets

Being able to view your current and forecast spend is, of course, important, but even more useful is the ability to alert you when thresholds are met. For this, we can use **Budgets**:

1.   From the current **Subscription** blade, click the **Budgets** option in the left-hand menu.

2.   Click **Add** on the top menu.

3.   Set the name, for example, MonthlyBudget.

4.   Set a budget amount, for example, 30. The budget amount is specified in your local currency.

5.   Click **Next**.

6.   Set an alert as an amount of your monthly budget. For example, if you want to be alerted when you have spent 80% of your budget, enter 80.

7.   Assign an action group if you set one up previously.

8.   Enter a recipient email for whoever should be alerted when the budget alert is reached.

9.   Click **Create**.

Setting budgets doesn't prevent you from overspending, but it will alert you when your set thresholds are met.

Summary

In this chapter, we covered the *Implement and Monitor Azure Infrastructure* objective. We covered the various aspects of Azure Monitor and showed how you can use metrics to monitor all of your Azure resources and alerts to get notified when certain things are happening with your Azure resources.

We saw how Log Analytics and created queries so that we can get valuable data out of the logs. We looked at how Network Watcher can help you identify and resolve connectivity problems between devices. We covered how Activity Logs can highlight and notify you of security events. Finally, we also looked at how we can use the cost reporting and alerting features to help control costs.

In the next chapter, we will cover the second part of this exam objective. We will learn how to create and configure storage accounts.

Questions

Answer the following questions to test your knowledge of the information in this chapter. You can find the answers in the *Assessments* section at the end of this book:

1. Is Azure Log Analytics now a part of Azure Monitor?

    a) Yes

    b) No

2. Suppose that you want to create a query to retrieve specific log data from a VM. Do you need to write a SQL statement to retrieve this?

    a) Yes

    b) No

3. Are action groups used to enable metrics for Azure Monitor?

    a) Yes

    b) No

4. If you need to confirm connectivity between two endpoints on a specific port, what monitoring tool could you use?

    a) IP flow verify

    b) Next Hop

    c) Packet capture

5. What Azure feature would you use to create a monetary consumption alert from a resource group?

    a) Budgets from Resource Group

    b) Budgets from Subscription

    c) Budgets from Azure Monitor

Further reading

You can check out the following links for more information about the topics that were covered in this chapter:

- Azure Monitor overview: `https://docs.microsoft.com/en-us/azure/azure-monitor/overview`

- Azure Resource logs overview: `https://docs.microsoft.com/en-us/azure/azure-monitor/platform/diagnostic-logs-overview`

- Overview of log queries in Azure Monitor: `https://docs.microsoft.com/en-us/azure/azure-monitor/log-query/log-query-overview`

- Create custom views by using View Designer in Azure Monitor: `https://docs.microsoft.com/en-us/azure/azure-monitor/platform/view-designer`

- Azure Network Watcher:`https://docs.microsoft.com/en-us/azure/network-watcher/network-watcher-monitoring-overview`

- Azure Security Monitoring: `https://docs.microsoft.com/en-us/azure/security/fundamentals/management-monitoring-overview`

2
Creating and Configuring Storage Accounts

In the previous chapter, we covered the first part of this book's objective by covering how to analyze resource utilization and consumption in Azure. We covered how to monitor different Azure resources using Azure Monitor, and how to use Azure Log Analytics to query logs.

This chapter will introduce a new objective in terms of implementing and managing storage. In this chapter, we are going to cover the different types of storage accounts, and you will learn which types are available for storing your data in Azure. We will also cover how to install Azure Storage Explorer, which can be used to manage data inside Azure Storage accounts. We are going to look at how to secure data using **Shared Access Signatures** (**SAS**) and how to implement storage replication to keep data safe.

The following topics will be covered in this chapter:

- Understanding Azure Storage accounts
- Creating and configuring a storage account
- Installing and using Azure Storage Explorer

- Configuring network access to the storage account
- SAS tokens and access keys
- Implementing Azure Storage replication and failover

Technical requirements

This chapter will use Azure PowerShell (`https://docs.microsoft.com/en-us/powershell/azure/install-az-ps`) for examples.

The source code for our sample application can be downloaded from `https://github.com/PacktPublishing/Microsoft-Azure-Architect-Technologies-Exam-Guide-AZ-303/tree/master/Ch02`.

Understanding Azure Storage accounts

Azure offers a variety of types of storage accounts that can be used to store all sorts of files in Azure. You can store files, documents, and datasets, but also blobs and **Virtual Hard Disks (VHDs)**. There is even a type of storage account specifically for archiving. In the next section, we are going to look at the different types of storage accounts and storage account replication types that Azure has to offer.

Storage account types

Azure Storage offers three different account types, which can be used for blob, table, file, and queue storage.

General-Purpose v1

The **General-Purpose v1 (GPv1)** storage account is the oldest type of storage account. It offers storage for page blobs, block blobs, files, queues, and tables, but it is not the most cost-effective storage account type. It is the only storage account type that can be used for the classic deployment model. It doesn't support the latest features, such as access tiers.

Blob storage

The Blob storage account offers all of the features of StorageV2 accounts, except that it only supports block blobs (and append blobs). Page blobs are not supported. It offers access tiers, which consist of hot, cool, and archive storage, and which will be covered later in this chapter, in the *Access tiers* section.

General-Purpose v2 (GPv2)

StorageV2 is the newest type of storage account, and it combines V1 storage with Blob storage. It offers all of the latest features, such as access tiers for Blob storage, with a reduction in costs. Microsoft recommends using this account type over the V1 and Blob storage account types.

V1 storage accounts can easily be upgraded to V2.

> **Important note**
> For more information on pricing and billing for these different account types, you can refer to the following pricing page: https://azure. microsoft.com/en-us/pricing/details/storage/.

Storage replication types

Data that is stored in Azure is always replicated to ensure durability and high availability. This way, it is protected from unplanned and planned events, such as network or power outages, natural disasters, and terrorism. It also ensures that, during these types of events, your storage account still meets the SLA. Data can be replicated within the same data center, across zonal data centers within the same region, and across different regions. Depending on the type of storage account, these replication types include **Locally Redundant Storage (LRS)**, **Zone-Redundant Storage (ZRS)**, **Geo-Redundant Storage (GRS)**, **Geo-Zone-Redundant Storage (GZRS)**, and **Read-Access Geo-Redundant Storage (RA-GRS)**, and will be covered in more detail in the upcoming sections.

> **Important note**
> You choose a replication type when you create a new storage account. Storage accounts can be created inside the Azure portal, as well as from PowerShell or the CLI.

Locally redundant storage

LRS is the cheapest option and replicates the data three times within the same data center. When you make a write request to your storage account, it will be synchronously written during this request to all three replicas. The request is committed when the data is completely replicated. With LRS, the data will be replicated across multiple update domains and fault domains within one storage scale unit.

Zone-redundant storage

ZRS replicates three copies across two or three data centers. The data is written synchronously to all three replicas, in one or two regions. It also replicates the data three times inside the same data center where the data resided, just like LRS. ZRS provides high availability with synchronous replication across three Azure availability zones.

Geo-redundant storage

GRS replicates the data three times within the same region, like ZRS, and replicates three copies to other regions asynchronously. Using GRS, the replica isn't available for read or write access unless Microsoft initiates a failover to the secondary region. In the case of a failover, you'll have read and write access to that data after the failover has completed.

Geo-zone-redundant storage

Together with maximum durability, this option provides high availability. Data is replicated synchronously across three Azure availability zones. Then, the data is replicated asynchronously to the secondary region.

Read access to the secondary region can be enabled as well. GZRS is specially designed to provide at least 99.99999999999999% (16 9s) durability of objects over a given year.

At the time of writing this book, GZRS and RA-GRS is only available in the following regions:

- Asia Southeast
- Europe North
- Europe West
- Japan East
- UK South
- US Central
- US East
- US East 2
- US West 2

Read-access geo-redundant storage

RA-GRS provides geo-replication across two regions, with read-only access to the data in the secondary location. This will maximize the availability of your storage account. When you enable RA-GRS, your data will be available on a primary and a secondary endpoint for your storage account as well. The secondary endpoint will be similar to the primary endpoint, but it appends the secondary suffix to it. The access keys that are generated for your storage account can be used for both endpoints.

Now that we have covered the different storage replication types that are set when you create a storage account, we can look at the different storage accounts that Azure has to offer.

Azure Blob storage

Azure Blob storage offers unstructured data storage in the cloud. It can store all kinds of data, including documents, VHDs, images, and audio files. There are two types of blobs that you can create. One type is page blobs, which are used for the storage of disks. So, when you have a VHD that needs to be stored and attached to your **Virtual Machine (VM)**, you will have to create a page blob. The maximum size of a page blob is 8 TB.

The other type is block blobs, which basically covers all of the other types of data that you can store in Azure, such as files and documents. The maximum size of a block blob is 200 GB. However, there is also a third blob type, an append blob, but this one is used internally by Azure and can't be used to store actual files. There are a couple of ways that you can copy blobs to your Blob storage account. You can use the Azure portal (only one at a time) or Azure Storage Explorer, or you can copy your files programmatically using .NET, PowerShell, or the CLI, or by calling the REST API.

Access tiers

Blob storage accounts use access tiers to determine how frequently the data is accessed. Based on this access tier, you will get billed. Azure offers three storage access tiers: hot, cool, and archive.

Hot access tier

The hot access tier is most suitable for storing data that's accessed frequently and data that is in active use. For instance, you would store images and style sheets for a website inside the hot access tier. The storage costs for this tier are higher than for the other access tiers, but you pay less for accessing the files.

Cool access tier

The cool access tier is the most suitable for storing data that is not accessed frequently (less than once in 30 days). Compared with the hot access tier, the cool tier has lower storage costs, but you pay more for accessing the files. This tier is suitable for storing backups and older content that is not viewed often.

Archive

The archive storage tier is set on the blob level and not on the storage level. It has the lowest costs for storing data and the highest cost for accessing data compared to the hot and cool access tiers. This tier is for data that will remain in the archive for at least 180 days, and it will take a couple of hours of latency before it can be accessed. This tier is most suitable for long-term backups or compliance and archive data. A blob in the archive tier is offline and cannot be read (except for the metadata), copied, overwritten, or modified.

> **Important note**
> The archive tier is not currently supported for ZRS, GZRS or RA-GRS accounts.

Azure file storage

With Azure Files, you can create file shares in the cloud. You can access your files using the **Server Message Block (SMB)** or **Network File System (NFS)** protocols, which is an industry standard and can be used on Linux, Windows, and macOS devices. Azure Files can also be mounted as if it is a local drive on these same devices, and can be cached for fast access on Windows Server using Azure File Sync.

File shares can be used across multiple machines, which makes them suitable for storing files or data that are accessed from multiple machines, such as tools for development machines, configuration files, or log data. Azure File Share is part of the Azure Storage client libraries and offers an Azure Storage REST API, which can be leveraged by developers in their solutions.

Azure disk storage

The disks that are used for VMs are stored in Azure Blob storage as page blobs. Azure stores two disks for each VM: the actual operating system (VHD) of the VM, and a temporary disk that is used for short-term storage. This data is erased when the VM is turned off or rebooted.

There are two different performance tiers that Azure offers: standard disk storage and premium disk storage.

Standard disk storage

Standard disk storage offers HDD and SSD drives to store the data on, and it is the most cost-effective storage tier that you can choose. It can only use LRS or GRS to support high availability for your data and applications.

Premium disk storage

With premium disk storage, your data is stored on SSDs. Not all Azure VM series can use this type of storage. It can only be used with DS, DSv2, GS, LS, or FS series Azure VMs. It offers high-performance and low-latency disk support.

Ultra disk storage

For high IOPS, high throughput, and consistent low-latency disk storage for Azure IaaS VMs, Azure offers Azure Ultra Disks. This type of disk offers a number of additional benefits, including the ability to dynamically change the performance of the disk, along with your workloads, without the need to restart your VMs. Ultra disks are well suited for data-intensive workloads, such as top-tier databases, SAP HANA, and transaction-heavy workloads. Ultra disks can only be used as data disks, and premium SSDs as OSes are then recommended.

> **Tip**
> **IOPS**, or **input/output operations per second**, is a measure of how fast data can be written to and read from a disk

Unmanaged versus managed disks

Managed disks automatically handle storage account creation for you. With unmanaged disks, which are the traditional disks used for VMs, you need to create a storage account manually, and then select that storage account when you create the VM. With managed disks, this task is handled for you by Azure. You select the disk type and the performance tier (standard or premium), and the managed disk is created. It also handles scaling automatically for you.

Managed disks are recommended by Microsoft over unmanaged disks.

Now that we have covered all of the background information that you need to know in relation to the different storage accounts, we are going to create a new storage account.

Creating and configuring a storage account

Before you can upload any data or files to Azure Storage, a storage account needs to be created. This can be done using the Azure portal, PowerShell, the CLI, ARM templates, or Visual Studio.

In this demonstration, we are going to create a storage account with the Azure CLI:

1. First, we need to log in to the Azure account:

    ```
    az login
    ```

2. If necessary, select the right subscription:

    ```
    az account set --subscriptionid "********-****-****-****-
    ************"
    ```

3. Create a resource group:

    ```
    az group create --name PacktPubStorageAccount --location
    eastus
    ```

4. Create a storage account. The account name should be unique, so replace this with your own account name:

    ```
    az storage account create --resource-group
    PacktPubStorageAccount --name packtpubstorage --location
    eastus --sku Standard_GRS
    ```

> **Tip**
> In this demonstration, we created a new storage account using the Azure CLI. If you are new to storage accounts, I highly recommend creating a storage account from the Azure portal as well. That way, you will see all of the available storage account types, storage replication types, and access tiers that you can choose from and the different performance tiers (standard or premium), and how these are all connected. You can refer to the following tutorial on creating a storage account from the Azure portal: https://docs.microsoft.com/en-us/azure/storage/common/storage-quickstart-create-account?tabs=azure-portal.

Now that we have created a new storage account, we can install the Azure Storage Explorer tool.

Installing and using Azure Storage Explorer

Azure Storage Explorer is a standalone application that can be used to easily work with the different types of data that are stored in an Azure Storage account. You can upload, download, and manage files, queues, tables, blobs, Data Lake Storage, and Cosmos DB entities using Azure Storage Explorer. Aside from that, you can also use the application to configure and manage **Cross-Origin Resource Sharing** (**CORS**) rules for your storage accounts. This application can be used on Windows, Linux, and macOS devices.

To install the application, you have to perform the following steps:

1. Navigate to `https://azure.microsoft.com/en-us/features/storage-explorer/` to download the application.

2. Once it has been downloaded, install the application.

3. When it is installed, open the application. You will be prompted to connect to your Azure environment. There are a couple of options to choose from. You can add an Azure account by connecting to your Azure environment using your administrator credentials, using a shared access signature (which will be covered later in this chapter), and using a storage account name and key, and you can select the **Attach to a local emulator** option if you so desire. For this demonstration, keep the default option selected and click on **Sign in...**:

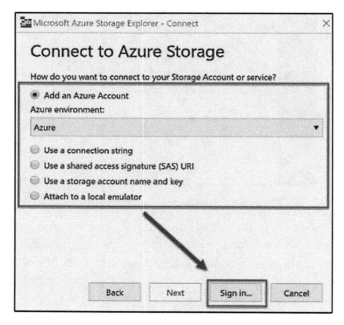

Figure 2.1 – Connecting to Azure Storage

4. Provide your credentials and log in.

5. All of your subscriptions will be added to the left-hand pane. Once this is done, click on **Apply**:

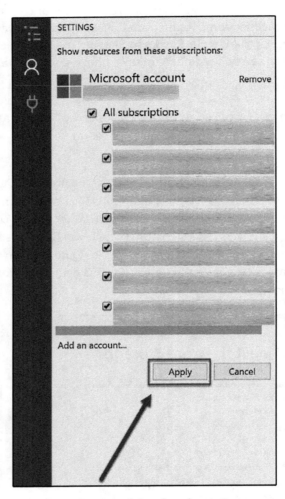

Figure 2.2 – Applying the subscriptions

6. You can now drill down to the subscription and the storage account that we created in the first demonstration from the left-hand pane. Select the storage account. From there, you can access the blob containers, file shares, queues, and tables:

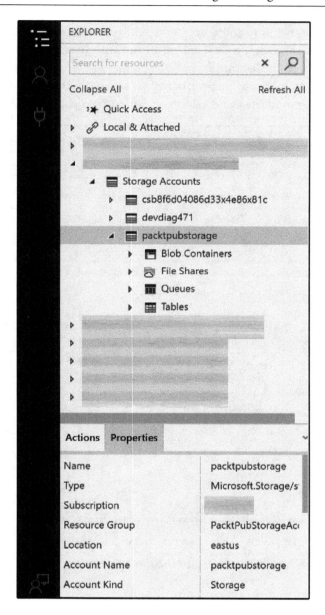

Figure 2.3 – Storage account settings

7. To add some files to a blob container, we need to create a blob container in the storage account. Therefore, right-click on **Blob Containers** in the left-hand menu and select **Create Blob Container**. Call the container `packtblobcontainer`. Now, you can upload files to that container. Click on the **Upload** button in the top menu, click on **Upload files**, and select some files from your local computer:

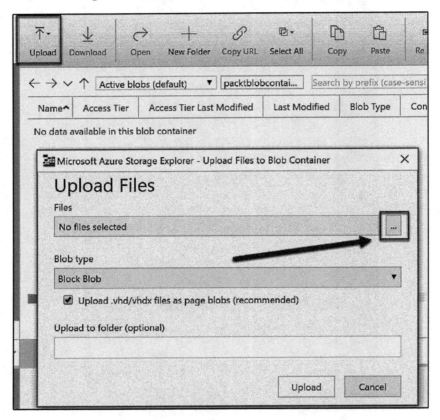

Figure 2.4 – Uploading files to the blob container

You will see that the files will be uploaded to the blob container.

> **Tip**
>
> If you navigate to the overview blade of the storage account in the Azure portal, you will see a button in the top menu – **Open in explorer**. This will open Azure Storage Explorer, which can then be used to easily manage all of the data that resides in the storage account.

Now that we have installed the Azure Storage Explorer tool and uploaded some files to a blob container, we can configure network access to the storage account.

Configuring network access to the storage account

You can secure your storage account to a specific set of supported networks. For this, you have to configure network rules so that only applications that request data over the specific set of networks can access the storage account. When these network rules are effective, the application needs to use proper authorization on the request. This authorization can be provided by Azure Active Directory credentials for blobs and queues, with an SAS token or a valid account access key, which we cover in the *SAS tokens and access keys* section.

In the following demonstration, we are going to configure network access to the storage account that we created in the previous step. You can manage storage accounts through the Azure portal, PowerShell, or CLIv2. We are going to set this configuration from the Azure portal. Therefore, we have to perform the following steps:

1. Navigate to the Azure portal by opening `https://portal.azure.com`.

2. Go to the storage account that we created in the previous step.

3. From the overview blade, in the left-hand menu, **select Firewalls and virtual networks**.

4. To grant access to a virtual network with a new network rule, under **Virtual Networks,** there are two options to choose from: **All networks**, which allows traffic from all networks (both virtual and on-premises) and the internet to access the data, and **Selected networks**. If you select this option, you can configure which networks are allowed to access the data from the storage account. Select **Selected networks**. Then, you can select whether you want to add an existing virtual network or create a new one. For this demonstration, click on **+ Add new virtual network**:

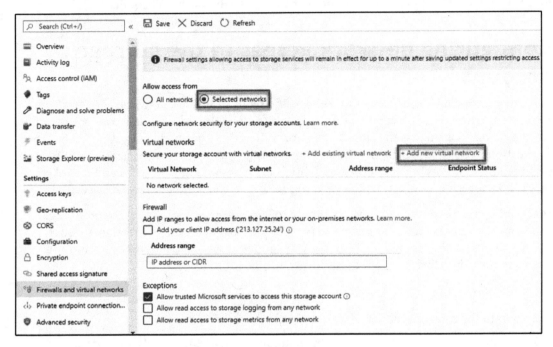

Figure 2.5 – Creating a new network

5. A new blade will open, where you will have to specify the network configuration. Specify the configuration that's shown in the following screenshot and click on **Create**:

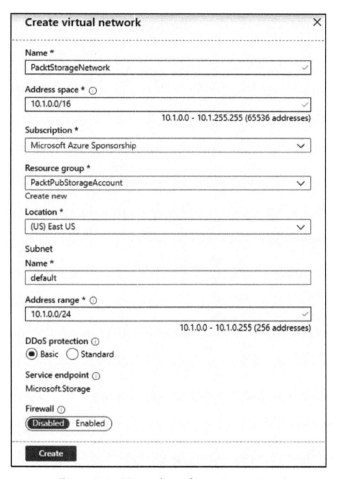

Figure 2.6 – Network configuration settings

6.   The virtual network will be added to the overview blade. This storage account is now secure and can be accessed only from applications and other resources that use this virtual network. In this same blade, you can also configure the firewall and only allow certain IP ranges from the internet or your on-premises environment:

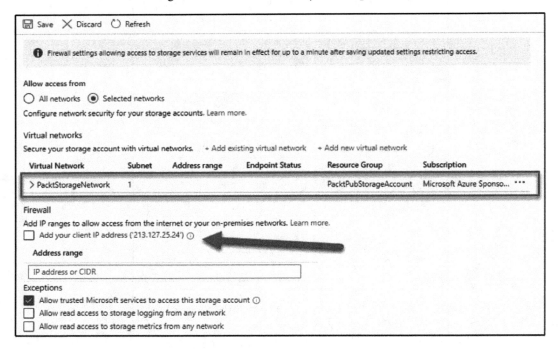

Figure 2.7 – IP ranges

This concludes this demonstration. In the next demonstration, we are going to generate and manage SAS.

SAS tokens and access keys

By using an SAS, you can provide a way to grant limited access to objects and data that are stored inside your storage account to the clients that connect to it. Using an SAS, you don't have to expose your access keys to the clients.

When you create a storage account, primary and secondary access keys are created. Both of these keys can grant administrative access to your account and all of the resources within it. Exposing these keys can also open your storage account to negligent or malicious use. SAS provides a safe alternative to this that will allow clients to read, write, and delete data in your storage account according to the permissions you've explicitly granted without the need for an account key.

In the next section, we're going to look at how to manage our access keys and how to generate an SAS for our storage account.

Managing access keys

To manage access keys, perform the following steps:

1. Navigate to the Azure portal by opening `https://portal.azure.com`.

2. Again, go to the storage account that we created in the previous step.

3. Once the overview blade is open, under **Settings**, select **Access keys**.

4. Here, you can see both of the access keys that were generated for you when the storage account was created. The reason that Azure created two access keys for you is that if you regenerate a new key, all of the SAS that you created for this key will no longer work. You can then let applications access that data using the secondary key, and once the key is regenerated, you can share the new key with your clients. You can generate new keys by clicking on the buttons next to both keys:

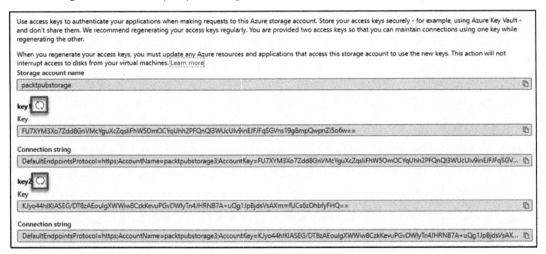

Figure 2.8 – Access keys

5. There is also a connection string provided for each key that can be used by client applications to access the storage account.

In the next section, we're going to generate an SAS for the access keys.

Generating an SAS

In this demonstration, we are going to generate an SAS for our blob store. An SAS access key is always in the following format:

```
?sv=<version>&ss=<allowed services>&srt=<allowed resource
types>&sp=<Allowed permissions>&se=<end date and
time>&st=<start date and time>&spr=<protocol>&sig=<random
signature>
```

To generate an SAS, perform the following steps:

1. Navigate to the Azure portal by opening `https://portal.azure.com`.

2. Again, go to the storage account that we created in the previous step.

3. Once the overview blade is open, under **Settings**, select **Shared access signature**:

A shared access signature (SAS) is a URI that grants restricted access rights to Azure Storage resources. You can provide a shared access signature to clients who should not be trusted with your storage account key but whom you wish to delegate access to certain storage account resources. By distributing a shared access signature URI to these clients, you grant them access to a resource for a specified period of time.

An account-level SAS can delegate access to multiple storage services (i.e. blob, file, queue, table). Note that stored access policies are currently not supported for an account-level SAS.

Learn more

Allowed services ⓘ

☑ Blob ☑ File ☑ Queue ☑ Table

Allowed resource types ⓘ

☑ Service ☑ Container ☑ Object

Allowed permissions ⓘ

☑ Read ☑ Write ☑ Delete ☑ List ☑ Add ☑ Create ☑ Update ☑ Process

Start and expiry date/time ⓘ

Start

| 12/17/2019 | 📅 | 5:21:28 PM |

End

| 12/18/2019 | 📅 | 1:21:28 AM |

(UTC+01:00) Sarajevo, Skopje, Warsaw, Zagreb ⌄

Allowed IP addresses ⓘ

for example, 168.1.5.65 or 168.1.5.65-168.1.5.70

Allowed protocols ⓘ

◉ HTTPS only ○ HTTPS and HTTP

Signing key ⓘ

key1 ⌄

Generate SAS and connection string

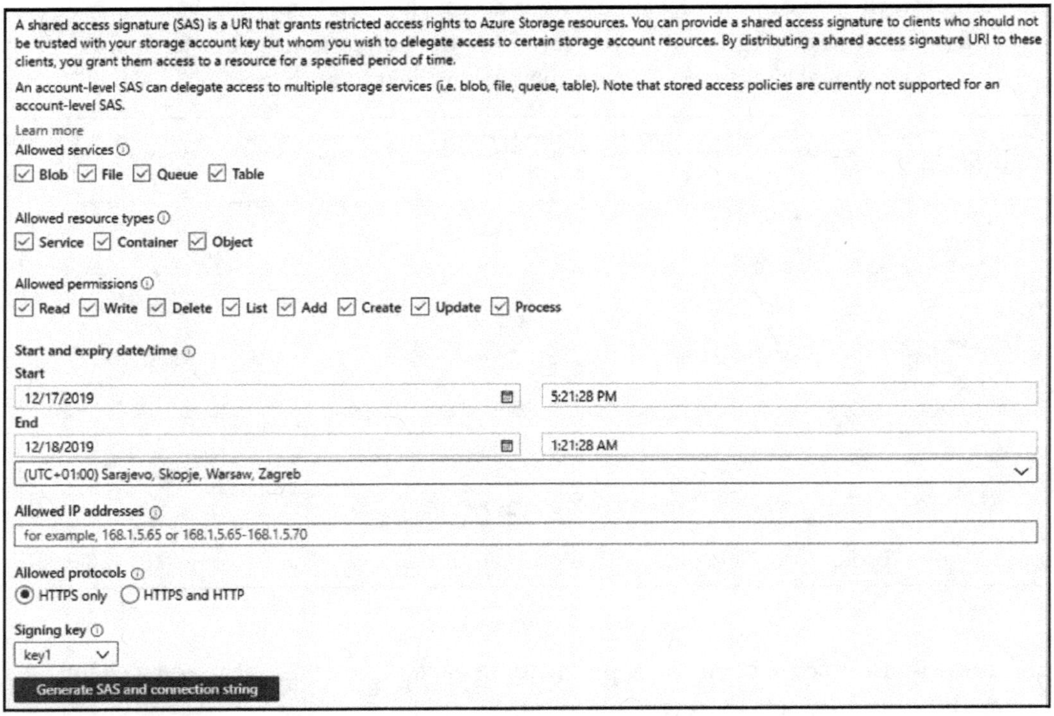

Figure 2.9 – Selecting Shared access signature

4. To only allow the Blob storage to be accessed, disable the file, queue, and table. Keep the default permissions, and then select an expiration date and time. You can also set the allowed protocols here. At the bottom of the screen, you can apply these permissions to the different keys. Keep **key1** selected and click on **Generate SAS and connection string**:

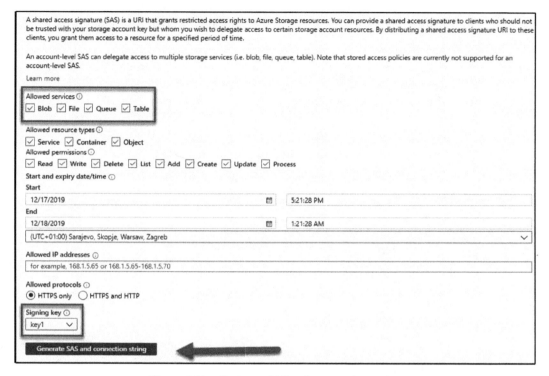

Figure 2.10 – Generate SAS and connection string

5. You can now use this token to request the data from the Blob storage.

In this section, we looked at the different ways in which we can provide access to a storage account. Next, we are going to look at how to implement Azure Storage replication.

Implementing Azure Storage replication and failover

The data in Azure is always replicated to ensure durability and high availability. Azure Storage copies your data so that it is protected from planned and unplanned events, including transient hardware failures, network or power outages, and massive natural disasters. We have already covered the different replication types that Azure offers for your storage accounts.

Storage replication can be set during the creation of the storage account. You can change the type of replication later as well by using the Azure portal, PowerShell, or the CLI. To change this in the Azure portal, you have to perform the following steps:

1. Navigate to the Azure portal by opening `https://portal.azure.com`.

2. Go to the storage account that we created in the previous step.

3. Under **Settings**, select **Configuration**. In this blade, under **Replication**, you can change the type of replication:

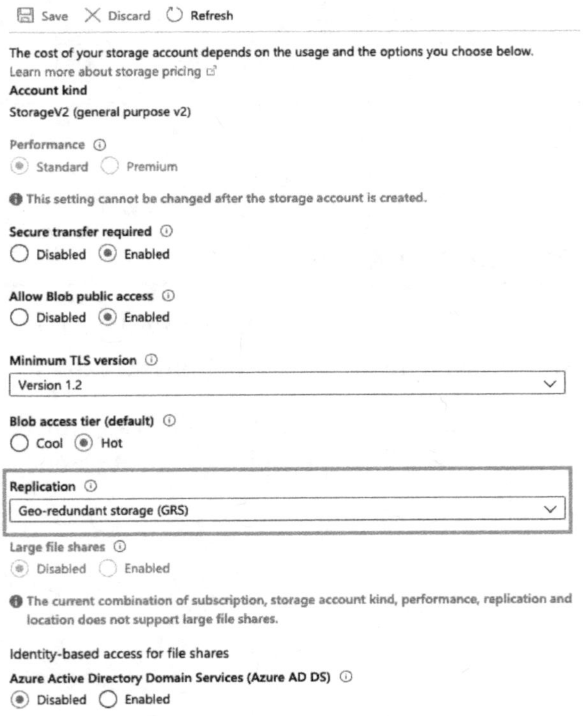

Figure 2.11 – Setting global replication on a storage account

When GRS is enabled after creation, as opposed to when you initially create the account, it can take some time for all your data to replicate over to the paired region.

In the event of a region outage that contains your storage account, failover is not automatic. You must manually switch your secondary replica to be the primary. Note that because data is copied from one region to the other asynchronously, that is, the data is copied to the secondary region after it has been written to the primary, there is still a chance therefore that some data could be lost if the outage happens before the data has had chance to replicate.

To perform a failover from the primary region to the secondary region, perform the following steps:

1. Navigate to the Azure portal by opening `https://portal.azure.com`.

2. Go to the storage account that we created in the previous step.

3. Under **Settings**, select **Geo Replication**. You will see where your replicas are located, as shown in the following screenshot:

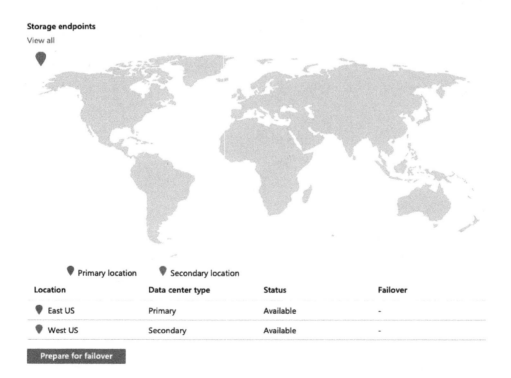

Figure 2.12 – Geo-replica failover

4.   To initiate the failover, click the **Prepare for failover** button.

5.   You will be prompted to confirm – type yes and click **Failover**.

Failover will now be initiated and will take a few minutes.

Summary

In this chapter, we covered the second part of the *Implement and Monitor Azure Infrastructure* objective. We covered the different types of storage that are available to us in Azure and when we should use them. We also covered how we can manage our data using Azure Storage Explorer and how we can secure our data using SAS. Finally, we covered how to replicate data across regions and initiate a failover.

In the next chapter, we'll cover the third part of this exam objective. In that chapter, we will cover how to implement and manage VMs.

Questions

Answer the following questions to test your knowledge of the information in this chapter. You can find the answers in the *Assessments* section at the end of this book:

1.   Can the Azure Storage Explorer application only be used on Windows devices?

a) Yes

b) No

2.   Can you configure storage accounts to be accessed from specific virtual networks and not from on-premises networks?

a) Yes

b) No

3.   Can you only set the type of replication for your storage accounts during the creation of the storage account?

a) Yes

b) No

Further reading

You can check out the following links for more information about the topics covered in this chapter:

- Azure Storage documentation: `https://docs.microsoft.com/en-us/azure/storage/`

- Getting started with Storage Explorer: `https://docs.microsoft.com/en-us/azure/vs-azure-tools-storage-manage-with-storage-explorer?tabs=windows`

- Configuring Azure Storage firewalls and virtual networks: `https://docs.microsoft.com/en-us/azure/storage/common/storage-network-security`

- Azure Storage redundancy: `https://docs.microsoft.com/en-us/azure/storage/common/storage-redundancy`

3
Implementing and Managing Virtual Machines

In the previous chapter, we covered the different types of storage that are available in Azure and when you should use them. We also covered how to install and use **Azure Storage Explorer** to manage your data.

This chapter proceeds with the third part of the Implement and Monitor Azure Infrastructure objective. In this chapter, we are going to cover **Virtual Machines** (**VMs**) in Azure, and the different VM sizes that are available for both Azure and Linux. VMs are still considered a core component of many solutions, and therefore a good understanding is essential.

You will learn how you can create and configure VMs for Windows and Linux. We will also cover high availability and what actions you can take to configure your VMs for high availability. You will also learn how to how to automate your deployment using scale sets, how to deploy and modify **Azure Resource Manager** (**ARM**) templates and DevOps, how to configure Azure Disk Encryption for VMs, and how to use dedicated hosts for additional VM security and isolation.

The following topics will be covered in this chapter:

- Understanding VMs
- Understanding Availability Sets
- Provisioning VMs
- Understanding VM scale sets
- Modifying and deploying ARM templates
- Deploying resources with Azure DevOps
- Configuring Azure Disk Encryption for VMs
- Using Azure Dedicated Host

Technical requirements

This chapter will use Azure PowerShell (`https://docs.microsoft.com/en-us/powershell/azure`) and Visual Studio Code (`https://code.visualstudio.com/download`) for the examples we'll cover.

You will also need an Azure DevOps account, which you can get at `https://azure.microsoft.com/en-us/services/devops`.

The source code for this chapter can be downloaded from `https://github.com/PacktPublishing/Microsoft-Azure-Architect-Technologies-Exam-Guide-AZ-303/tree/master/Ch03`.

Understanding VMs

You can run both Windows VMs as well as Linux VMs in Azure. VMs come in all sorts of sizes and a variety of prices, ranging from VMs with a small amount of memory and processing power for general purposes, to large VMs that can be used for **Graphics Processing Unit** (**GPU**)-intensive and high-performance computing workloads.

To create a VM, you can choose from several predefined images. There are images available for operating systems such as Windows Server or Linux, as well as predefined applications, such as SQL Server images, and complete farms that consist of multiple VMs that can all be deployed at once. An example of a farm is a three-tier SharePoint farm.

VMs can be created and managed either from the Azure portal, PowerShell, or the CLI, and they come in the following series and sizes.

VM series and sizes

At the time of writing this book, the following VM series and sizes are available:

Series	Type	Description
B, Dsv3, Dv3, Dasv4, Dav4, DSv2, Dv2, Av2, DC, DCv2, Dv4, Dsv4, Ddv4, Ddsv4	General-purpose	These VMs have a balanced CPU-to-memory ratio and are ideal for testing and development scenarios. They are also suitable for small and medium databases and web servers with low-to-medium traffic.
F, Fs, Fsv2	Compute-optimized	These VMs have a high CPU-to-memory ratio and are suitable for web servers with medium traffic, application servers, and network appliances for nodes in batch processing.
Esv3, Ev3, Easv4, Eav4, Ev4, Esv4, Edv4, Edsv4, Mv2, M, DSv2, Dv2	Memory-optimized	These VMs have a high memory-to-CPU ratio and are suitable for relational database servers, medium-to-large caches, and in-memory analytics.
Lsv2	Storage-optimized	These VMs have high disk throughput and I/O and are suitable for big data, SQL, and NoSQL databases.
NC, NCv2, NCv3, NCasT4_v3 (Preview), ND, NDv2 (Preview), NV, NVv3, NVv4	GPU	These VMs are targeted for heavy graphics rendering and video editing, deep learning applications, and machine learning model training. These VMs are available with single or multiple GPUs.
HB, HBv2, HC, H	High-performance compute	These are the fastest VMs available. They offer the most powerful CPU with optional high-throughput network interfaces (Remote Direct Memory Access (RDMA)).

For the exam, you will not need to know the individual VM sizes, but you will be expected to know the different families and their use cases.

> **Important note**
> VM series are updated constantly. New series, types, and sizes are added and removed frequently. To stay up to date with these changes, you can refer to the documentation for VM sizes, at `https://docs.microsoft.com/en-us/azure/virtual-machines/sizes`.

Managed disks

Azure managed disks are the default disks selected when you create a VM in the Azure portal. They handle storage for your VMs completely. Previously, you would have had to manually create storage accounts to store VM hard disks, and when your VM needed to scale up, you had to add additional storage accounts to make sure you didn't exceed the limit of 20,000 **Input/Output Operations Per Second (IOPS)** per account.

With managed disks, this burden is now handled for you by Azure. You can now create 10,000 VM disks inside a subscription, which can result in thousands of VMs inside a subscription, without the need to copy disks between storage accounts.

There are a number of different disk types that can be provisioned, depending on the VM series you choose. The different types and their specifications are listed in the following table:

	Ultra disk	Premium SSD	Standard SSD	Standard HDD
Disk type	SSD	SSD	SSD	HDD
Scenario	I/O-intensive	Production and performance	Web servers, lightly used, applications and dev/test	Backup, non-critical
Max disk size	65,536 GB	32,767 GB	32,767 GB	32,767 GB
Max throughput	2,000 Mb/s	900 Mb/s	750 Mb/s	500 Mb/s
Max IOPS	160,000	20,000	6,000	2,000

As with the VM families, the important points to remember are around the use cases for each disk type.

Sizes and disk types cover the basic configuration elements of a VM. Next, we will look at some of the advanced features available, starting with **Availability Sets**.

Understanding Availability Sets

To create a reliable infrastructure, adding your VMs to an Availability Set is key. Several scenarios can have an impact on the availability of your Azure VMs. These are as follows:

- **Unplanned hardware maintenance events**: When hardware is about to fail, Azure fires an unplanned hardware maintenance event. Live migration technology is used, which predicts the failure and then moves the VM, the network connections, memory, and storage to different physical machines, without disconnecting the client. When your VM is moved, the performance is reduced for a short time because the VM is paused for 30 seconds. Network connections, memory, and open files are still preserved.

- **Unexpected downtime**: The VM is down when this event occurs because Azure needs to heal your VM. A hardware or physical infrastructure failure often causes this event to happen.

- **Planned hardware maintenance events**: This type of event is a periodic update from Microsoft in Azure to improve the platform. Most of these updates don't have a significant impact on the uptime of VMs, but some of them may require a reboot or restart.

To provide redundancy during these types of events, you can group two or more VMs in an Availability Set. By leveraging Availability Sets, VMs are distributed across multiple isolated hardware nodes in a cluster. This way, Azure can ensure that, during an event or failure, only a subset of your VMs is impacted and your overall solution will remain operational and available. This way, the 99.95% uptime promised by the Azure **Service Level Agreement** (**SLA**) can still be met during outages and other failures.

> **Tip**
> VMs can only be assigned to an Availability Set during initial deployment.

When you place your VMs in an Availability Set, Azure guarantees to spread them across **fault domains** and **update domains**. We will look at what this means next and how it protects your VMs.

Fault domains and update domains

By default, Azure will assign 3 fault domains and 5 update domains (which can be increased to a maximum of 20) to the Availability Set.

When spreading your VMs over fault domains, your VMs sit across three different racks in the Azure data center, each with its own networking and power. So, in the case of an event or failure of the underlying platform, only one rack gets affected and the other VMs remain accessible, as depicted in the following diagram:

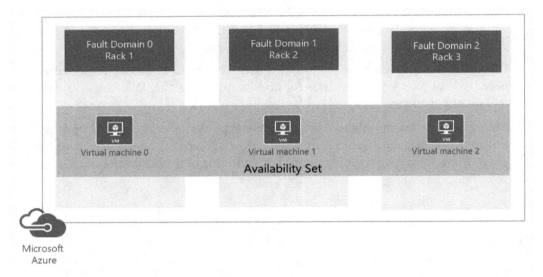

Figure 3.1 – VMs spread over three fault domains

Update domains are useful in the case of an OS or host update. When you spread your VMs across multiple update domains, one domain will be updated and rebooted while the others remain accessible, as depicted in the following diagram:

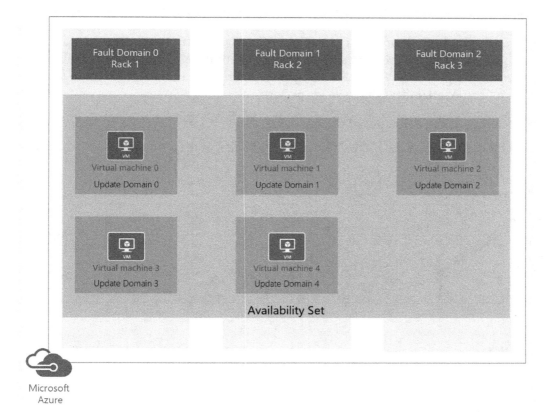

Figure 3.2 – VMs spread over five update domains and three fault domains

In the next section, we are going to create a new Windows VM in the Azure portal.

Understanding how to provision VMs

In the upcoming demonstrations, we are going to deploy a Windows Server VM from both the Azure portal and PowerShell.

Important note

Deploying Linux machines is quite similar to deploying Windows machines. We are not going to cover how to deploy Linux machines. For more information about how to deploy Linux machines in Azure, you can refer to the *Further reading* section at the end of this chapter

---

Deploying a Windows VM from the Azure portal

In this demonstration, we are going to deploy a Windows VM from the Azure portal. We are going to set up networking and storage and select an SKU size (RAM or CPU, for example) for this VM. We are also going to configure high availability for this VM by placing it in an **Availability Set**. To do so, perform the following steps:

1. Navigate to the Azure portal by opening `https://portal.azure.com`.

2. In the left menu, click **Virtual machines**, and then, in the top menu, click **+ Add** as follows:

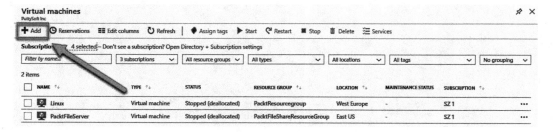

Figure 3.3 – Creating a new VM

3. We are going to create a Windows VM, so in the **Basics** blade, add the following values:

    a) **Subscription**: Choose a subscription.

    b) **Resource group**: Type `PacktVMGroup`.

    c) **Virtual machine name**: Type `PacktWindowsVM`.

    d) **Region**: Choose a region.

    e) **Availability options**: Here, select **Availability set**.

    f) **Availability set**: Create a new one and call it `PacktWindowsAS`. Keep the default fault domains and update the domains for this VM.

    g) **Image**: Choose **Windows Server Datacenter 2019**.

    h) **Azure Spot instance**: Select **No**.

    i) **Size**: Here, you can choose between the different sizes. Click **Change size** and select **Standard DS1 v2**.

    j) **Administrator account**: Provide a username and a password.

    k) **Inbound port rules**: Select **Allow selected ports** and enable **Remote Desktop Protocol (RDP)**. You will need this to log in to the server after creation.

l) **Save money**: If you already have a valid Windows Server Datacenter license, you get a discount on this VM.

4. Click **Next: Disks**.

5. Here, you can select the disk type. Keep the default as follows, which is **Premium SSD**:

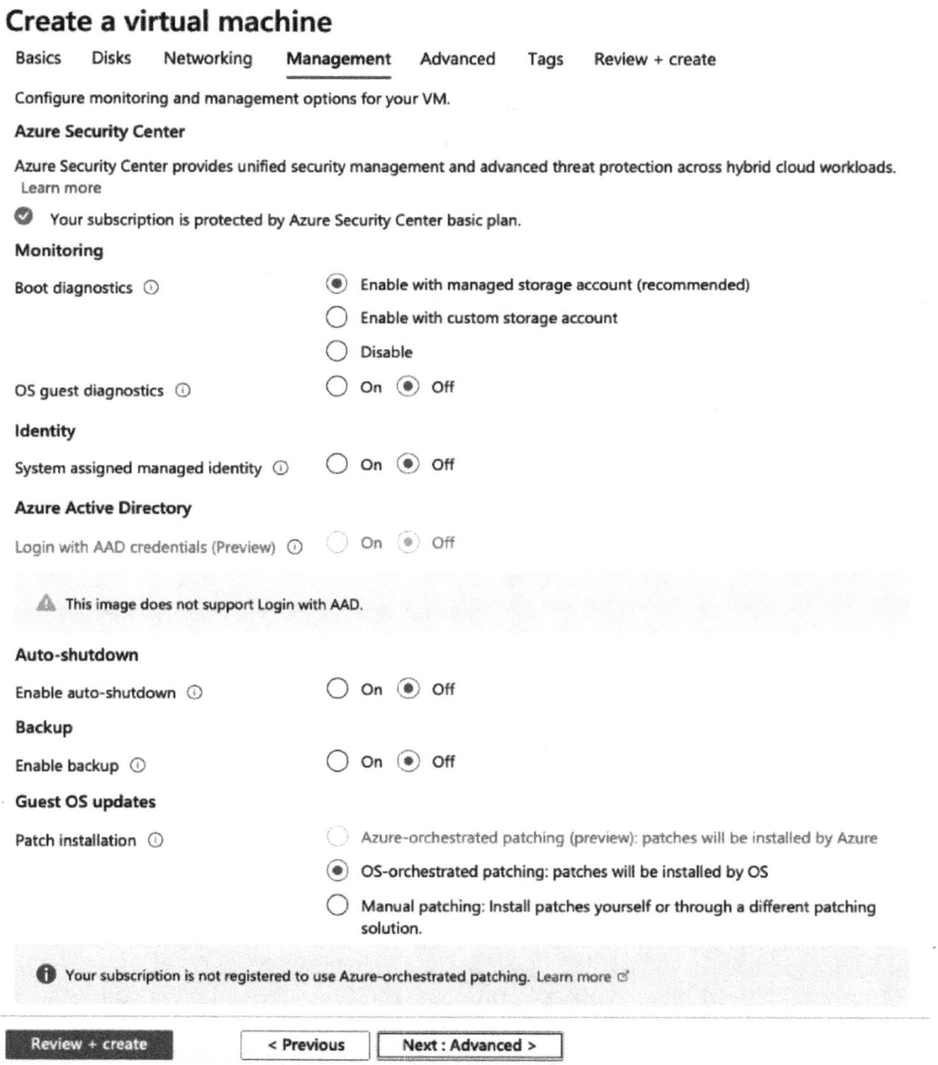

Create a virtual machine

Basics   Disks   Networking   **Management**   Advanced   Tags   Review + create

Configure monitoring and management options for your VM.

Azure Security Center

Azure Security Center provides unified security management and advanced threat protection across hybrid cloud workloads.
Learn more

✅ Your subscription is protected by Azure Security Center basic plan.

Monitoring

Boot diagnostics ⓘ
- ◉ Enable with managed storage account (recommended)
- ◯ Enable with custom storage account
- ◯ Disable

OS guest diagnostics ⓘ   ◯ On ◉ Off

Identity

System assigned managed identity ⓘ   ◯ On ◉ Off

Azure Active Directory

Login with AAD credentials (Preview) ⓘ   ◯ On ◉ Off

⚠ This image does not support Login with AAD.

Auto-shutdown

Enable auto-shutdown ⓘ   ◯ On ◉ Off

Backup

Enable backup ⓘ   ◯ On ◉ Off

Guest OS updates

Patch installation ⓘ
- ◯ Azure-orchestrated patching (preview): patches will be installed by Azure
- ◉ OS-orchestrated patching: patches will be installed by OS
- ◯ Manual patching: Install patches yourself or through a different patching solution.

ⓘ Your subscription is not registered to use Azure-orchestrated patching. Learn more ◰

[Review + create]   [< Previous]   [Next : Advanced >]

Figure 3.4 – Select disk type

6. Click **Next: Networking**.

7. In the **Networking** blade, you can configure the virtual network. You can keep the default values for this machine as follows:

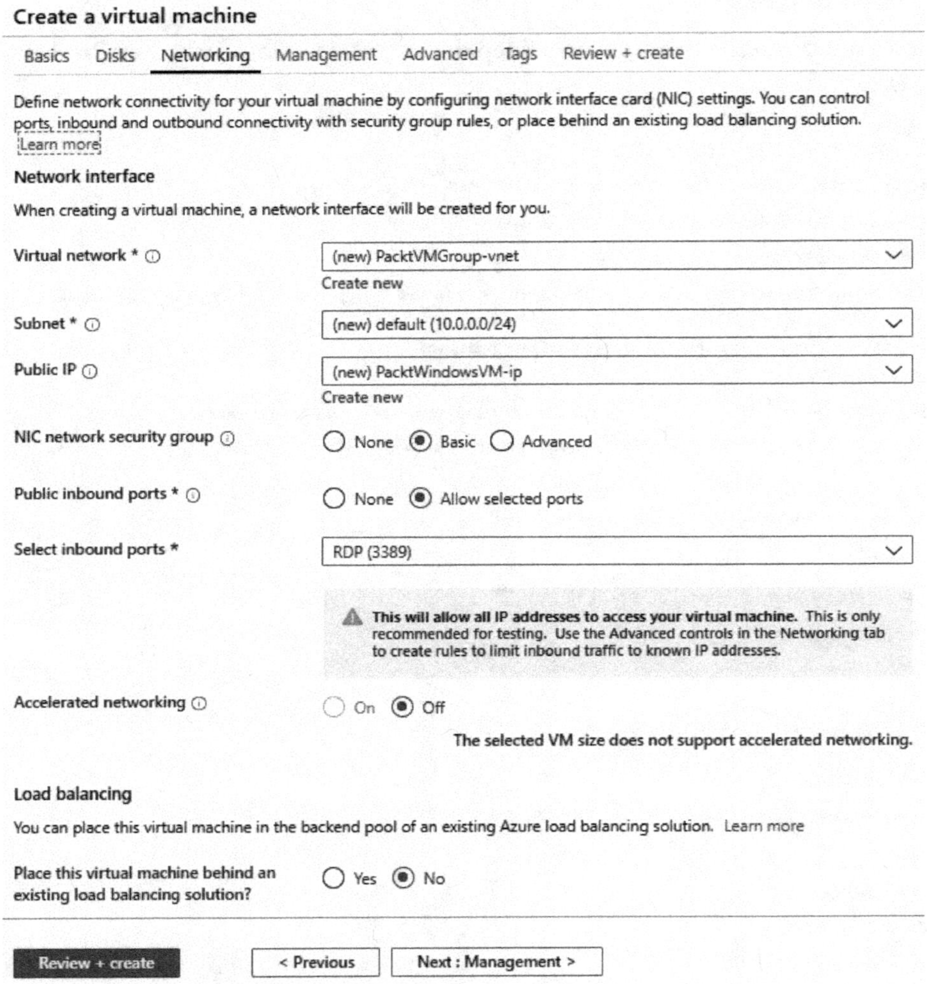

Create a virtual machine

Basics   Disks   Networking   Management   Advanced   Tags   Review + create

Define network connectivity for your virtual machine by configuring network interface card (NIC) settings. You can control ports, inbound and outbound connectivity with security group rules, or place behind an existing load balancing solution. Learn more

Network interface

When creating a virtual machine, a network interface will be created for you.

Virtual network * ⓘ	(new) PacktVMGroup-vnet ∨
	Create new
Subnet * ⓘ	(new) default (10.0.0.0/24) ∨
Public IP ⓘ	(new) PacktWindowsVM-ip ∨
	Create new
NIC network security group ⓘ	○ None ⦿ Basic ○ Advanced
Public inbound ports * ⓘ	○ None ⦿ Allow selected ports
Select inbound ports *	RDP (3389) ∨

⚠ **This will allow all IP addresses to access your virtual machine.** This is only recommended for testing. Use the Advanced controls in the Networking tab to create rules to limit inbound traffic to known IP addresses.

Accelerated networking ⓘ	○ On ⦿ Off
	The selected VM size does not support accelerated networking.

Load balancing

You can place this virtual machine in the backend pool of an existing Azure load balancing solution. Learn more

Place this virtual machine behind an existing load balancing solution?	○ Yes ⦿ No

[Review + create]   [< Previous] [Next : Management >]

Figure 3.5 – Set up networking for VM

8. Click **Next: Management**.

9. In the **Management** blade, you can configure monitoring, and create and select a storage account for monitoring. You can also assign a system-assigned managed identity, which can be used to authenticate to various Azure resources, such as Azure Key Vault, without storing any credentials in code. You can also enable auto-shutdown here, as follows:

Create a virtual machine

Basics   Disks   Networking   **Management**   Advanced   Tags   Review + create

Configure monitoring and management options for your VM.

Azure Security Center

Azure Security Center provides unified security management and advanced threat protection across hybrid cloud workloads.
Learn more

✓ Your subscription is protected by Azure Security Center standard plan.

Monitoring

Boot diagnostics ⓘ   ● On   ○ Off

OS guest diagnostics ⓘ   ○ On   ● Off

Diagnostics storage account * ⓘ
| (new) packtvmgroupdiag156 ⌄ |
Create new

Identity

System assigned managed identity ⓘ   ○ On   ● Off

Azure Active Directory

Login with AAD credentials (Preview) ⓘ   ○ On   ◉ Off

⚠ This image does not support Login with AAD.

Auto-shutdown

Enable auto-shutdown ⓘ   ● On   ○ Off

Shutdown time ⓘ
| 7:00:00 PM |

Time zone ⓘ
| (UTC) Coordinated Universal Time ⌄ |

Notification before shutdown ⓘ   ○ On   ● Off

Backup

Enable backup ⓘ   ○ On   ● Off

| Review + create | | < Previous | | Next : Advanced > |

Figure 3.6 – Set up management features

10. We can now create the VM. Click **Review + create** and the settings will be validated.
    After that, click **Create** to actually deploy the VM.

We have now deployed a Windows VM, placed it in an Availability Set, and looked at the
networking, storage, and monitoring features and capabilities for this VM. In the next
section, we are going to deploy a Windows Server VM from PowerShell.

Deploying a Windows VM from PowerShell

In the next demonstration, we are going to create two Windows Server VMs from PowerShell and place them in an Availability Set. To do so, you have to perform the following steps:

1. First, we need to log in to the Azure account, as follows:

    ```
    Connect-AzAccount
    ```

2. If necessary, select the right subscription, as follows:

    ```
    Select-AzSubscription -SubscriptionId "********-****-
    ****-****-
    ***********"
    ```

3. Create a resource group for the Availability Set, as follows:

    ```
    New-AzResourceGroup -Name PacktVMResourceGroup -Location
    EastUS
    ```

4. Then, we can create an Availability Set for the VMs, as follows:

    ```
    New-AzAvailabilitySet `
    -Location "EastUS" `
    -Name "PacktVMAvailabilitySet" `
    -ResourceGroupName PacktVMResourceGroup `
    -Sku aligned `
    -PlatformFaultDomainCount 2 `
    -PlatformUpdateDomainCount 2
    ```

5. We have to set the administrator credentials for the VMs, as follows:

    ```
    $cred = Get-Credential
    ```

6. We can now create the two VMs inside the Availability Set, as follows:

```
for ($i=1; $i -le 2; $i++)
{
New-AzVm `
-ResourceGroupName PacktVMResourceGroup `
-Name "PacktVM$i" `
-Location "East US" `
-VirtualNetworkName "PacktVnet" `
-SubnetName "PacktSubnet" `
-SecurityGroupName "PacktNetworkSecurityGroup" `
-PublicIpAddressName "PacktPublicIpAddress$i" `
-AvailabilitySetName "PacktVMAvailabilitySet" `
-Credential $cred
}
```

In the last two demonstrations, we created VMs inside an Availability Set from the Azure portal and PowerShell. In the next section, we are going to cover scale sets.

Understanding VM scale sets

VM scale sets are used for deploying multiple VMs at once without the need for manual actions or using scripts. You can then manage them all at once from a single place. VM scale sets are typically used to build large-scale infrastructures, where keeping all of your VMs in sync is key. The maintenance of VMs, including keeping them in sync, is handled by Azure. VM scale sets use Availability Sets under the hood. VMs inside a scale set are automatically spread over the fault and update domains by the underlying platform. VM scale sets use Azure autoscale by default. You can, however, add or remove instances yourself instead of using autoscale.

You can configure the autoscale settings to scale out (add instances) or in (reduce the number of instances) based on CPU thresholds. For example, you can create a rule to add additional instances if the CPU threshold is over 80% for 5 minutes. You can then have the scale set scaled back if the CPU threshold drops below 30% for 5 minutes.

When creating a scale set, a couple of artifacts are created for you automatically. As well as the number of VMs you have specified being added to the set, Azure Load Balancer and Azure autoscale are added, along with a virtual network and a public IP address, as shown in the following screenshot:

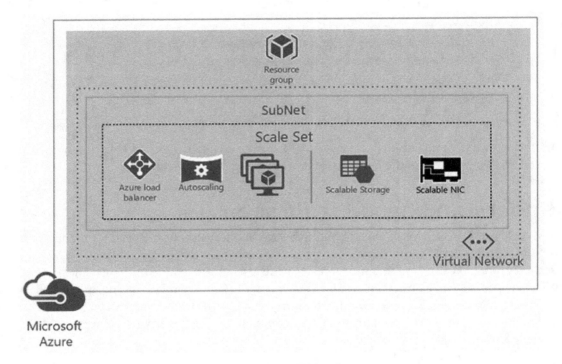

Figure 3.7 – Azure VM scale set

In the next section, we are going to deploy and configure scale sets.

Deploying and configuring scale sets

To create a VM scale set from the Azure portal, take the following steps:

1. Navigate to the Azure portal by opening `https://portal.azure.com`.

2. Click on **Create a resource** and type `Scale Set` into the search bar. Select **Virtual machine scale set**.

   On the next screen, click **Create** and add the following settings for the scale set to be created:

   a) **Virtual machine scale set name**: `PacktScaleSet`.

   b) **Operating system disk image: Windows Server 2016 Datacenter**.

   c) **Subscription**: Select a subscription.

   d) **Resource group**: `PacktVMGroup`.

   e) **Location: (US) East US**.

   f) **Availability zone: None**.

   g) **Username:** `SCPacktUser`.

   h) **Password**: Fill in a password.

   i) **Instance count: 2.**

   j) **Instance size: Standard DS1 v2.**

   k) **Deploy as low priority: No.**

   l) **Use managed disks: Yes.**

3. If you scroll down, you can configure the autoscale settings, choose between the different load balancing settings, and configure the networking and monitoring capabilities, as follows:

AUTOSCALE

Autoscale ⓘ	◯ Disabled ⦿ Enabled
Minimum number of VMs * ⓘ	1
Maximum number of VMs * ⓘ	10

Scale out

CPU threshold (%) * ⓘ	75
Number of VMs to increase by * ⓘ	1

Scale in

CPU threshold (%) * ⓘ	25
Number of VMs to decrease by * ⓘ	1

NETWORKING

Microsoft Azure Application Gateway is a dedicated virtual appliance providing application delivery controller (ADC) as a service.
Azure Load Balancer allows you to scale your applications and create high availability for your services.
Learn more about load balancer differences

Resources	Optimal for	Supported Protocols	SSL offloading	RDP to instance
Application Gateway	Web-based traffic	HTTP/HTTPS/WebSoc...	Supported	Not supported
Load balancer	Stream-based traffic	Any	Not supported	Supported

Choose Load balancing options	◯ Application Gateway ◯ Load balancer ⦿ None

Configure virtual networks

Virtual network * ⓘ	(new) PacktVMGroup-vnet ⌄
	Create new
Subnet *	(new) default (10.0.0.0/24) ⌄
Public IP address per instance ⓘ	◯ On ⦿ Off
Accelerated networking ⓘ	◯ On ⦿ Off

Figure 3.8 – Scale set configuration settings

4. Scroll down again and keep the default settings for network security ports and so on.

5. Click **Create**. The scale set with the specified number of VMs in it is now deployed.

In the next section of this chapter, we are going to cover how to automate the deployment of VMs using ARM templates.

Modifying and deploying ARM templates

ARM templates define the infrastructure and configuration of your Azure solution. Azure is managed by an API, which is called the **Azure Resource Manager** or **ARM** API. You can use this API to deploy infrastructure as code and configure your Azure environment. This API can be called from various tooling and resources; you can do it using the Azure portal, PowerShell, or the CLI, or by calling the API directly, or by creating ARM templates.

You can create an ARM template in JSON format and use this to repeatedly deploy your solution across your Azure environment in a consistent state. The template is processed by ARM like any other request, and it will parse the template and convert the syntax into REST API operations for the appropriate resource providers. The REST API uses the resources section inside the template to call the resource-specific APIs. An example of a resource provider is `Microsoft.Storage/storageAccounts`.

> **Tip**
>
> Microsoft offers various predefined ARM templates that can be downloaded and deployed. You can download and deploy the quickstart templates straight from GitHub, or download them and make the necessary adjustments: `https://github.com/Azure/azure-quickstart-templates`.

In the next section, we are going to modify an ARM template in the Azure portal.

Modifying an ARM template

In this demonstration, we are going to create an ARM template of a storage account in the Azure portal. We are going to modify this template so that it will generate a storage account name automatically. We will then deploy this template again and use it to create a new storage account from the Azure portal. Therefore, you have to take the following steps:

1. Navigate to the Azure portal by opening `https://portal.azure.com`.

2. Select **Create a resource**, then **Storage**, and then select **Storage account**. Create a new storage account.

3. Add the following values:

    a) **Subscription**: Pick a subscription.

    b) **Resource group**: Create a new one and call it `PacktARMResourceGroup`.

    c) **Storage account name**: Type `packtarm`.

d) **Location**: Select **(US) East US**.

e) **Performance**: Select **Standard**.

f) **Account kind**: Select **StorageV2 (general purpose v2)**.

g) **Replication**: Select **Read-access geo-redundant storage (RA-GRS)**.

h) **Access tier**: Select **Hot**.

4. Click **Review + create**. Do not select **Create**.

5. Then, in the next step, select **Download a template for automation**:

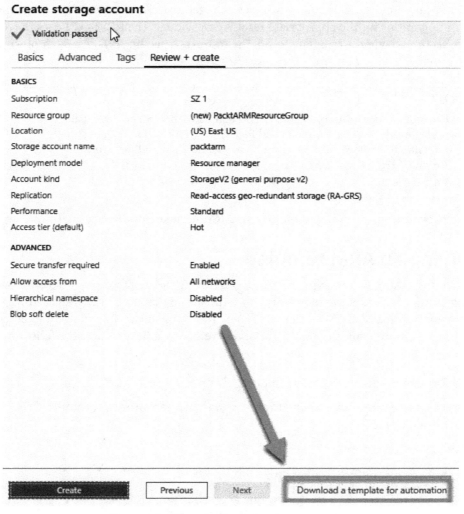

Figure 3.9 – Download template for automation

The editor will be opened and the generated template will be displayed. The main pane shows the template. It has six top-level elements: `schema`, `contentVersion`, `parameters`, `variables`, `resources`, and `output`. There are also six parameters. `storageAccountName` is highlighted in the following screenshot. In the template, one Azure resource is defined. The type is `Microsoft.Storage/storageAccounts`.

6.   Select **Download** from the top menu:

Figure 3.10 – Main ARM template

7. Open the downloaded ZIP file and then save `template.json` to your computer. In the next section, you will use a template deployment tool to edit the template.

8. Select **Parameters** from the top menu and look at the values. We will need this information later during the deployment:

Figure 3.11 – ARM template parameters

9. The Azure portal can be used for the basic editing of ARM templates. More complex ARM templates can be edited using Visual Studio Code, for instance. We are going to use the Azure portal for this demonstration. Therefore, select **+ Create a resource**; then, in the search box, type `Template Deployment` (deploy using custom templates). Then select **Create**.

10. In the next blade, you have different options for loading templates. For this demonstration, select **Build your own template in the editor**:

Figure 3.12 – Template options

11. Select **Load file**, and then follow the instructions to load the `template.json` file that we downloaded in the last section. Make the following changes:

a) Remove the `storageAccountName` parameter.

b) Add a new variable:

```
"storageAccountName":
"[concat(uniqueString(subscription().subscriptionId),
'storage')]",
```

c) Replace `"name": "[parameters('storageAccountName')]"` with `"name": "[variables('storageAccountName')]"`:

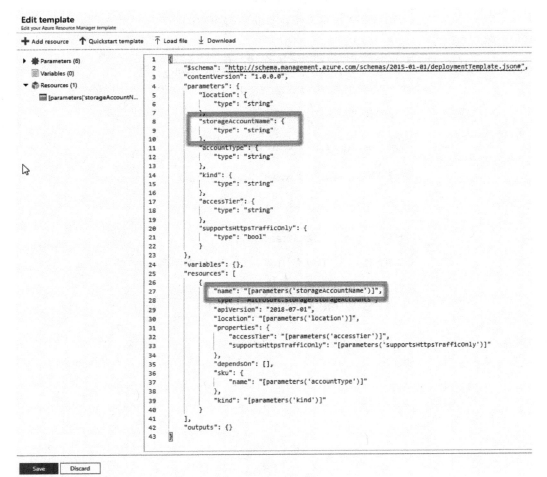

Figure 3.13 – Make changes to the highlighted sections

d) The code of the template will look as follows:

The `schema` and `parameters` sections are as follows:

```
{
"$schema":
"http://schema.management.azure.com/schemas/2015-01-01
/deploymentTemplate.json#", "contentVersion": "1.0.0.0",
"parameters": {
"location": {
"type": "string"


},
"accountType": { "type": "string"
},
"kind": {
"type": "string"
},
"accessTier": { "type": "string"
},
"supportsHttpsTrafficOnly": { "type": "bool"
}
},
```

The `variables` and `resources` sections are as follows:

```
"variables": { "storageAccountName":
"[concat(uniqueString(subscription().subscriptionId),
'storage')]"
},
"resources": [
{
"name": "[variables('storageAccountName')]",
"type": "Microsoft.Storage/storageAccounts",
"apiVersion": "2018-07-01",
"location": "[parameters('location')]", "properties": {
"accessTier": "[parameters('accessTier')]",
"supportsHttpsTrafficOnly":
```

```
"[parameters('supportsHttpsTrafficOnly')]"
},
"dependsOn": [], "sku": {
"name": "[parameters('accountType')]"
},
"kind": "[parameters('kind')]"
}
],
"outputs": {}
}
```

e) Then select **Save**.

12. In the next screen, fill in the values for creating the storage account. You will see that the parameter for filling in the storage account name has been removed. This will be generated automatically. Fill in the following values:

a) **Resource group**: Select the resource group name you created in the previous section.

b) **Location**: Select **East US**.

c) **Account type**: Select **Standard_LRS**.

d) **Kind**: Select **StorageV2**.

e) **Access Tier**: Select **Hot**.

f) **Https Traffic Only Enabled**: Select **true**.

g) **I agree to the terms and conditions stated above**: Select this checkbox.

13. Select **Purchase**.

14. The ARM template will now be deployed. After deployment, go to the **Overview** blade of the resource group. You will see that the storage account name has automatically been generated for you:

Figure 3.14 – Storage account name

> **Tip**
>
> For more information about the syntax and structure of ARM templates, you can refer to the documentation at `https://docs.microsoft.com/ en-us/azure/azure-resource-manager/resource-group- authoring-templates`.

We have now modified an ARM template in the Azure portal and created a new storage account using the modified ARM templates. In the next demonstration, we are going to save a deployment as an ARM template.

Saving a deployment as an ARM template

For this demonstration, we are going to save a deployment as an ARM template from the Azure portal. We are going to export the template for the first VM we created via the portal.

Once downloaded, you can then make changes to it, and redeploy it in Azure using PowerShell or code. The generated ARM template consists of a large amount of code, which makes it very difficult to make changes to it. For saving a deployment as an ARM template, take the following steps:

1. Navigate to the Azure portal by opening `https://portal.azure.com`.

2. Open the resource group that we created in the previous demonstration, and under **Settings**, select **Export template** as follows:

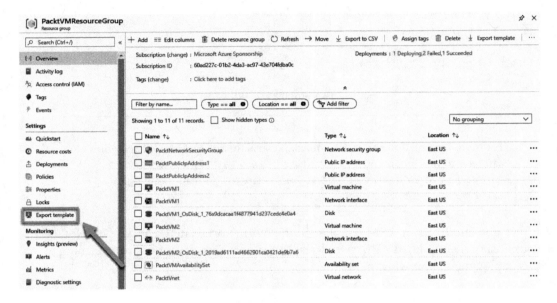

Figure 3.15 – Export template

3. The template is generated for you based on the settings that we made during the creation of the different resources. You can download the template and redeploy it from here. You can also download the scripts for the CLI, PowerShell, .NET, and Ruby, and create different resources using these programming languages. Select **Download** from the top menu, as follows:

Figure 3.16 – Download template

The template is downloaded as a ZIP file to your local filesystem.

4. You can now extract the template files from the ZIP file and open them in Visual Studio Code. If you don't have this installed, you can use the download link provided at the beginning of this chapter or use Notepad or some other text-editing tool. The ZIP file contains three different deployment files, created in different languages. There is one each for PowerShell, the CLI, and Ruby. It also consists of a `DeploymentHelper.cs` file, a `parameters.json` file, and a `template.json` file.

5. In Visual Studio Code, you can make all of the modifications to the parameters and template files that are needed. If you then want to deploy the template again to Azure, use one of the deployment files inside the container. In the case of PowerShell, right-click on `deploy.ps1` and select **Run with PowerShell**. Fill in the subscription ID, provide the resource group name and deployment name and log in using your Azure credentials. This will start the deployment.

We have now saved a deployment as an ARM template. In the next section, we will look at an alternative method for deploying resources – Azure DevOps.

Deploying resources with Azure DevOps

ARM templates provide new ways of working; by codifying our infrastructure in files, we can leverage software development patterns and principles. **Infrastructure as Code (IaC)** is a pattern whereby you define and store an entire environment in configuration scripts.

By storing your templates and configurations in a *code repository* such as **Azure DevOps**, you can control who has access to change the files and see a full audit trail of those changes.

With IaC and DevOps, you are also able to align another pattern called **Immutable Infrastructure**. Immutable infrastructure states that modifications are never applied directly to the components, such as a VM; instead, any changes are made in code and then redeployed.

Azure DevOps supports this through automation – pipelines that can be configured to automatically deploy new code as soon as updates are committed and approved.

This section will use the ARM templates we just downloaded to create an automated deployment pipeline.

Setting up your first DevOps project

The first step is to set up a DevOps organization and create a project, as follows:

1. Navigate to Azure DevOps by opening `https://azure.microsoft.com/en-us/services/devops`.

2. Click **Start free**.

3. You will be prompted to accept the license terms, so click the checkbox and then click **Continue**.

4. You will now be prompted to pick a name – either accept the default or enter your own. It will need to be unique. You can also specify the region in which all your DevOps code will be stored. See the following screenshot for an example. Click **Continue**:

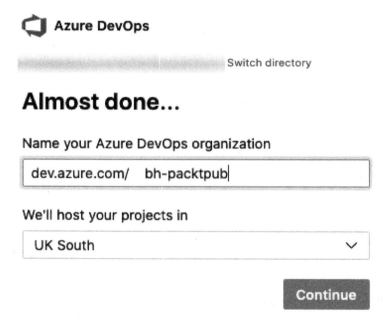

Figure 3.17 – Signing up for Azure DevOps

5. You will be taken to your organization. Now create a new project by giving it a name and clicking **Create project**. You also have the option of making the project public or private as follows:

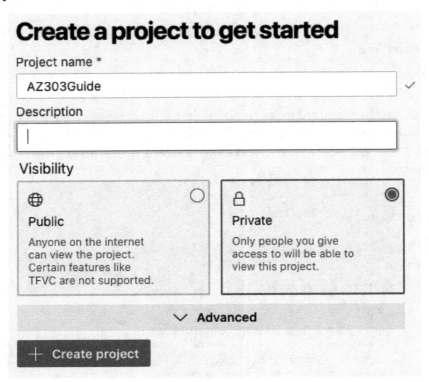

Create a project to get started

Project name *

AZ303Guide   ✓

Description

|

Visibility

🌐   ○   🔒   ◉
Public   Private
Anyone on the internet   Only people you give
can view the project.   access to will be able to
Certain features like   view this project.
TFVC are not supported.

                        ⌄ Advanced

+ Create project

Figure 3.18 – Creating a DevOps project

6. On the left-hand menu, click **Repos**, then **Files**. Your repository is empty – we can commit individual files to it, but for simplicity, we will import the **Packt Publishing AZ303** GitHub repository.

7. Under **Import a repository**, click **Import**, then paste in the URL for GitHub repo mentioned in the previous step: `https://github.com/PacktPublishing/Microsoft-Azure-Architect-Technologies-Exam-Guide-AZ-303.git`. Lastly, click **Import** as follows:

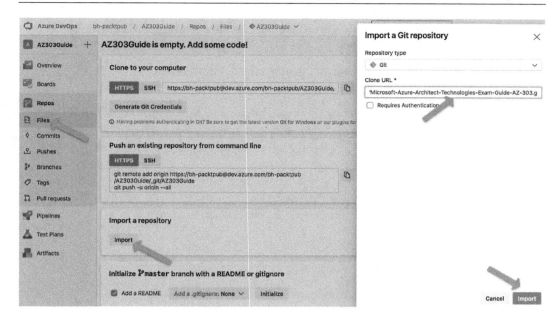

Figure 3.19 – Importing into a repository

For this exercise, we only really need three files. While still on the **Files** view, you will now see a list of chapter folders corresponding to this book's chapters. Expand **Chapter03** by clicking on the yellow folder icon, then click on the `azuredeploy.json` file inside that folder.

This file is an ARM template for building a simple **Linux VM**. You can view and even edit the file within DevOps. Now select the `azuredeploy.parameters.json` file – this is the parameters file where we can set the username, password, and server name of the machine we wish to deploy. You can change these parameters if you want to by clicking the **Edit** button. Click **Commit** once you have finished making your changes.

We now have the files we need, but before we can create our deployment pipeline, we must create a service connection.

Creating a service connection

A service connection is a configuration item that will grant access from your Azure subscription to your Azure DevOps project. Let's set it up as follows:

1. At the bottom of the screen, click **Project Settings**.
2. On the left-hand menu, click **Service Connections**.
3. Click **Create Service Connection**.

4. In the **New Service Connection** window, select **Azure Resource Manager** as the type and click **Next**.

5. On the next window, for **Authentication method**, choose the default option – **Service Principal (automatic)**.

6. In the next window, set the scope to **Subscription** and select your subscription from the drop-down list. Enter a name for your connection, such as `PacktPubServiceConnection` shown in the following screenshot. Make a record of this name as you will need it later on:

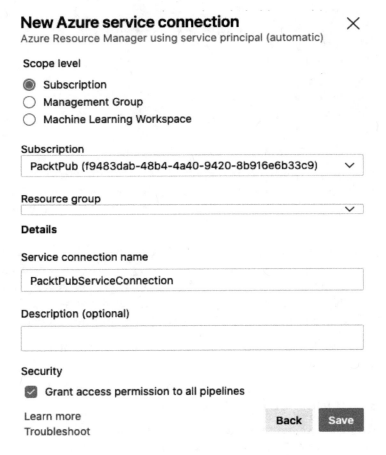

Figure 3.20 – Creating a service connection

7. Click **Save**.

Once the project has been set up and your code is ready, we can now build the deployment pipeline.

Creating the pipeline

Follow these steps to create the pipeline:

1. On the left-hand menu, click **Pipelines**, then **Create Pipeline**.

2. On the next screen, choose the top option, **Azure Repos Git (YAML)**.

3. Select the repository that is in your project (in this example, AZ303Guide) – it will be the same name as the project you created in *step 5* of *Setting up your first DevOps project*.

4. Select **Existing Azure Pipelines YAML file**.

5. In the **Path** drop-down box, choose /Chapter03/pipeline.yaml as follows:

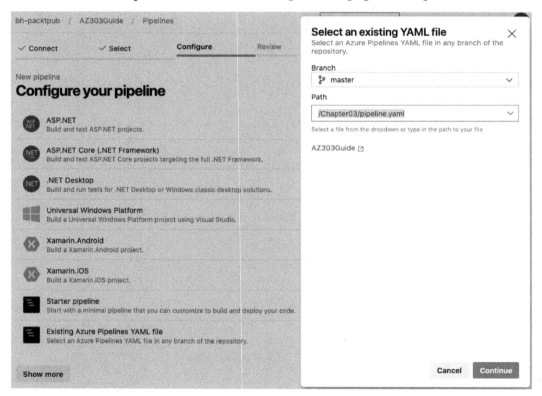

Figure 3.21 – Creating a YAML pipeline

6. Click **Continue**.

7. The **YAML** file will be shown – you will need to update the `azureSubscription` entry with the name of the **service connection** you created earlier. (In our example, this was `PacktPubServiceConnection`.) The setting is near the bottom of the file as follows:

Figure 3.22 – Editing the YAML pipeline

8. Click **Save and run**.

The deployment will start, and you will be taken to a screen that will track the progress as follows:

Figure 3.23 – Deployment report

DevOps pipelines provide a great way to manage and record deployments. They are well suited to modern patterns such as **Agile**, **Infrastructure as Code**, and **Immutable Infrastructure**.

This section has shown how to create automated pipelines that make deploying our infrastructure quicker and repeatable. Next, we will configure Azure Disk Encryption for VMs to enhance our VMs' security.

Configuring Azure Disk Encryption for VMs

Azure Disk Encryption for VMs can help you to meet your organizational security and compliance commitments by encrypting the disks of your VMs in Azure. For Windows VMs, it uses the **BitLocker** feature and, for Linux VMs, it uses the **DM-Crypt** feature to encrypt the OS and data disks. Azure Disk Encryption is available for Windows and Linux VMs with a minimum of 2 GB of memory, and for Standard VMs and VMs with Azure Premium Storage; however, it is not available for Basic, A-series, or generation 2 VMs.

> **Tip**
> For more information about the prerequisites of Azure Disk Encryption, you can refer to the documentation at `https://docs.microsoft.com/en-us/azure/security/azure-security-disk-encryption-prerequisites`.

It uses **Azure Key Vault** to help to control and manage the disk encryption keys and secrets. Azure Disk Encryption also ensures that disks that are stored in Azure Storage are encrypted at rest.

You will get a **High Severity** alert in **Azure Security Center** if you have VMs that are not encrypted. From there, you will get the recommendation to encrypt these VMs, and you can take action to encrypt the VMs from the Azure Security Center directly.

In the next demonstration, we are going to encrypt the data disk of one of the VMs that we created in the *Deploying a Windows VM from PowerShell* section. However, before we can encrypt the disk, we first need to create an Azure Key Vault. We are going to create this from the Azure portal.

Creating an Azure Key Vault

To create the Key Vault from the Azure portal, take the following steps:

1. Navigate to the Azure portal by opening `https://portal.azure.com`.
2. From the left menu, click **Create a resource**. In the search box, type `Key Vault` and create a new Key Vault.

3. Fill in the following values:

    a) **Subscription**: Select the same subscription that you used in the *Deploying a Windows VM from PowerShell* section.

    b) **Resource group**: Select the resource group name you created in the previous exercise, which is `PacktVMResourceGroup`.

    c) **Key Vault Name**: Type `PacktEncryptionVault`.

    d) **Region**: Select **East US**.

    e) **Pricing Tier**: Select **Standard**.

4. Click **Review + create** and then **Create**.

Now that the Key Vault is in place, we can encrypt the disk.

Encrypting the disk

Disks can only be encrypted using PowerShell and the CLI. In the next demonstration, we are going to encrypt the disk of *PacktVM1* using PowerShell.

To encrypt the disk, take the following steps:

1. First, we need to log in to the Azure account, as follows:

    ```
    Connect-AzAccount
    ```

2. If necessary, select the right subscription, as follows:

    ```
    Select-AzSubscription -SubscriptionId "********-****-
    ****-****-************"
    ```

3. Set some parameters, as follows:

    ```
    $ResourceGroupName = 'PacktVMResourceGroup'
    $vmName = 'PacktVM1'
    $KeyVaultName = 'PacktEncryptionVault'
    ```

4. Then, retrieve the Key Vault, as follows:

    ```
    $KeyVault = Get-AzKeyVault -VaultName $KeyVaultName -
    ResourceGroupName $ResourceGroupName
    $diskEncryptionKeyVaultUrl = $KeyVault.VaultUri
    $KeyVaultResourceId = $KeyVault.ResourceId
    ```

5. Then, encrypt the disk, as follows:

```
Set-AzVMDiskEncryptionExtension `
-ResourceGroupName $ResourceGroupName `
-VMName $vmName `
-DiskEncryptionKeyVaultUrl $diskEncryptionKeyVaultUrl `
-DiskEncryptionKeyVaultId $KeyVaultResourceId
```

It will take approximately 10 minutes before the disk is encrypted. This concludes this demonstration.

Disk Encryption provides a greater level of protection for your VMs that should satisfy most security requirements. However, VMs will run on shared hardware, and for some organizations, the fact that data could be processed on the same physical CPU as another customer's VM is not permitted. For companies that require physically separate hardware, Microsoft offers **Azure Dedicated Host**.

Azure Dedicated Host

Standard VMs in Azure are automatically placed and run on physical host servers. Therefore, your VM will be running on the same hardware as VMs from other customers at any one time. Although isolated by the virtualization later, any processing performed by your VM may share CPU and RAM space with that of other customers.

Azure dedicated hosts are physical servers that you can request. When you create a dedicated host, you gain control over the physical components and set your maintenance routines for things such as OS patching and reboots.

Dedicated hosts are typically required for compliance reasons, such as when your corporate policies restrict the use of shared infrastructure and demand hardware isolation. Once you have created your dedicated host, you can then provision virtual hosts on top. Billing is per physical host and is the same regardless of how many virtual servers you run on it – you do not get billed separately for the virtual server's compute and RAM usage.

Software is billed separately and is dependent on the number of virtual cores you are using, so although the number of virtual servers you host won't affect the hardware costs, it will affect the license costs. This can be mitigated with **Azure Hybrid Benefit**, which allows you to bring your own licenses (for example, an Enterprise Agreement).

Because your VMs run on a host dedicated to you, any hardware failure will affect your VMs. To ensure protection against such events, you can create multiple dedicated hosts in a **host group**, with each server in a different availability zone.

> **Important note**
> Azure subscriptions have quotas; by default, you are limited to 4 cores for trial subscriptions and 10 cores for pay as you go. Therefore, to create a dedicated host, you will need to request an increase of this quota to at least 64 cores, as that is the smallest host you can provision.

Let's walk through creating a host.

Implementing a dedicated host

Let's follow these steps to implement a dedicated host:

1. Navigate to the Azure portal by opening `https://portal.azure.com`.

2. Click the menu burger and click **Create a resource**.

3. Search for `Dedicated Hosts`, then click **Create** when presented.

4. Fill in the details for your host, as shown in the screenshot that follows:

    a) **Subscription**: Select your subscription.

    b) **Resource group**: Click **Create new**, type in `dedicatedhostsresourcegroup`, and click **OK**.

    c) **Name**: Enter a name, such as `PacktPubDedicatedHost`.

    d) **Location: East US**.

    e) **Host group**: Click **Create new host group**, enter the following information, then click **Add**:

    　　i) **Host group name**: `PacktPubHostGroup`.

    　　ii) **Availability zone**: Leave this empty.

    　　iii) **Fault domain count: 1**.

If you have licenses, select the options to state that you have either Windows Server or SQL Server licenses to receive hybrid benefits:

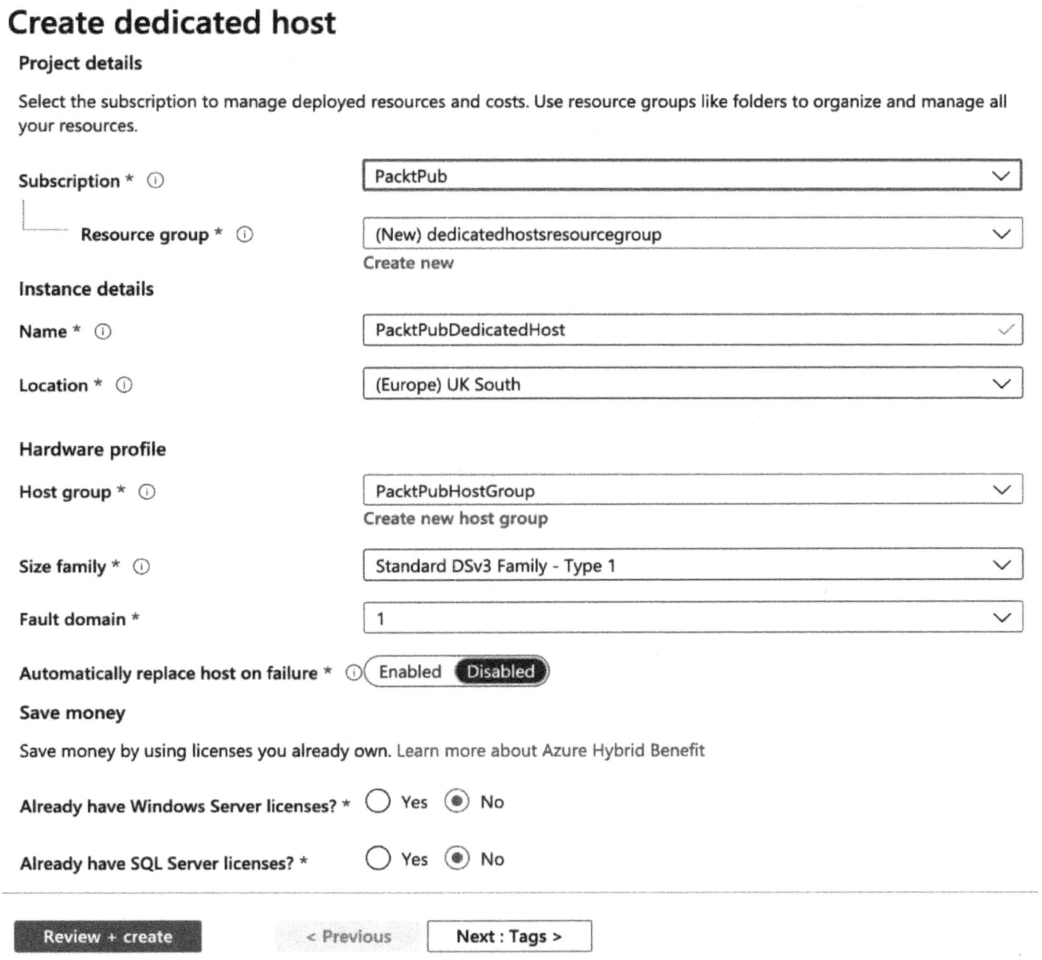

Create dedicated host

Project details

Select the subscription to manage deployed resources and costs. Use resource groups like folders to organize and manage all your resources.

Subscription * ⓘ | PacktPub   ∨ |

Resource group * ⓘ | (New) dedicatedhostsresourcegroup   ∨ |
                                     Create new

Instance details

Name * ⓘ | PacktPubDedicatedHost   ✓ |

Location * ⓘ | (Europe) UK South   ∨ |

Hardware profile

Host group * ⓘ | PacktPubHostGroup   ∨ |
                                     Create new host group

Size family * ⓘ | Standard DSv3 Family - Type 1   ∨ |

Fault domain * | 1   ∨ |

Automatically replace host on failure * ⓘ (Enabled **Disabled**)

Save money

Save money by using licenses you already own. Learn more about Azure Hybrid Benefit

Already have Windows Server licenses? *   ◯ Yes   ⦿ No

Already have SQL Server licenses? *   ◯ Yes   ⦿ No

[Review + create]   < Previous   [Next : Tags >]

Figure 3.24 – Creating a dedicated host

5. Click **Review + create**.

6. Click **Create**.

It can take a few minutes to provision a new host. Next, we will look at how to deploy VMs to the dedicated host.

Creating VMs on a dedicated host

Once deployment is complete, perform the following steps:

1. Navigate to the Azure portal by opening `https://portal.azure.com`.

2. Go to the **Subscriptions** blade and select your subscription.

3. Click **Resource Groups** on the left-hand menu and select the `dedicatedhostsresourcegroup` resource group you created earlier.

4. Select the `PacktPubDedicatedHost` host.

5. You will see a list of VM types you can create – make a note of these, as you have to select a VM of this type on the next screen. On the left-hand menu, click **Instances**.

6. On the top menu, click **+Add VM**.

7. Fill in the details as per the steps earlier for creating a normal VM, however, ensure the **Resource group** and **Location** are the same as for your dedicated host, and that you choose a **VM Size** that is compatible with your host (See *step 5*).

8. Work through the screens as presented until you get to the **Advanced** section. The first half is the same as a normal VM – that is, you can select extensions to install, however, there is now a new **Host** section – see the following screenshot for an example. Enter the following details:

    a) **Host group**: Select the host group you created.

    b) **Host**: Select the host you created.

    c) **VM Generation: Gen 1**:

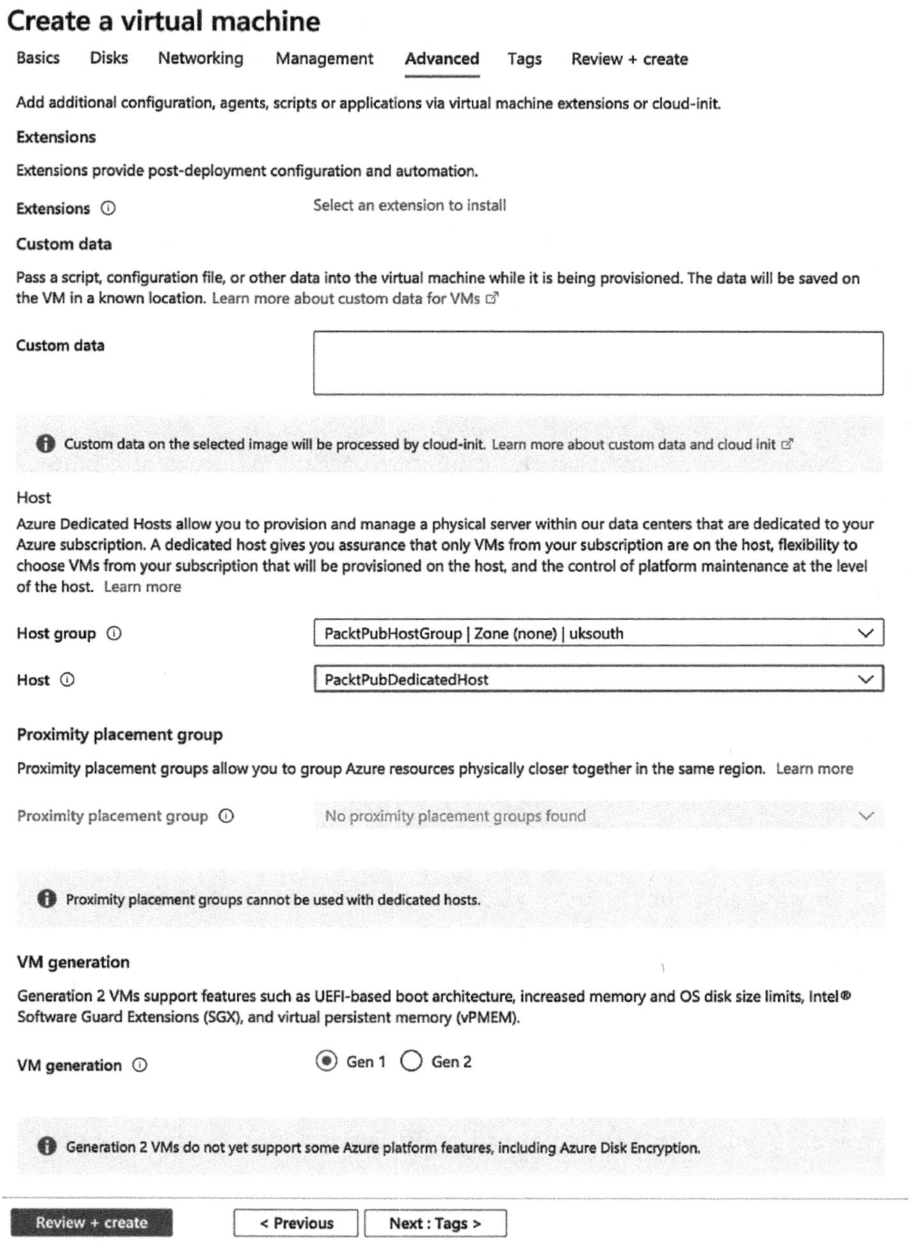

Create a virtual machine

Basics   Disks   Networking   Management   **Advanced**   Tags   Review + create

Add additional configuration, agents, scripts or applications via virtual machine extensions or cloud-init.

Extensions

Extensions provide post-deployment configuration and automation.

Extensions ⓘ   Select an extension to install

Custom data

Pass a script, configuration file, or other data into the virtual machine while it is being provisioned. The data will be saved on the VM in a known location. Learn more about custom data for VMs ☐

Custom data

> ⓘ Custom data on the selected image will be processed by cloud-init. Learn more about custom data and cloud init ☐

Host

Azure Dedicated Hosts allow you to provision and manage a physical server within our data centers that are dedicated to your Azure subscription. A dedicated host gives you assurance that only VMs from your subscription are on the host, flexibility to choose VMs from your subscription that will be provisioned on the host, and the control of platform maintenance at the level of the host. Learn more

Host group ⓘ   PacktPubHostGroup | Zone (none) | uksouth   ∨

Host ⓘ   PacktPubDedicatedHost   ∨

Proximity placement group

Proximity placement groups allow you to group Azure resources physically closer together in the same region. Learn more

Proximity placement group ⓘ   No proximity placement groups found   ∨

> ⓘ Proximity placement groups cannot be used with dedicated hosts.

VM generation

Generation 2 VMs support features such as UEFI-based boot architecture, increased memory and OS disk size limits, Intel® Software Guard Extensions (SGX), and virtual persistent memory (vPMEM).

VM generation ⓘ   ◉ Gen 1   ○ Gen 2

> ⓘ Generation 2 VMs do not yet support some Azure platform features, including Azure Disk Encryption.

[Review + create]   [< Previous]   [Next : Tags >]

Figure 3.25 – Advanced VM settings

9. Click **Review + create**.

10. Click **Create**.

Creating and using dedicated hosts is as easy as standard VMs; however, you gain far greater control and visibility of the hardware. Although the compute resource is dedicated to you, note that the network traffic still uses the same shared networking as all other Azure services.

> **Important note**
> Dedicated hosts are relatively expensive; therefore, ensure you delete the host once you have finished with it.

Summary

This chapter covered the third part of the *Implement and Monitor Azure Infrastructure* objective by covering how to provision and configure VMs for Windows and Linux. You learned about the various aspects and components that are created when you deploy a VM in Azure. We also covered how to automate the deployment of VMs using scale sets and ARM templates, and how this can be done using PowerShell or through automated pipelines using DevOps. We also covered how to configure Azure Disk Encryption for VMs and use Azure Dedicated Hosts to enhance the security and compliance standards of your infrastructure.

In the next chapter, we will continue with this objective by covering how to implement and manage virtual networking.

Questions

Answer the following questions to test your knowledge of the information in this chapter. You can find the answers in the *Assessments* section at the end of this book:

1. Can you use VM scale sets to automate the deployment of multiple VMs?

    a) Yes.

    b) No.

2. Can you use Availability Sets for spreading VMs across update and fault domains?

    a) Yes.

    b) No.

3. Do you have to define resource providers in your ARM templates to deploy the various resources in Azure?

   a) Yes.

   b) No.

4. The compute costs of VMs running on dedicated hosts are billed separately to the host.

   a) True.

   b) False.

Further reading

You can check out the following links for more information about the topics that were covered in this chapter:

- Quickstart: Create a Linux VM in the Azure portal: `https://docs.microsoft.com/en-us/azure/virtual-machines/linux/quick-create-portal`

- VM scale sets documentation: `https://docs.microsoft.com/en-us/azure/virtual-machine-scale-sets/`

- Manage the availability of Windows VMs in Azure: `https://docs.microsoft.com/en-us/azure/virtual-machines/windows/manage-availability`

- Understand the structure and syntax of ARM templates: `https://docs.microsoft.com/en-us/azure/azure-resource-manager/resource-group-authoring-templates`

- Deploy resources with ARM templates and Azure PowerShell: `https://docs.microsoft.com/en-us/azure/azure-resource-manager/resource-group-template-deploy`

- Define resources in ARM templates: `https://docs.microsoft.com/en-us/azure/templates/`

- Using Azure DevOps: `https://azure.microsoft.com/en-gb/services/devops/`

- Understanding Azure Dedicated Host: `https://azure.microsoft.com/en-gb/services/virtual-machines/dedicated-host/`

4
Implementing and Managing Virtual Networking

In the previous chapter, we covered the third part of the *Implement and Monitor Azure Infrastructure objective*. We covered **virtual machines** (**VMs**) in Azure, as well as the different VM sizes that are available for both Azure and Linux. We also learned how to provision VMs and how to create and deploy **Azure Resource Manager** (**ARM**) templates.

This chapter introduces the fourth part of this objective. In this chapter, we are going to focus on implementing and managing virtual networking in Azure. You will learn about the basics of Azure virtual networking, including private and public IP addresses, and learn how to configure subnets, **Virtual Networks** (**VNets**), and public and private IP addresses.

Many **Infrastructure as a Service** (**IaaS**) components in Azure, such as virtual machines, use VNets, and many **Platform as a Service** (**PaaS**) components enable VNet integration to provide more secure communications. It is therefore important to understand how communications flow over VNets and the different ways in which they can be configured.

The following topics will be covered in this chapter:

- Understanding Azure VNet
- Understanding IP addresses
- Configuring subnets and VNets
- Configuring private and public IP addresses
- User-defined routes

Technical requirements

The code example for this chapter uses Azure PowerShell. For more details, visit `https://docs.microsoft.com/en-us/powershell/azure/`.

The source code for our sample application can be downloaded from `https://github.com/PacktPublishing/Microsoft-Azure-Architect-Technologies-Exam-Guide-AZ-303/tree/master/Ch04`.

Understanding Azure VNets

An Azure VNet is a virtual representation of a traditional network that's hosted in the cloud. It is totally software based, whereas traditional networks use cables, routers, and more. VNets provide a secure and isolated environment, and they connect Azure resources to each other. By default, the different resources can't be reached from outside of the VNet. However, you can connect multiple VNets to each other or connect a VNet to your on- premises network. All the Azure resources that are connected to each other inside the same VNet must reside in the same region and subscription.

When you create a VNet, one subnet is automatically created for you. You can create multiple subnets inside the same VNet (with a maximum of 3,000 subnets per VNet). Connecting multiple VNets together is called VNet peering. A maximum of 500 peerings are allowed per virtual network.

Each subnet reserves five Ipv4 addresses for internal use, which are as follows:

- `x.x.x.0`: Network address
- `x.x.x.1`: Reserved for the default gateway
- `x.x.x.2, x.x.x.3`: Reserved to map Azure DNS IPs to the VNet space
- `x.x.x.255`: Network broadcast address

Therefore, the smallest subnet that can be used in Azure is the /29 subnet, which consists of eight addresses, of which only three are useable. The largest is /8, which consists of 16 million addresses.

> **Tip**
>
> For more information on subnetting, please refer to the *Subnet Mask Cheat Sheet*: https://bretthargreaves.com/html-cheatsheet/.

All devices connected to a VNet require an IP address; however, there are different types of IP addresses available, which we will examine next.

Understanding IP addresses

A VNet in Azure can have private and public IP addresses. Private IP addresses are only accessible from within the VNet, though public IP addresses can be accessed from the internet as well. You can access private IP addresses from a VPN gateway or an ExpressRoute connection. Both private and public IP addresses can be static or dynamic, but when you create a new VNet, the IP address is static by default. You can change the IP address to *static* from the Azure portal, PowerShell, or **command-line interface (CLI)**. The following are the two states of an IP address:

- **Dynamic**: Dynamic IP addresses are assigned by Azure automatically and are selected from the configured subnet's address range from the virtual network where the Azure resource resides. The IP address is assigned to the Azure resource upon creation or start. The IP address will be released when the resource is stopped and deallocated (when you stop the VM from the Azure portal, the VM is deallocated automatically) and added back to the pool of available addresses inside the subnet by Azure.

- **Static**: Static IP addresses (private and public) are preassigned and will remain the same until you delete the assignment. You can select a static private IP address manually. They can only be assigned to non-internet-facing connections, such as an internal load balancer. You can assign a private static IP address to a connection on your on-premises network or to an ExpressRoute circuit. Public static IP addresses are created by Azure automatically and can be assigned to internet-facing connections such as an external load balancer.

Public IP addresses

Public IP addresses can be used for internal communication between Azure services and external communication over the internet. Azure now supports dual-stacked IPV4/IPV6 addressing; however, the subnets for IPV6 must be /64 in size to ensure future compatibility.

When an Azure resource is started or created, Azure will assign the public IP address to the network interface of the VNet. When an outbound connection is initiated, Azure will map the private IP address to the public IP addresses, which is also known as **Source Network Address Translation (SNAT)**.

Azure assigns the public IP address to the network interface when the Azure resource is started or created. When an outbound connection is initiated, Azure will map the private IP address to the public IP address (SNAT). Returning traffic to the resource is allowed as well. Public IP addresses are typically used for VMs, internet-facing load balancers, VPN gateways, and application gateways. A default maximum of 10 dynamic public IP addresses and 10 static public IP addresses is allowed per basic subscription. The default limits vary by subscription offer (such as Pay-As-You-Go, CSP, or Enterprise Agreement); however, they can be extended by contacting support.

Private IP addresses

Private IP addresses are typically used for VMs, internal load balancers, and application gateways. Because a VPN is always internet-facing, it cannot have a private IP address. You can have a maximum of 65,536 private IP addresses per VNet, and you can create multiple VNets (with a maximum number of 1,000 VNets per subscription).

> **Important note**
> These limits are based on the default limits from the following page:
> `https://docs.microsoft.com/en-us/azure/azure-resource-manager/management/azure-subscription-service-limits#networking-limits`. You can open a support request to raise these limits.

Now that we have some background information about the various networking aspects in Azure, we can configure a VNet with a subnet.

Configuring VNets and subnets

In this section, we are going to create and configure a VNet and a subnet from the Azure portal. We created both of these in earlier demonstrations, for instance, when we created VMs in *Chapter 3, Implementing and Managing Virtual Machines*. Now, we are going to cover this topic in more detail.

Here, we are going to configure a VNet and a subnet using PowerShell. Therefore, we have to follow these steps:

1. First, we need to log into our Azure account, as follows:

   ```
   Connect-AzAccount
   ```

2. If necessary, select the right subscription, as follows:

   ```
   Select-AzSubscription -SubscriptionId "********-****-
   ****-****-
   ***********"
   ```

3. Create a resource group for the VNet as follows:

   ```
   New-AzResourceGroup -Name PacktVNetResourceGroup
   -Location EastUS
   ```

4. Next, we can create the VNet, as follows:

   ```
   $virtualNetwork = New-AzVirtualNetwork `
   -ResourceGroupName PacktVNetResourceGroup `
   -Location EastUS `
   -Name PacktVirtualNetwork `
   -AddressPrefix 10.0.0.0/16
   ```

5. Then, we can create the subnet, as follows:

   ```
   $subnetConfig = Add-AzVirtualNetworkSubnetConfig `
   -Name default `
   -AddressPrefix 10.0.0.0/24 `
   -VirtualNetwork $virtualNetwork
   ```

6. Finally, we can associate the subnet with the virtual network, as follows:

   ```
   $virtualNetwork | Set-AzVirtualNetwork
   ```

Now, we have created a VNet and a subnet via PowerShell. We will use this for further demonstration purposes in this chapter. In the next section, we are going to configure both a private and a public IP address in PowerShell and associate them with this VNet.

Configuring private and public IP addresses

In this section, we are going to configure both a private and a public IP address. When we created the VNet, a private IP address was created for us automatically by Azure. However, we are going to create another in this demonstration and associate it, along with the public IP address, to a **network interface card** (**NIC**). To configure a private and public IP address from PowerShell, you have to perform the following steps:

1. In the same PowerShell window, add the following code to retrieve the VNet and subnet configuration:

```
$vnet = Get-AzVirtualNetwork -Name PacktVirtualNetwork -
ResourceGroupName PacktVNetResourceGroup
$subnet = Get-AzVirtualNetworkSubnetConfig -Name default
- VirtualNetwork $vnet
```

2. Next, create a private and a public IP address and assign them to the configuration, as follows:

```
$publicIP = New-AzPublicIpAddress `
-Name PacktPublicIP `
-ResourceGroupName PacktVNetResourceGroup `
-AllocationMethod Dynamic `
-Location EastUS

$IpConfig = New-AzNetworkInterfaceIpConfig `
-Name PacktPrivateIP `
-Subnet $subnet `
-PrivateIpAddress 10.0.0.4 `
-PublicIPAddress $publicIP `
-Primary
```

3. Then, create a network interface and assign the configuration to it, as follows:

```
$NIC = New-AzNetworkInterface `
-Name PacktNIC `
-ResourceGroupName PacktVNetResourceGroup `
-Location EastUS `
-IpConfiguration $IpConfig
```

Now, we have configured an NIC, a public and a private IP address, and associated them with the VNet that we created in the previous section. Next, we will look at how we can control the flow of traffic between networks.

User-defined routes

When you create subnets, Azure creates system routes that enable all the resources in a subnet so that they can communicate with each other. Every subnet has a default system route table, which contains the following minimum routes:

- **Local VNet**: This is a route for resources that reside in the VNet. For these routes, there is no next hop address. If the destination IP address contains the local VNet prefix, traffic is routed there.

- **On-premises**: This is a route for defined on-premises address spaces. For this route, the next hop address will be the VNet gateway. If the destination IP address contains the on-premises address prefix, traffic is routed there.

- **Internet**: This route is for all the traffic that goes over the public internet, and the internet gateway is always the next hop address. If the destination IP address doesn't contain the VNet or on-premises prefixes, traffic is routed to the internet using **Network Address Translation (NAT)**.

You can override these system routes by creating **User-Defined Routes** (**UDRs**). This way, you can force traffic to follow a particular route. For instance, you have a network that consists of two subnets and you want to add a VM that is used as a **Demilitarized Zone** (**DMZ**) and has a firewall installed on it. You only want traffic to go through the firewall and not between the two subnets. To create UDRs and enable IP forwarding, you have to create a routing table in Azure. When this table is created and there are custom routes in there, Azure prefers the custom routes over the default system routes.

> **Important note**
>
> A demilitarized zone, or **DMZ**, is a network that has been firewalled off from, but is still connected to, your internal network in order to provide a public-facing service. In this way, you can still send traffic between your internal network and the DMZ, but it is more controlled and provides protection for your internal devices in the event the DMZ is compromised.

Creating UDRs

To create UDRs, follow these steps:

1. Navigate to the Azure portal by going to `https://portal.azure.com/`.

2. Click **Create a resource**, type `Route Table` into the search bar, and create a new one.

3. Add the following values, as shown in the following screenshot:

    a) **Name**: `PacktRoutingTable`.

    b) **Subscription**: Select a subscription.

    c) **Resource Group**: Create a new one and call it `PacktRoutingTable`.

    d) **Location**: **(US) East US**:

Figure 4.1 – Creating a new route table

4. Click **Create**.

5. A new and empty route table will be created. After creation, open the **Overview** blade of the route table. To add custom routes, click **Routes** in the left menu, as follows:

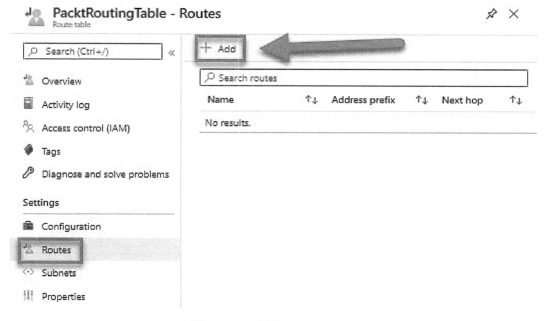

Figure 4.2 – Adding a new route

6. In this example, we want all the internet traffic to go through the firewall. To do this, add the following values, as shown in the following screenshot:

a) **Name**: `DefaultRoute`.

b) **Address prefix**: `0.0.0.0/0`.

c) **Next hop type**: **Virtual appliance**; this is the firewall.

d) **Next hop address**: `10.1.1.10`. This will be the internal IP address of the firewall:

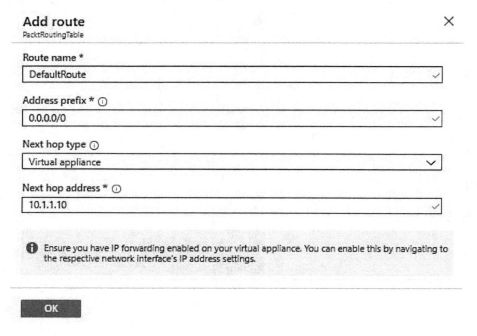

Figure 4.3 – Adding a route to the table

7. Click **OK**. The route will be created for you.

> **Tip**
>
> For more detailed instructions on how to create UDRs and virtual appliances, refer to the following tutorial: `https://docs.microsoft.com/en-us/azure/virtual-network/tutorial-create-route-table-portal`.

In this section, we created a custom route table and added a route to it that routes all the traffic to a firewall. This concludes this chapter.

Summary

In this chapter, we completed the fourth part of the *Implement and Monitor Azure Infrastructure* objective by covering the basics of Azure virtual networking, including private and public IP addresses. You are now able to configure subnets, VNets, and public and private IP addresses with the knowledge gained from this chapter.

In the next chapter, we will continue with the fifth part of the *Implement and Monitor Azure Infrastructure* objective by covering how to create connectivity between virtual networks.

Questions

Answer the following questions to test your knowledge of the information in this chapter.

You can find the answers in the *Assessments* section at the end of this book:

1. There is a maximum of 120 dynamic public IP addresses and 10 static public IP addresses per subscription.

   a) Yes

   b) No

2. Can you create custom route tables to adjust the routing between the different resources inside your VNets?

   a) Yes

   b) No

3. Can you assign IPv6 addresses in Azure?

   a) Yes

   b) No

Further reading

Check out the following links if you want to find out more about the topics that were covered in this chapter:

- Azure networking limits: https://docs.microsoft.com/en-us/azure/azure-resource-manager/management/azure-subscription-service-limits#networking-limits

- Quickstart: Creating a virtual network using the Azure portal: `https://docs.microsoft.com/en-us/azure/virtual-network/quick-create-portal`

- Tutorial: Routing network traffic with a route table using the Azure portal: `https://docs.microsoft.com/en-us/azure/virtual-network/tutorial-create-route-table-portal`

5
Creating Connectivity between Virtual Networks

In the previous chapter, we covered the fourth part of the *Implementing and Monitoring Azure Infrastructure* objective. We also previously covered virtual networking in Azure and how to implement and manage it. By doing this, you learned about the basics of virtual networking, including private and public IP addresses, and more.

In this chapter, you will learn how to create connectivity between virtual networks. We are going to cover how to create and configure **Virtual Network** (**VNet**) peering, how to create and configure VNet-to-VNet, how to verify VNet connectivity, and the differences between VNet peering, VNet-to-VNet connections, and when to use what type of connection.

The following topics will be covered in this chapter:

- Understanding VNet peering
- Creating and configuring VNet peering
- Understanding VNet-to-VNet
- Creating and configuring VNet-to-VNet
- Verifying VNet connectivity
- VNet peering versus VNet-to-Vnet

Let's get started!

Technical requirements

This chapter will use Azure PowerShell (`https://docs.microsoft.com/en-us/powershell/azure/install-az-ps`) for the examples provided.

The source code for our sample application can be downloaded from `https://github.com/PacktPublishing/Microsoft-Azure-Architect-Technologies-Exam-Guide-AZ-303/tree/master/Ch05`.

Understanding VNet peering

VNet peering is a mechanism that seamlessly connects two VNets in the same region through Azure's backbone infrastructure. Once peered, the VNets appear as one for connectivity purposes, just like routing traffic between **virtual machines** (**VMs**) that are created in the same VNet. The VMs that reside in the peered VNets communicate with each other using private IP addresses. VNet peering is the easiest and most effective way to connect two VNets.

Azure supports the following types of peering:

- **VNet peering**: This is used for connecting VNets in the same Azure region.
- **Global VNet peering**: This is used for connecting VNets across different Azure regions.

The network traffic between peered VNets is private. The traffic is kept on the Microsoft backbone network completely, so there is no need to use any additional gateways or to route traffic over the public internet. There is also no encryption required in the communication between the peered VNets. It uses a low-latency, high-bandwidth connection between the resources in the different virtual networks.

You can use VNet peering to connect VNets that are created through the Resource Manager and the classic deployment model. This allows you to transfer data across Azure regions and subscriptions.

> **Important note**
> The other way to connect VNets is to set up VNet-to-VNet connections. This requires you to deploy gateways in each of the connected VNets, which are both connected by a tunnel. This limits the connection speeds to the bandwidth of the gateway.

Creating and configuring VNet peering

In the following demonstration, we are going to create and configure VNet peering from the Azure portal. We need two VNets for this. Here, we are going to use the VNet that we created in the first demonstration, along with the resource group that we created for it, in the previous chapter, and then create an additional VNet that has a different address space than the first VNet. Note that you can't use overlapping address spaces when you peer two VNets together.

To create the VNet and set up VNet peering from the Azure portal, go through the following steps:

1. Navigate to Azure portal by going to `https://portal.azure.com/`.

2. Click **Create a resource | Networking | Virtual network**. Create a new VNet.

3. Add the following values:

    a) **Subscription**: Choose your subscription.

    b) **Resource group**: **Create new** – `PacktVNetResourceGroup`.

    c) **Name**: `PacktVNetPeering`.

    d) **Location**: **East US**.

4. Click **Next: IP Addresses**, then enter the following values:

    a) **IPV4 Address space**: Leave as `10.1.0.0/16`.

    b) **Add Ipv6 Address space**: Leave unticked.

    c) **Subnet**: Leave as `default`.

    d) **Address range**: Leave as `10.1.0.0/24`.

5. Next, click **Next: Security** and leave all the default values as-is.

6. Click **Review + Create**.

7. Click **Create**.

8. With that, the VNet has been created. Now, open the **VNet Overview** blade of the VNet that we created in the previous demonstration of this chapter, which is called `PacktVirtualNetwork`, as follows:

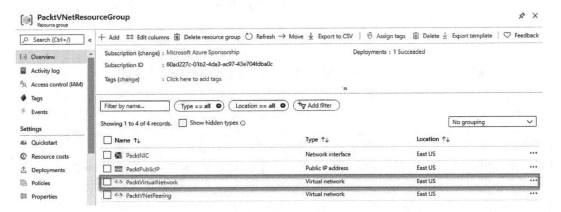

Figure 5.1 – VNet Overview blade

9. Then, under **Settings**, select **Peerings**. Click **Add** at the top of the menu, as shown in the following screenshot:

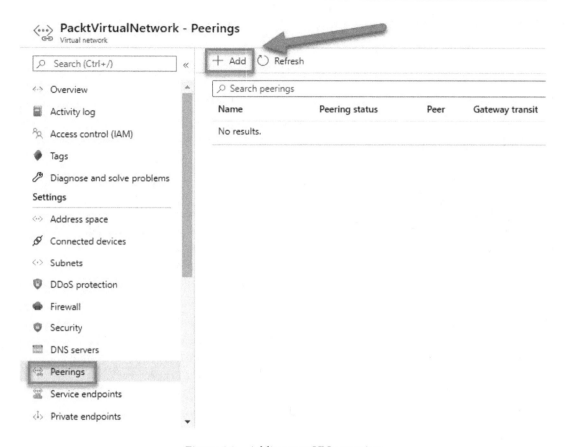

Figure 5.2 – Adding new VNet peering

10. In the **Add peering** blade, under **This virtual network**, add the following values:

   a) **Peering link name**: SourceToDestinationPeering.

   b) **Traffic to remote virtual network**: Leave as **Allow (default)**.

   c) **Traffic forwarded from remote virtual network**: Leave as **Allow (default)**.

   d) **Virtual network gateway**: Leave as **None (default)**.

11. Under **Remote virtual network**, add the following values:

   a) **Peering link name**: DestinationToSourcePeering.

   b) **Virtual network deployment model: Resource Manager**.

   c) **Subscription**: Choose your subscription.

   d) **Virtual Network**: PacktVNETPeering.

e) **Traffic to remote virtual network**: Leave as **Allow (default)**.

f) **Traffic forwarded from remote virtual network**: Leave as **Allow (default)**.

g) **Virtual network gateway**: Leave as **None (default)**.

12. Finally, click **Add**.

We have now configured VNet peering from Azure portal. In the next section, we are going to look at VNet-to-VNet connections.

Understanding VNet-to-VNet

A VNet-to-VNet connection is a simple way to connect multiple VNets together. Connecting a virtual network to another virtual network is similar to creating a site-to-site IPSec connection to an on-premises environment. Both connection types use the Azure **virtual private network (VPN)** gateway. This VPN gateway provides a secure IPSec/IKE tunnel, and each side of the connection communicates in the same way. The only difference is in the way the local network gateway is configured.

The local network gateway address space is automatically created and populated when a VNet-to-VNet connection is created. When the address space of one VNet has been updated, this results in the other VNet automatically routing its traffic to the updated address space. This makes it more efficient and quicker to create a VNet-to-VNet connection instead of a **Site-to-Site (S2S)** connection.

There are a couple of reasons why you might want to set up a VNet-to-VNet connection, as follows:

- **Cross-region geo-redundancy and geo-presence**: Using a VNet-to-VNet connection, you can set up your own geo-replication or synchronization without going over internet-facing endpoints. You can set up highly available workloads with geo-redundancy across multiple Azure regions using Azure Load Balancer and Azure Traffic Manager.

- **Regional multi-tier applications with isolation or administrative boundaries**: In the case of isolation or administrative requirements, you can set up multi-tier applications with multiple networks in the same region. Then those networks can be connected.

You can combine VNet-to-VNet communication with multisite configurations. Network topologies that combine inter-virtual network connectivity with cross-premises connectivity can be established using these configurations, as shown in the following diagram:

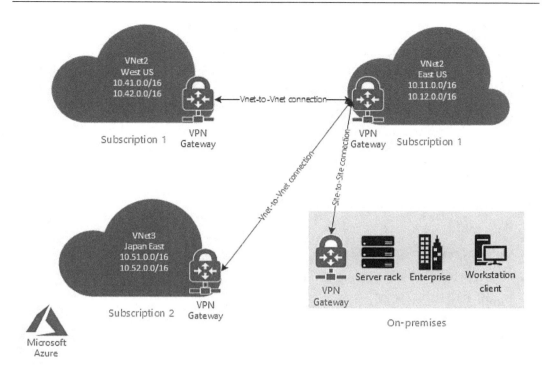

Figure 5.3 – Combined VNet-to-VNet communication

In the next section, we'll learn how to create a VNet-to-VNet connection.

Creating and configuring VNet-to-VNet

In this demonstration, we are going to create a VNet-to-VNet connection using PowerShell.

In the following steps, we will create two VNets, along with their gateway configurations and subnets. We will then create a VPN connection between the two VNets. We are going to set up a connection in the same Azure subscription. In the first step, we are going to plan the IP address ranges for the network configurations. To do this, we need to make sure that none of the VNet or local ranges overlap.

Planning IP ranges

We are going to use the following values for the IP ranges:

The values for **PacktVNet1** are as follows:

- **VNet Name**: `PacktVNet1`
- **Resource Group**: `PacktResourceGroup1`
- **Location**: **East US**
- **PacktVNet1**: `10.11.0.0/16` and `10.12.0.0/16`
- **Frontend**: `10.11.0.0/24`
- **Backend**: `10.12.0.0/24`
- **Gateway Subnet**: `10.12.255.0/27`
- **Gateway Name**: `PacktVNet1Gateway`
- **Public IP**: `PacktVNet1GWIP`
- **VPN Type**: `RouteBased`
- **Connection**: `VNet1toVNet2`
- **Connection Type**: `VNet2VNet`

The values for **TestVNet2** are as follows:

- **VNet Name**: `PacktVNet2`
- **Resource Group**: `PacktResourceGroup2`
- **Location**: **West US**
- **PacktVNet2**: `10.41.0.0/16` and `10.42.0.0/16`
- **Frontend**: `10.41.0.0/24`
- **Backend**: `10.42.0.0/24`
- **Gateway Subnet**: `10.42.255.0/27`
- **Gateway Name**: `PacktVNet2Gateway`
- **Public IP**: `PacktVNet2GWIP`
- **VPN Type**: `RouteBased`
- **Connection**: `VNet2toVNet1`
- **Connection Type**: `VNet2VNet`

Now that we have planned our IP configurations, we can create the VNets.

Creating PacktVNet1

To create `PacktVNet1`, perform the following steps:

1. First, we need to log into our Azure account:

   ```
   Connect-AzAccount
   ```

2. If necessary, select the right subscription:

   ```
   Select-AzSubscription -SubscriptionId "********-****-
   ****-****-************"
   ```

3. Define the variables for the first VNet:

   ```
   $RG1 = "PacktResourceGroup1"
   $Location1 = "East US"
   $VNetName1 = "PacktVNet1"
   $FESubName1 = "FrontEnd"
   $BESubName1 = "Backend"
   $VNetPrefix01 = "10.11.0.0/16"
   $VNetPrefix02 = "10.12.0.0/16"
   $FESubPrefix1 = "10.11.0.0/24"
   $BESubPrefix1 = "10.12.0.0/24"
   $GWSubPrefix1 = "10.12.255.0/27"
   $GWName1 = "PacktVNet1Gateway"
   $GWIPName1 = "PacktVNet1GWIP"
   $GWIPconfName1 = "gwipconf1"
   $Connection01 = "VNet1toVNet2"
   ```

4. Create a resource group:

   ```
   New-AzResourceGroup -Name $RG1 -Location $Location1
   ```

5. Now, we need to create the subnet configurations for `PacktVNet1`. For this demonstration, we are going to create a VNet, called `PacktVNet1`, and three subnets, called `FrontEnd`, `Backend`, and `GatewaySubnet`. It is important to name your gateway subnet `GatewaySubnet`; otherwise, the gateway creation process will fail.

It is recommended that you create a gateway subnet using a /27. This includes more addresses that will accommodate possible additional configurations that you may want to make in the future:

```
$fesub1 = New-AzVirtualNetworkSubnetConfig -Name
$FESubName1 -AddressPrefix $FESubPrefix1

$besub1 = New-AzVirtualNetworkSubnetConfig -Name
$BESubName1 -AddressPrefix $BESubPrefix1

$gwsub1 = New-AzVirtualNetworkSubnetConfig -Name
"GatewaySubnet" -AddressPrefix $GWSubPrefix1
```

6. Create PacktVNet1:

```
New-AzVirtualNetwork -Name $VNetName1 `
-ResourceGroupName $RG1 `
-Location $Location1 `
-AddressPrefix $VNetPrefix01,$VNetPrefix02 `
-Subnet $fesub1,$besub1,$gwsub1
```

7. Request a public IP address to be allocated to the gateway that you will create for PacktVNet1. Note that you cannot specify the IP address that you want to use. It's dynamically allocated to your gateway:

```
$gwpip1 = New-AzPublicIpAddress `
-Name $GWIPName1 `
-ResourceGroupName $RG1 `
-Location $Location1 `
-AllocationMethod Dynamic
```

8. Create the gateway configuration. The gateway configuration defines the subnet and the public IP address that will be used:

```
$vnet1 = Get-AzVirtualNetwork `
-Name $VNetName1 `
-ResourceGroupName $RG1
$subnet1 = Get-AzVirtualNetworkSubnetConfig `
-Name "GatewaySubnet" `
-VirtualNetwork $vnet1
$gwipconf1 = New-AzVirtualNetworkGatewayIpConfig `
-Name $GWIPconfName1 `
-Subnet $subnet1 `
```

```
-PublicIpAddress $gwpip1
```

9. Create the gateway for `PacktVNet1`. VNet-to-VNet configurations require a `RouteBased` setting for `VpnType`:

```
New-AzVirtualNetworkGateway `
-Name $GWName1 `
-ResourceGroupName $RG1 `
-Location $Location1 `
-IpConfigurations $gwipconf1 `
-GatewayType Vpn `
-VpnType RouteBased `
-GatewaySku VpnGw1
```

> **Important note**
> Creating a gateway can take 45 minutes or more, depending on the selected gateway SKU.

With that, we have created our first VNet. In the next section, we are going to create `PacktVNet2`.

Creating PacktVNet2

To create `PacktVNet2`, perform the following steps:

1. Define the variables for the first VNet:

```
$RG2 = "PacktResourceGroup2"
$Location2 = "West US"
$VNetName2 = "PacktVNet2"
$FESubName2 = "FrontEnd"
$BESubName2 = "Backend"
$VnetPrefix11 = "10.41.0.0/16"
$VnetPrefix12 = "10.42.0.0/16"
$FESubPrefix2 = "10.41.0.0/24"
$BESubPrefix2 = "10.42.0.0/24"
$GWSubPrefix2 = "10.42.255.0/27"
$GWName2 = "PacktVNet2Gateway"
$GWIPName2 = "PacktVNet1GWIP"
```

```
$GWIPconfName2 = "gwipconf2"
$Connection2 = "VNet2toVNet1"
```

2. Create a resource group:

```
New-AzResourceGroup -Name $RG2 -Location $Location2
```

3. Create subnet configurations for PacktVNet2:

```
$fesub2 = New-AzVirtualNetworkSubnetConfig -Name
$FESubName2 -AddressPrefix $FESubPrefix2
```

```
$besub2 = New-AzVirtualNetworkSubnetConfig -Name
$BESubName2 -AddressPrefix $BESubPrefix2
```

```
$gwsub2 = New-AzVirtualNetworkSubnetConfig -Name
"GatewaySubnet" -AddressPrefix $GWSubPrefix2
```

4. Create PacktVNet2:

```
New-AzVirtualNetwork `
-Name $VnetName2 `
-ResourceGroupName $RG2 `
-Location $Location2 `
-AddressPrefix $VnetPrefix11,$VnetPrefix12 `
-Subnet $fesub2,$besub2,$gwsub2
```

5. Request a public IP address:

```
$gwpip2 = New-AzPublicIpAddress `
-Name $GWIPName2 `
-ResourceGroupName $RG2 `
-Location $Location2 `
-AllocationMethod Dynamic
```

6. Create the gateway configuration:

```
$vnet2 = Get-AzVirtualNetwork `
-Name $VnetName2 `
-ResourceGroupName $RG2
$subnet2 = Get-AzVirtualNetworkSubnetConfig `
-Name "GatewaySubnet" `
-VirtualNetwork $vnet2
```

```
$gwipconf2 = New-AzVirtualNetworkGatewayIpConfig `
-Name $GWIPconfName2 `
-Subnet $subnet2 `
-PublicIpAddress $gwpip2
```

7. Create the gateway:

```
New-AzVirtualNetworkGateway -Name $GWName2 `
-ResourceGroupName $RG2 `
-Location $Location2 `
-IpConfigurations $gwipconf2 `
-GatewayType Vpn `
-VpnType RouteBased `
-GatewaySku VpnGw1
```

Wait for the gateway to be created. Once it's been created, we can create connections between the VNets.

Creating connections

To create connections between our VNets, perform the following steps:

1. First, get a reference to the two gateways:

```
$vnet1gw = Get-AzVirtualNetworkGateway -Name $GWName1 -
ResourceGroupName $RG1
$vnet2gw = Get-AzVirtualNetworkGateway -Name $GWName2 -
ResourceGroupName $RG2
```

2. Next, we need to create a connection. Make sure that the keys match:

```
New-AzVirtualNetworkGatewayConnection `
-Name $Connection01 `
-ResourceGroupName $RG1 `
-VirtualNetworkGateway1 $vnet1gw `
-VirtualNetworkGateway2 $vnet2gw `
-Location $Location1 `
-ConnectionType Vnet2Vnet `
-SharedKey 'AzurePacktGateway'


New-AzVirtualNetworkGatewayConnection `
```

```
-Name $Connection02 `
-ResourceGroupName $RG2 `
-VirtualNetworkGateway1 $vnet2gw `
-VirtualNetworkGateway2 $vnet1gw `
-Location $Location2 `
-ConnectionType Vnet2Vnet `
-SharedKey 'AzurePacktGateway'
```

In this demonstration, we configured a VNet-to-VNet connection. To do this, we created two VNets, both with a virtual network gateway. We also set up the connections between the gateways. In the next section, we are going to learn how to verify the network connectivity of your VNet-to-VNet connection.

Verifying your virtual network's connectivity

To verify whether the VNet-to-VNet connection has been successfully set up, perform the following steps:

1. In PowerShell, use the following code to verify the network connection:

```
Get-AzVirtualNetworkGatewayConnection -Name $Connection01
- ResourceGroupName $RG1
```

```
Get-AzVirtualNetworkGatewayConnection -Name $Connection02
- ResourceGroupName $RG2
```

2. If the connection status is **Connected**, then this means the connection was successful:

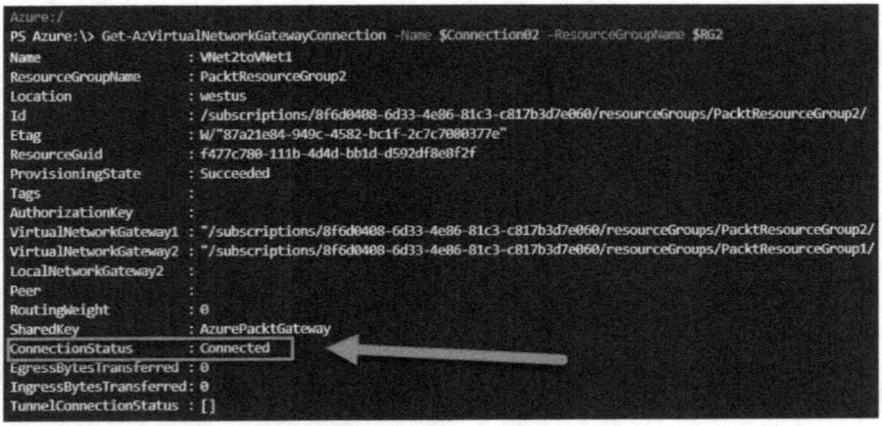

Figure 5.4 – Verifying the connection

In this demonstration, we verified our VNet-to-VNet connection. In the next section, we will look at the differences between VNet peering and VNet-to-VNet connections.

VNet peering versus VNet-to-VNet connections

VNet peering and VNet-to-VNet both offer ways to connect VNets. But based on your specific scenario and needs, you might want to pick one over the other:

- **VNet peering**: This offers high-bandwidth, low-latency connections, which are useful in cross-region data replication and database failover scenarios. The traffic remains on the Microsoft backbone and is completely private; that's why customers with strict data security requirements prefer to use VNet peering, since public internet is not involved. There are also no extra hops because a gateway isn't used here, which ensures low-latency connections. You can peer up to 500 VNets to one VNet. The ingress and egress traffic is charged using VNet peering. In region/cross-region scenarios, VNet peering is recommended.

- **VPN gateways**: These provide a limited bandwidth connection and are useful in scenarios where encryption is needed, but bandwidth restrictions are tolerable. These bandwidth limitations vary, based on the type of gateway, and range from 100 Mbps to 10 Gbps. In these scenarios, customers are also not as latency-sensitive. Each VNet can only have one VPN gateway, and the gateway and egress are charged. A public IP address is bound to the gateway.

The following table shows some of the most important differences between them:

	Vnet Peering	**VPN Gateway**
Cross-Region Support?	Yes, via global VNet peering	Yes
Limits	Up to 500 VNets	Only one VPN gateway
Pricing	Ingress/egress charges	Gateway and egress charged
Encryption	Software encryption recommended	Yes – using an IPSec/IKE policy
Bandwidth Limitations?	None	100 Mbps – 1.25 Gbps, depending on SKU
Private?	Yes	No – public IP involved
Example Use Cases	Data replication, database failover	Encryption-specific scenarios where latency is not a factor

In this section, we covered the scenarios where VNet peering and VPN gateways are the most suitable. This concludes this chapter.

Summary

In this chapter, we covered the fifth part of the *Implement and Monitor Azure Infrastructure* objective by covering how to create connectivity between virtual networks in Azure. With the knowledge you've gained from this chapter, you can now create and configure VNet peering, create and configure VNet-to-VNet connections, and verify virtual network connectivity. You now also understand the main differences between VNet peering and VNet-to-VNet connections.

In the next chapter, we will continue with this objective by covering how to manage Azure **Active Directory (AD)**.

Questions

Answer the following questions to test your knowledge of the information provided in this chapter. You can find the answers in the *Assessments* section at the end of this book:

1. When you use VNet peering, do you have to create an Azure VPN gateway to connect both of the VNets to each other?

    a) Yes

    b) No

2. Is a VNet-to-VNet connection most suitable in scenarios where you don't want to use a public IP address?

    a) Yes

    b) No

3. VNet peering doesn't have any bandwidth limitations.

    a) Yes

    b) No

Further reading

Check out the following links for more information about the topics that were covered in this chapter:

*   Virtual network peering: `https://docs.microsoft.com/en-us/azure/virtual-network/virtual-network-peering-overview`.

*   What is a VPN gateway?: `https://docs.microsoft.com/en-us/azure/vpn-gateway/vpn-gateway-about-vpngateways`.

- Configuring a VNet-to-VNet VPN gateway connection using the Azure portal: `https://docs.microsoft.com/en-us/azure/vpn-gateway/vpn-gateway-howto-vnet-vnet-resource-manager-portal`.

6
Managing Azure Active Directory (Azure AD)

In the previous chapter, we covered the fifth part of the *Implement and Monitor Azure Infrastructure* objective. We covered how to create connectivity between virtual networks in Azure, as well as how to create and configure VNet peering, how to create and configure VNet-to-VNet, and how to verify network connectivity. Finally, we learned about the differences between VNet peering and VNet-to-VNet.

This chapter covers the sixth part of this objective. In this chapter, we are going to cover how to create and manage users and groups in **Azure Active Directory** (**Azure AD**). You will learn how to manage this from the Azure portal and how to perform bulk updates inside your Azure AD tenant. You will learn how to configure a self-service password reset for your users in order to reduce user management overhead. By doing so, we will cover Conditional Access policies. We are also going to cover Azure AD join and how you can manage devices that have been registered or joined in Azure AD. To conclude this chapter, we will add a custom domain to Azure AD.

The following topics will be covered in this chapter:

- Understanding Azure AD
- Creating and managing users and groups
- Adding and managing guest accounts
- Performing bulk user updates
- Configuring a self-service password reset
- Understanding Conditional Access policies and security defaults
- Working with Azure AD join
- Adding custom domains

Understanding Azure AD

Azure AD offers a directory and identity management solution from the cloud. It offers traditional username and password identity management, as well as roles and permissions management. On top of that, it offers more enterprise-grade solutions, including **Multi-Factor Authentication** (**MFA**) and application monitoring, solution monitoring, and alerting.

Azure AD can easily be integrated with your on-premises Active Directory to create a hybrid infrastructure.

Azure AD offers the following pricing plans:

- **Free**: This offers the most basic features, such as support for up to 500,000 objects, **single sign-on** (**SSO**), Azure B2B for external users, support for Azure AD Connect synchronization, self-service password change, groups, MFA (app authenticator only), and standard security reports.
- **Basic**: This offers no object limit, has an SLA of 99.9%, a self-service password reset, company branding features, and support for the application proxy.
- **Premium P1**: This offers advanced reporting, full MFA, conditional access, MDM auto-enrollment, cloud app discovery, and Azure AD Connect Health.
- **Premium P2**: This offers identity protection and privileged identity management.

> **Tip**
>
> For a detailed overview of the different pricing plans and all the features that are offered for each plan, you can refer to the following pricing page: `https://azure.microsoft.com/en-us/pricing/details/active-directory/`.
>
> Note that Azure AD Premium is part of the enterprise mobility and security suite.

In the next section, we are going to create and manage users and groups inside an Azure AD tenant.

Creating and managing users and groups

In this section, we are going to create and manage users and groups in the Azure portal. You can also use PowerShell and its CLI to create users.

> **Important note**
>
> We aren't going to create an Azure AD tenant in this section since it's assumed that you already have one. If you need to create an Azure AD tenant, you can refer to the following tutorial: `https://docs.microsoft.com/en-us/azure/active-directory/develop/quickstart-create-new-tenant`.
>
> You can also create multiple Azure AD tenants in one subscription. These directories can be used for development and testing purposes, for example.

Creating users in Azure AD

We will begin by creating a couple of users in our Azure AD tenant from the Azure portal. To do this, perform the following steps:

1. Navigate to the Azure portal by going to `https://portal.azure.com`.

2. In the left menu, select **Azure Active Directory.**

3. In the **Overview** blade of Azure AD, in the left menu, select **Users | All users**. Then, select **+ New user** from the top menu, as follows:

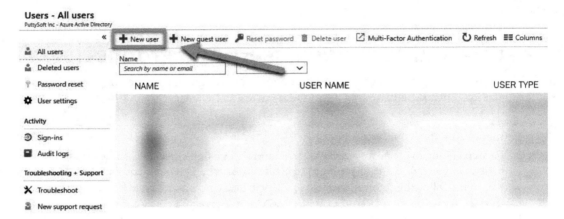

Figure 6.1 – Creating a new user

4. Next, we are going to create three users. Add the following values, which can be seen in the following screenshot:

a) **Name**: `PacktUser1`.

b) **Username**: The username is the identifier that the user enters to sign in to Azure AD. Use the domain name that has been configured for you and add this to the end of the username. In my case, this is `PacktUser1@sjoukjezaal.com`.

c) **Profile**: Here, you can create a new profile for your user. Add the **First name**, **Last name**, **Job title**, and **Department** boxes. After that, click **Create**, as follows:

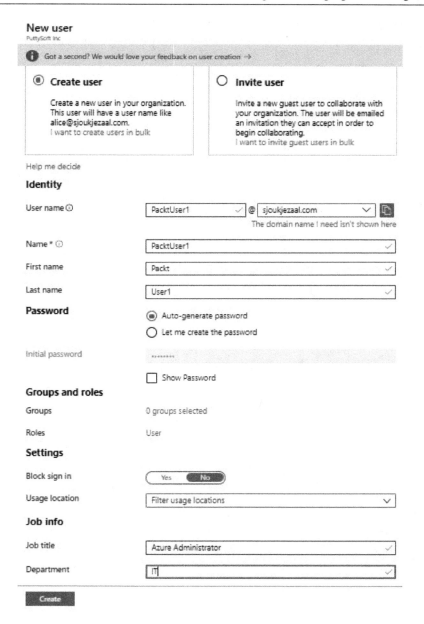

Figure 6.2 – Creating a profile for your user

d) **Group**: You can also add your user to a group from here. We are going to do this in the next section, so you can skip this part for now.

e) **Roles**: Here, you can assign the user to an RBAC role. Select **Directory readers** from the list, as follows:

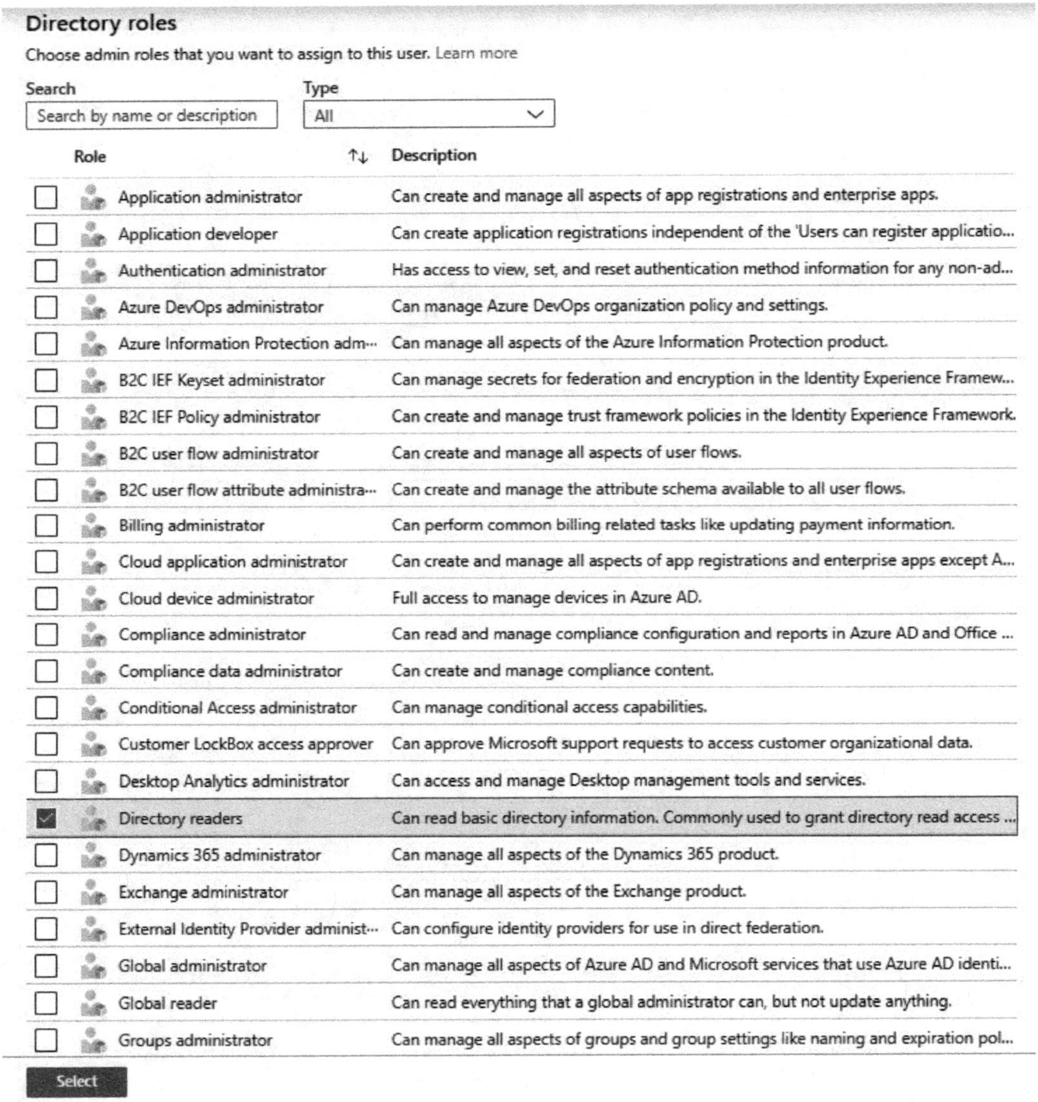

Directory roles

Choose admin roles that you want to assign to this user. Learn more

Search	Type
Search by name or description	All ⌄

	Role ↑↓	Description
☐	Application administrator	Can create and manage all aspects of app registrations and enterprise apps.
☐	Application developer	Can create application registrations independent of the 'Users can register applicatio...
☐	Authentication administrator	Has access to view, set, and reset authentication method information for any non-ad...
☐	Azure DevOps administrator	Can manage Azure DevOps organization policy and settings.
☐	Azure Information Protection adm···	Can manage all aspects of the Azure Information Protection product.
☐	B2C IEF Keyset administrator	Can manage secrets for federation and encryption in the Identity Experience Framew...
☐	B2C IEF Policy administrator	Can create and manage trust framework policies in the Identity Experience Framework.
☐	B2C user flow administrator	Can create and manage all aspects of user flows.
☐	B2C user flow attribute administra···	Can create and manage the attribute schema available to all user flows.
☐	Billing administrator	Can perform common billing related tasks like updating payment information.
☐	Cloud application administrator	Can create and manage all aspects of app registrations and enterprise apps except A...
☐	Cloud device administrator	Full access to manage devices in Azure AD.
☐	Compliance administrator	Can read and manage compliance configuration and reports in Azure AD and Office ...
☐	Compliance data administrator	Can create and manage compliance content.
☐	Conditional Access administrator	Can manage conditional access capabilities.
☐	Customer LockBox access approver	Can approve Microsoft support requests to access customer organizational data.
☐	Desktop Analytics administrator	Can access and manage Desktop management tools and services.
☑	Directory readers	Can read basic directory information. Commonly used to grant directory read access ...
☐	Dynamics 365 administrator	Can manage all aspects of the Dynamics 365 product.
☐	Exchange administrator	Can manage all aspects of the Exchange product.
☐	External Identity Provider administ···	Can configure identity providers for use in direct federation.
☐	Global administrator	Can manage all aspects of Azure AD and Microsoft services that use Azure AD identi...
☐	Global reader	Can read everything that a global administrator can, but not update anything.
☐	Groups administrator	Can manage all aspects of groups and group settings like naming and expiration pol...

Select

Figure 6.3 – Selecting a directory role

5. Click **Select** and then **Create**.

6. Repeat these steps and create `PacktUser2` and `PacktUser3`.

Now that we have created a couple of users in our Azure AD tenant, we can add them to a group in Azure AD.

Creating groups in Azure AD

To create and manage groups from the Azure AD tenant in the Azure portal, perform the following steps:

1. Navigate to the Azure portal by going to `https://portal.azure.com`.

2. From the left menu, select **Azure Active Directory**.

3. In the **Overview** blade of Azure AD, in the left menu, select **Groups | All groups**. Then, select **+ New group** from the top menu, as follows:

Figure 6.4 – Creating a new group

4. Add the following values to create the new group:

    a) **Group type**: **Security**.

    b) **Group name**: `PacktGroup`.

    c) **Membership type**: Here, you can choose between three different values. The first is **Assigned** and is where you assign the members manually to the group; then, there's **Dynamic user**, which is where the group membership is determined based on certain user properties. Dynamic group membership eliminates the management overhead of adding and removing users. The final option is **Dynamic device**, which is where the group membership is determined based on certain device properties. Select the first option: **Assigned**.

5. Click the **Members** tab to add members to this group. Select the three user accounts that we created in the previous section, as follows:

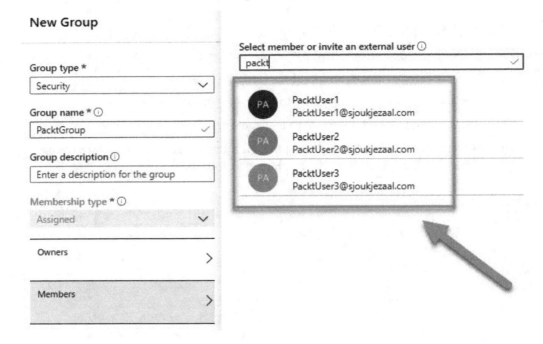

Figure 6.5 – Adding users to a group

6. Click **Select** to add the members and then **Create** to create the group.

> **Important note**
>
> We created the group as a security group. Security groups are specifically designed to allow you to grant access to resources in Azure. The other type of group is a Microsoft 365 group, which can be used to grant access to shared mailboxes, calendars, and SharePoint sites.

Now we have created a new group inside Azure AD and added the user accounts to it that we created in the previous section. In the next section, we are going to learn how to add and manage guest accounts.

Adding and managing guest accounts

You can also add guest accounts in Azure AD using Azure AD B2B. Azure AD B2B is a feature on top of Azure AD that allows organizations to work safely with external users. To be added to Azure B2B, external users don't need to have a Microsoft work or personal account that has been added to an existing Azure AD tenant. All sorts of accounts can be added to Azure B2B.

You don't have to configure anything in the Azure portal to use B2B. This feature is enabled by default for all Azure AD tenants. Perform the following steps to do this:

1. Adding guest accounts to your Azure AD tenant is similar to adding internal users to your tenant. When you go to the user's overview blade, you can choose **+ New guest user** from the top menu, as follows:

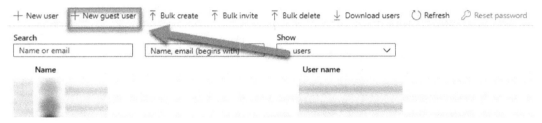

Figure 6.6 – Adding a guest user

2. Then, you can provide the same credentials as an internal user. You need to provide a name and an email address, as well as a personal message, which is sent to the user's inbox. This personal message includes a link so that you can log in to your tenant. You can also add the user to a group, as well as an RBAC role:

Figure 6.7 – External user properties

3. Click **Invite** to add the user to your Azure AD tenant and send out the invitation to the user's inbox.

4. To manage external users after creation, you can select them from the user overview blade. They will have a **USER TYPE** of **Guest**. Simply select the user from the list. Now, you'll be able to manage the settings that are displayed in the top menu for this user, as follows:

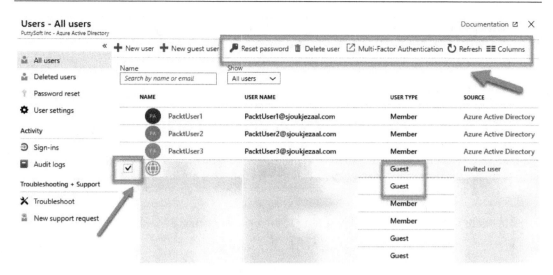

Figure 6.8 – Managing external users

In the next section, we are going to learn how to perform bulk user updates from the Azure portal.

Performing bulk user updates

Performing bulk user updates is similar to managing single users (internal and guest). The only property that can't be set for multiple users is resetting their passwords. This has to be done for a single user.

To perform a bulk user update, perform the following steps:

1. Go to the user's **Overview** blade again.

2. You can select multiple users in the **Overview** blade. From the top menu, select the property that you want to configure, as follows:

Figure 6.9 – Performing a bulk user update

This concludes how to perform bulk user updates. In the next section, we are going to cover how you can configure a self-service password reset for your users.

Configuring a self-service password reset

By enabling a self-service password for your users, they are able to change their passwords automatically without calling the help desk. This eliminates management overhead significantly.

A self-service password reset can easily be enabled from the Azure portal. To do so, perform the following steps:

1. Navigate to the Azure portal by going to `https://portal.azure.com`.

2. From the left menu, select **Azure Active Directory**.

3. In the Azure **AD Overview** blade, in the left menu, under **Manage**, select **Password reset**.

4. In the password reset **Overview** blade, you can enable self-service password resets for all your users by selecting **All** or, for selected users and groups, by selecting **Selected**. For demonstration purposes, enable it for all users and click **Save**, as follows:

Figure 6.10 – Enabling a self-service password reset for all users

5. Next, we need to set the different required authentication methods for your users. For this, under **Manage**, select **Authentication methods**.

6. In the next blade, we can set the number of authentication methods that are required to reset a password and what methods there are available for your users, as follows:

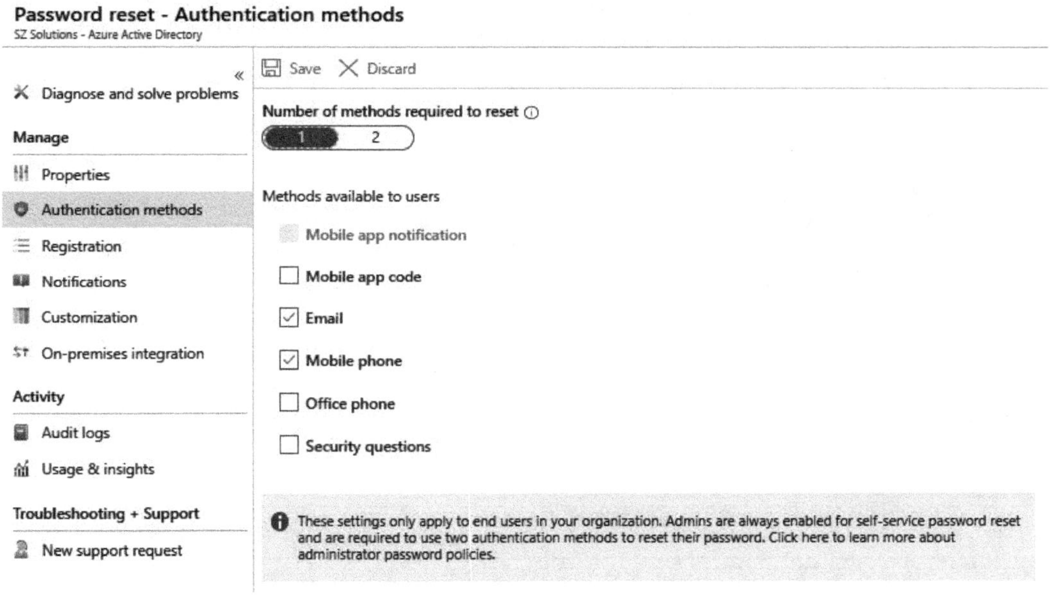

Figure 6.11 – Different authentication methods

7. Make a selection and then click **Save**.

Tip

If you want to test the self-service password reset after it's been configured, make sure that you use a user account without administrator privileges.

Now we have configured a self-service password reset for all our users inside our Azure AD tenant. In the next section, we are going to cover security defaults and Conditional Access policies in Azure AD.

Understanding Conditional Access policies and security defaults

Nowadays, modern security extends beyond the boundaries of an organization's network to include user and device identity. These identity signals can be used by organizations as part of their access control decisions. However, these require configuration and ongoing management, plus you must upgrade to the Azure AD Premium P1 tier to use it.

For some organizations, this may either involve too much effort (perhaps you are a small team) or possibly cost too much. Because security is so important, Microsoft offers security defaults as part of the free tier.

Security defaults

Security defaults are a set of built-in policies that protect your organization against common threats. Essentially, enabling this feature preconfigures your Active Directory tenant with the following:

- It requires all users to register for MFA.
- It requires users to use MFA when necessary (for example, in response to security events).
- It requires administrators to perform MFA.
- It blocks legacy authentication protocols.
- It protects privileged activities, such as accessing the Azure portal.

An important note to consider is that security defaults are completely free and do *not* require a premium AD license.

It's also important to understand that because this is a free tier, it's an all-or-nothing package. It's also worth noting that the MFA mechanism only works using security codes through a mobile app authenticator (such as Microsoft Authenticator) or a hardware token. In other words, you cannot use the text message or calling features.

Security defaults are enabled by default on new subscriptions, and therefore you don't need to do anything. However, if you do need to disable the feature (for example, to use Conditional Access policies), you can do so by performing the following steps:

1. Navigate to the Azure portal by going to `https://portal.azure.com`.
2. From the left menu, select **Azure Active Directory**.
3. On the left-hand menu, click **Properties**.

4. At the bottom of the page, click **Manage Security Defaults**.

5. A side window will appear with an option to enable or disable the security defaults.

Security defaults provide a basic level of security for your accounts. However, if you need granular and advanced control over how users gain access to your platform, you should consider Conditional Access policies.

Using Conditional Access policies

In their most basic form, Conditional Access policies are *if-then* statements. If a user wants to access a certain resource, they must complete a certain action. For instance, a guest user wants access to data that is stored in an Azure SQL database and is required to perform MFA to access it. This achieves administrators' two main goals: protecting the organization's assets and empowering users to be productive wherever and whenever. By implementing Conditional Access policies, you can apply the right access controls for all those different signals when needed to keep the organization's data and assets secure and enable different types of users and devices to easily get access to it. With Conditional Access policies, you have the choice to either block or grant access based on different signals.

The following common signals can be taken into account when policy decisions need to be made:

*   **User or group membership**: Administrators can get fine-grained control over access by targeting policies for specific users and groups.

*   **Device**: Policies and rules can be enforced for specific devices or platforms.

*   **Application**: Different Conditional Access policies can be triggered when users are trying to access specific applications.

*   **IP Location information**: Administrators can specify IP ranges and addresses to block or allow traffic from.

*   **Microsoft Cloud App Security (MCAS)**: User applications and sessions can be monitored and controlled in real time. This increases control and visibility over access and activities inside the cloud environment.

*   **Real-time and calculated risk detection**: The integration of signals with Azure AD Identity Protection allows Conditional Access policies to identify risky sign-in behavior. These risk levels can then be reduced, or access can be blocked, by enforcing Conditional Access policies that perform MFA or password changes.

> **Tip**
> Implementing a Conditional Access policy could come up as an exam question. For a complete walk-through on how to enable MFA for specific apps using a Conditional Access policy, you can refer to the following website: `https://docs.microsoft.com/en-us/azure/active-directory/conditional-access/app-based-mfa`.

In the next section, we are going to cover how we can join devices directly to Azure AD.

Working with Azure AD join

With Azure AD join, you are able to join devices directly to Azure AD without the need to join your on-premises Active Directory in a hybrid environment. While hybrid Azure AD join with an on-premises AD may still be preferred for certain scenarios, Azure AD join simplifies the addition of devices and modernizes device management for your organization. This can result in the reduction of device-related IT costs. Let's say your users are getting access to corporate assets through their devices. To protect these corporate assets, you want to control these devices. This allows your administrators to make sure that your users are accessing resources from devices that meet your standards for security and compliance.

Azure AD join is a good solution when you want to manage devices with a cloud device management solution, modernize your application infrastructure, and simplify device provisioning for geographically distributed users, and when your company is adopting Microsoft 365 as the productivity suite for your users.

Managing device settings

Azure AD offers you the ability to ensure that users are accessing Azure resources from devices that meet corporate security and compliance standards. Device management is the foundation for device-based conditional access and is where you can ensure that access to the resources in your environment is only possible from managed devices.

Device settings can be managed from the Azure portal. To manage your device settings, your device needs to be registered or joined to Azure AD.

To manage your device settings from the Azure portal, perform the following steps:

1. Navigate to the Azure portal by going to `https://portal.azure.com`.

2. From the left menu, select **Azure Active Directory**.

3. In the Azure AD **Overview** blade, under **Manage**, select **Devices**.

4. The **Device management** blade will open. Here, you can configure your device management settings, locate your devices, perform device management tasks, and review the device management-related audit logs.

5. To configure device settings, select **Device** settings from the left menu. Here, you can configure the following settings, which can be seen in the following screenshot:

   a) **Users may join devices to Azure AD**: Here, you can set which users can join their devices to Azure AD. This setting is only applicable to Azure AD join on Windows 10.

   b) **Users may register their devices with Azure AD**: This setting needs to be configured to allow devices to be registered with Azure AD. There are two options here: **None**, that is, devices are not allowed to register when they are not Azure AD joined or hybrid Azure AD joined, and **All**, that is, all devices are allowed to register. Enrollment with Microsoft Intune or **Mobile Device Management** (MDM) for Office 365 requires registration. If you have configured either of these services, **All** is selected and **None** is not available.

   c) **Require Multi-Factor Auth to join devices**: Here, you can specify that users are required to perform MFA when registering a device. Before you can enable this setting, MFA needs to be configured for users who register their devices.

   d) **Maximum number of devices per user**: This setting allows you to select the maximum number of devices that a user can have in Azure AD:

e) **Additional local administrators on all Azure AD joined devices**: Here, you can select the users who are granted local administrator permissions on a device. The users selected here are automatically added to the device administrator's role in Azure AD. Global administrators in Azure AD and device owners are granted local administrator rights by default (this is an Azure AD Premium option):

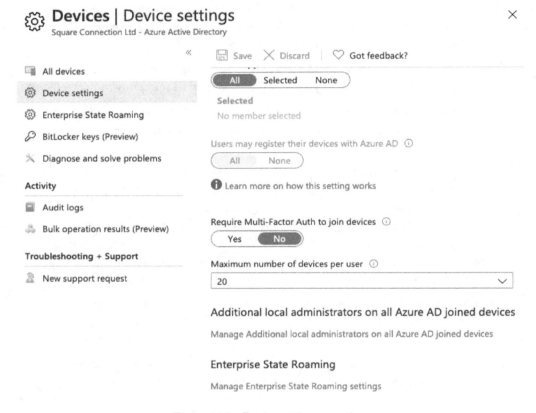

Figure 6.12 – Device settings overview

6. To locate your devices, under **Manage**, select **All devices**. Here, you will see all the joined and registered devices, as follows:

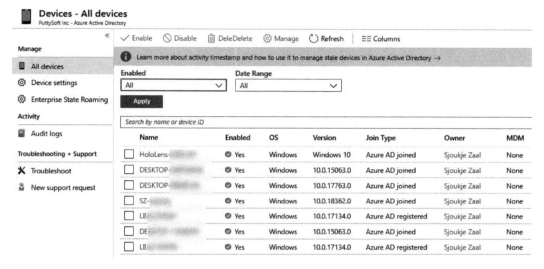

Figure 6.13 – Located devices

7. You can also select the different devices from the list to get more detailed information about the device in question. Here, global administrators and cloud device administrators can **Disable** or **Delete** the device, as follows:

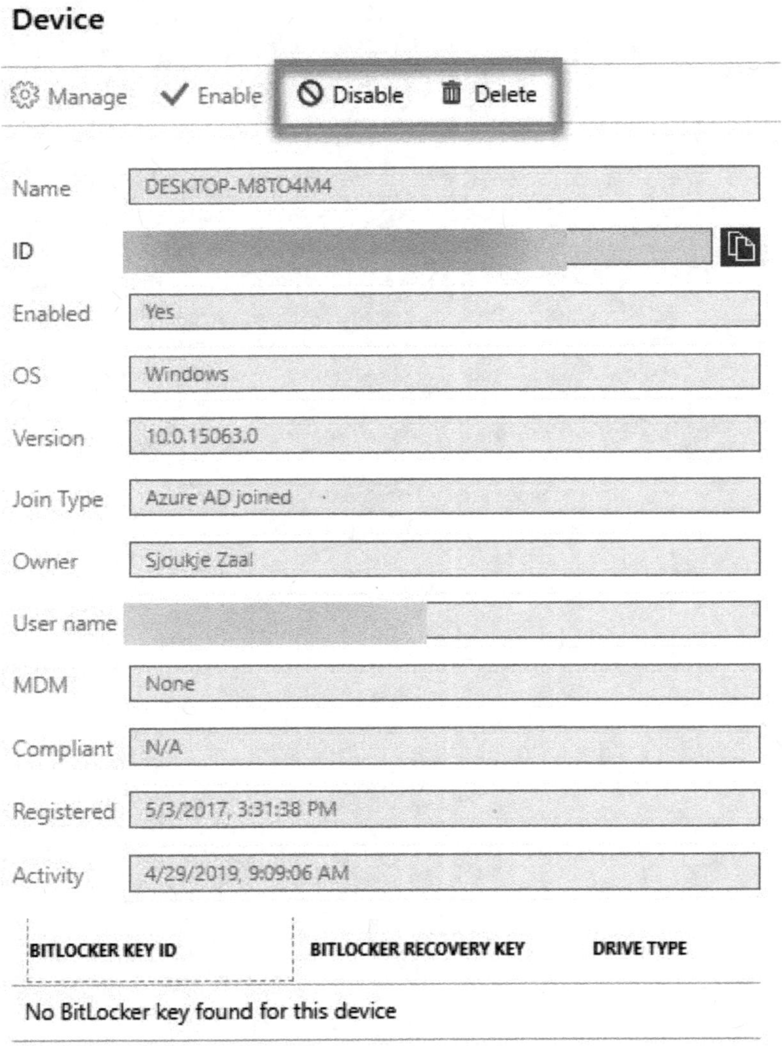

Device

⚙ Manage	✔ Enable	⦸ Disable	🗑 Delete

Name	DESKTOP-M8TO4M4
ID	▢
Enabled	Yes
OS	Windows
Version	10.0.15063.0
Join Type	Azure AD joined
Owner	Sjoukje Zaal
User name	
MDM	None
Compliant	N/A
Registered	5/3/2017, 3:31:38 PM
Activity	4/29/2019, 9:09:06 AM

BITLOCKER KEY ID	BITLOCKER RECOVERY KEY	DRIVE TYPE

No BitLocker key found for this device

Figure 6.14 – Device information

8. For audit logs, under **Activity**, select **Audit logs**. From here, you can view and download the different log files. You can also create filters to search through the logs, as follows:

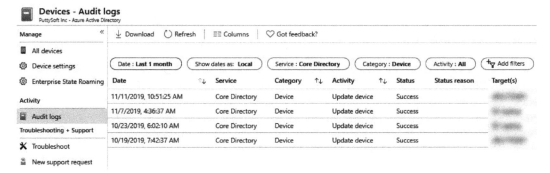

Figure 6.15 – Audit logs

Now we have looked at all the different management and configuration options for devices that are registered or joined to Azure AD. In the next section, we are going to learn how to add custom domains to Azure AD.

Adding custom domains

Azure creates an initial domain for every Azure AD tenant that is created in a subscription. This domain name consists of the tenant name, followed by `onmicrosoft.com` (`packtpub.onmicrosoft.com`). You cannot change or delete the initial domain name, but you can add custom domains to your Azure AD tenant.

This custom domain name can be registered at a third-party domain registrar and, after registration, can be added to the Azure AD tenant.

To add a custom domain to Azure AD from the Azure portal, perform the following steps:

1. Navigate to the Azure portal by going to `https://portal.azure.com`.
2. From the left menu, select **Azure Active Directory**.

3. In the Azure AD **Overview** blade, under **Manage**, select **Custom domain names**. To add a custom domain, select the **+ Add custom domain** button from the top menu, as follows:

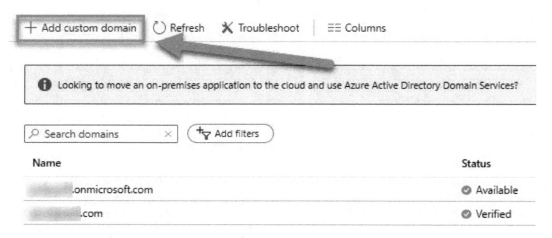

Figure 6.16 – Adding a custom domain

4. Type the custom domain name into the **Custom domain name** field (for example, `packtpub.com`) and select **Add domain**, as follows:

Figure 6.17 – Providing a custom domain name

5. After you've added your custom domain name to Azure AD, you need to create a **TXT** record inside the DNS settings of your domain registrar. Go to your domain registrar and add the Azure AD DNS information from your copied **TXT** file. Creating this **TXT** record for your domain *verifies* ownership of your domain name. After creating the **TXT** file, click **Verify**, as follows:

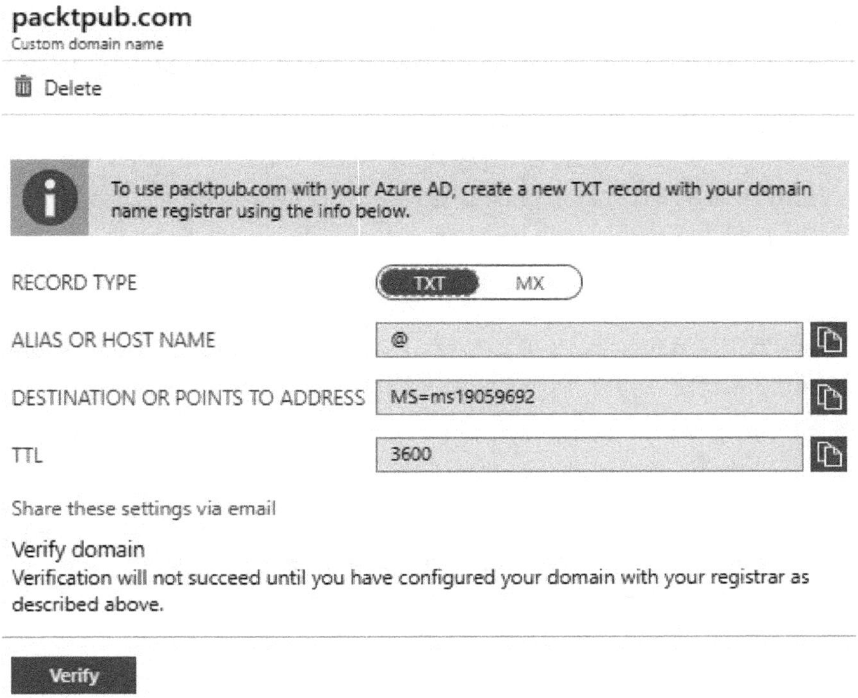

packtpub.com
Custom domain name

🗑 Delete

To use packtpub.com with your Azure AD, create a new TXT record with your domain name registrar using the info below.

RECORD TYPE	TXT MX
ALIAS OR HOST NAME	@
DESTINATION OR POINTS TO ADDRESS	MS=ms19059692
TTL	3600

Share these settings via email

Verify domain
Verification will not succeed until you have configured your domain with your registrar as described above.

Verify

Figure 6.18 – Verifying ownership of the domain

6. After you've verified your custom domain name, you can delete your verification **TXT** or **MX** file.

Now we have configured a custom domain for our Azure AD tenant. Your users can now use this domain name to log in to the various Azure resources to which they have access.

Summary

In this chapter, we covered the sixth part of the *Implement and Monitor Azure Infrastructure* objective. We covered the various aspects of Azure AD by learning how to add users and groups, how to add guest users, and how to manage devices in Azure AD. We also covered how to add custom domain names to our Azure AD tenant from the Azure portal. You have now understood how to set up and manage users effectively.

In the next chapter, we will learn how to implement and configure MFA.

Questions

Answer the following questions to test your knowledge of the information in this chapter. You can find the answers in the *Assessments* section at the end of this book:

1. If you want to create a guest user using PowerShell, you have to use the `New-AzureADMSInvitation` cmdlet.

   a) Yes

   b) No

2. If you want to use Azure AD join for your devices, you need to configure your on-premises AD environment in a hybrid environment, along with Azure AD.

   a) Yes

   b) No

3. When you add a custom domain to Azure AD, you need to verify it by adding a **TXT** record to the DNS settings of your domain registrar.

   a) Yes

   b) No

Further reading

You can check out the following links for more information about the topics covered in this chapter:

- Azure Active Directory documentation: `https://docs.microsoft.com/en-us/azure/active-directory/`

- Adding or deleting users using Azure Active Directory: `https://docs.microsoft.com/en-us/azure/active-directory/fundamentals/add-users-azure-active-directory`

- Azure Active Directory version 2 cmdlets for group management: `https://docs.microsoft.com/en-us/azure/active-directory/users-groups-roles/groups-settings-v2-cmdlets`

- Quickstart: Adding a guest user with PowerShell: `https://docs.microsoft.com/en-us/azure/active-directory/b2b/b2b-quickstart-invite-powershell`

- Quickstart: Self-service password reset: `https://docs.microsoft.com/en-us/azure/active-directory/authentication/quickstart-sspr`

- How to: Planning your Azure Active Directory join implementation: `https://docs.microsoft.com/en-us/azure/active-directory/devices/azureadjoin-plan`

- What is device management in Azure Active Directory? `https://docs.microsoft.com/en-us/azure/active-directory/devices/overview`

- Adding your custom domain name using the Azure Active Directory portal: `https://docs.microsoft.com/en-us/azure/active-directory/fundamentals/add-custom-domain`

7
Implementing Multi-Factor Authentication (MFA)

In the previous chapter, we covered how to manage **Azure Active Directory** (**Azure AD**). We've learned how to manage this from the Azure portal and how to perform bulk updates inside your Azure AD tenant. We also covered how to configure self-service password reset for your users to reduce user management overhead, how to configure Azure AD join, and how you can manage your devices that are registered or joined in Azure AD.

In this chapter, we'll take a deep dive into configuring Azure AD for **Multi-Factor Authentication** (**MFA**), and how it can provide an additional layer of security to the authentication process.

We'll also examine the different options for aligning capabilities to each organization's unique requirements, such as who must use MFA, what locations can bypass it, and sending alerts when fraudulent activity is detected.

The following topics will be covered in this chapter:

- Understanding Azure MFA
- Configuring user accounts for MFA
- Configuring verification methods
- Configuring trusted IPs
- Configuring fraud alerts
- Configuring bypass options

Understanding Azure MFA

MFA is a security feature that requires more than one method of authentication. You can use it to add an additional layer of security to the signing in of users. It enables two-step verification, where the user first signs in using something they know (such as a password) and then signs in with something they have (such as a smartphone or hardware token) or some human characteristic (such as biometrics).

Azure MFA maintains simplicity for users, as well as helps to keep data and applications safe by providing additional security and requiring a second form of authentication. It offers a variety of configuration methods set by an administrator that determines whether users are challenged for MFA or not.

Azure MFA is part of the following offerings:

- **Azure AD Free and Basic tiers**: MFA is enabled as part of the security defaults, but with limitations, such as only supporting mobile app verification codes and not text messages or phone calls.

- **Azure AD Premium P1 and P2 licenses**: With this license, you can use the Azure MFA service (cloud) and Azure MFA server (on-premises). The latter is most suitable in scenarios where an organization has **Active Directory Federation Services** (**AD FS**) installed and needs to manage infrastructure components.

- **Azure AD Global Administrators**: A subset of the MFA features is available for administrator accounts in Azure.

- **MFA for Office 365**: A subset of the MFA features is available for Office 365 users.

With Azure MFA, you can use the following verification methods:

Verification method	Description
Mobile app notification	A request for verification is sent to the user's smartphone. When necessary, the user will enter a PIN and then select Verify.
Mobile app verification code	The mobile app on the user's smartphone will display a verification code, which will refresh every 30 seconds. The user will select the most recent code and will enter it on the login page.
Voice call*	A call is made to the registered phone of the user. The user needs to enter a PIN for verification.
Text message*	A text message is sent to the user's mobile phone containing a six-digit code. The user needs to fill in this code on the login page.
Third-party tokens	The Azure MFA server can be configured to accept third-party security tokens.
App passwords*	Only in certain cases. Certain non-browser apps do not support MFA; if a user has been enabled for MFA and attempts to use non-browser apps, they are unable to authenticate. An app password allows users to continue to authenticate. If MFA is enforced through Conditional Access policies and not through per-user MFA, you cannot create app passwords. Applications that use Conditional Access policies to control access do not need app passwords.

*Not available in the Basic or Free tiers

In the upcoming sections, we will enable MFA for the Azure AD tenant, configure user accounts, configure fraud alerts, and configure bypass options.

> **Important note**
> For the demos in this chapter, I will use an Azure AD Premium P2 license.

Enabling MFA for an Azure AD tenant

The following are the four different options for enabling MFA for your users, data, and applications:

- **Security defaults**: Basic MFA settings are applied by default.

- **Using a Conditional Access policy**: You can use Conditional Access policies to enable MFA. This can be enabled at the user or application level. You can also enable MFA for security groups or for all external users using a Conditional Access policy. This is available for premium Azure AD licenses.

- **At the user level**: This option is covered in more detail in the next section of this chapter. This is the traditional method for enabling MFA. With this method, the user needs to perform MFA every time they sign in. It will override Conditional Access policies when these are set.

- **Using Azure AD Identity Protection**: With this option, you will create an Azure AD Identity Protection risk policy based on the sign-in risk for all of your cloud applications. This will also override Conditional Access policies, if they've been created. This option requires an Azure AD P2 license.

Configuring user accounts for MFA

Azure MFA is enabled in Azure AD at the user level. To enable MFA for a user account in Azure AD, take the following steps:

1. Navigate to the Azure portal by opening `https://portal.azure.com/`.

2. Go to **All services** in the left menu, then type `Azure Active Directory`, and open the Azure AD resource.

3. In the **Azure AD** blade, under **Manage**, select **Users**.

4. In the **All users** blade, select **Multi-Factor Authentication** in the top menu.

5. You then will be redirected to the **multi-factor authentication** portal. There, select a user and click **Enable** at the right side of the screen:

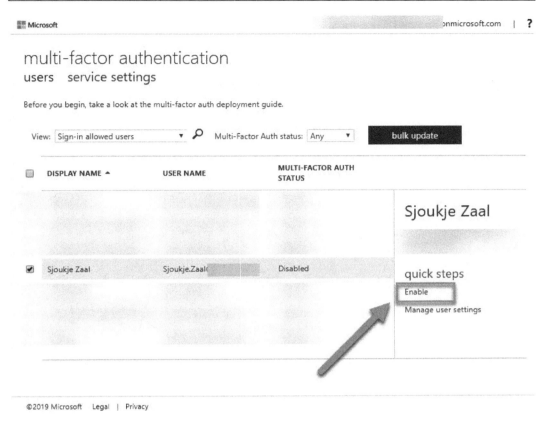

Figure 7.1 – Enabling MFA for a user

6. After clicking the link, you will receive the following warning:

About enabling multi-factor auth

Please read the deployment guide if you haven't already.

If your users do not regularly sign in through the browser, you can send them to this link to register for multi-factor auth: https://aka.ms/MFASetup

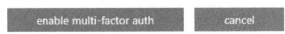

Figure 7.2 – MFA warning

7. Click **enable multi-factor auth** to activate MFA for this user.

Now that we have enabled MFA for the user, we can look at how to configure the verification methods.

Configuring verification methods

Verification methods are also configured in the Azure MFA portal, just as you enabled MFA for the user account in the previous step:

1. With the MFA portal still open, select **service settings** in the top menu:

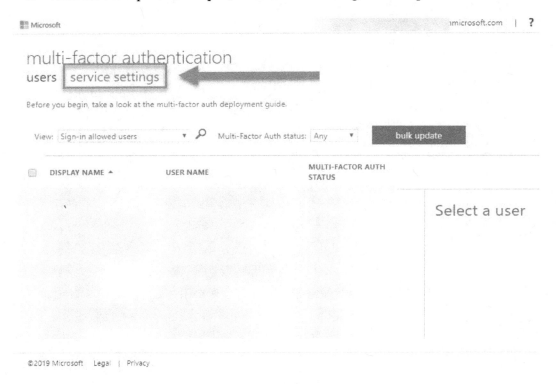

Figure 7.3 – MFA portal service settings

2. Under **verification options**, you can select the methods that you want to enable for your users. By default, all verification options are enabled:

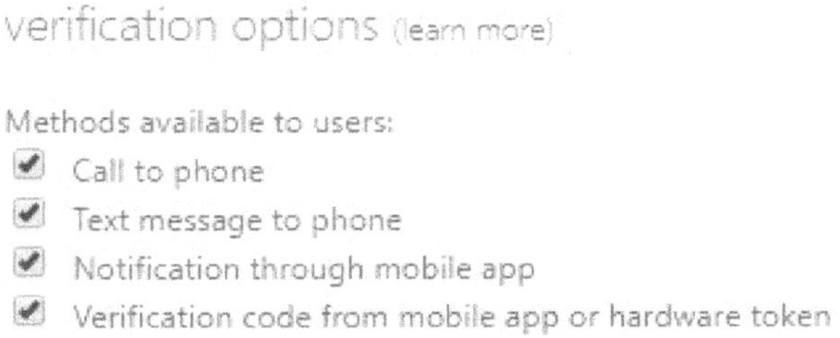

verification options (learn more)

Methods available to users:
- ☑ Call to phone
- ☑ Text message to phone
- ☑ Notification through mobile app
- ☑ Verification code from mobile app or hardware token

Figure 7.4 – MFA verification options

3.   If you want to disable any options, uncheck the checkbox and click the **Save** button.

We have seen how to configure the verification methods that users are allowed to use with MFA in Azure. In the next section, we are going to look at how to configure trusted IPs.

Configuring trusted IPs

Trusted IPs are used by administrators of an Azure AD tenant. This option will bypass the MFA for users that sign in from a trusted IP, such as the company intranet.

Trusted IPs can be configured from the **service settings** page from the MFA portal:

1.   With the **server settings** page still open from the previous demo, under **trusted ips**, check the checkbox that says **Skip multi-factor authentication for requests from federated users on my intranet**. Then, add an IP address or a range of IP addresses to the list:

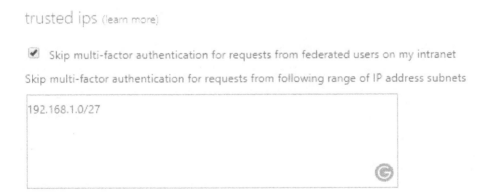

trusted ips (learn more)

☑ Skip multi-factor authentication for requests from federated users on my intranet

Skip multi-factor authentication for requests from following range of IP address subnets

192.168.1.0/27

Figure 7.5 – Trusted IP settings

2.   Click the **Save** button to save your settings.

In the next section, we are going to cover how to configure fraud alerts in the Azure portal.

Configuring fraud alerts

With the fraud alert feature, users can report fraudulent attempts to access their resources using their phone or the mobile app. This is an MFA server (on-premises) feature.

Fraud alerts are configured from the Azure portal, in the Azure AD settings:

1. Navigate to the Azure portal by opening `https://portal.azure.com`.

2. Select **All services** in the left menu, then type `Azure Active Directory` in the search bar, and open the settings:

    a) Under **Manage**, select **Security**.

    b) Then, in the **Security** blade, select **MFA**.

3. The **Getting started** blade is automatically opened. Under **Settings**, select **Fraud alert**.

4. The **Fraud alert** settings page is opened. Here, you can enable users to submit fraud alerts:

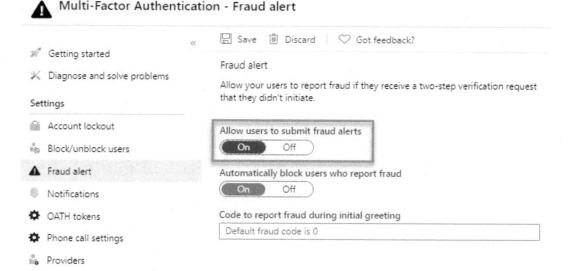

Figure 7.6 – Enabling submitting fraud alerts

5. Click the **Save** button to save the settings.

We've now seen how we can allow users to submit fraud alerts. In the next section, we will cover how to configure bypass options.

Configuring bypass options

With the one-time bypass feature, users can authenticate one time and bypass MFA. This setting is temporary, and after a specified number of seconds, it will expire automatically. This can be a solution in cases when a phone or mobile app doesn't receive a phone call or notification.

This setting is also configured from the Azure AD settings in the Azure portal, as follows:

1. Navigate to the Azure portal by opening `https://portal.azure.com`.

2. Select **All services** in the left menu, then type `Azure Active Directory` in the search bar, and open the settings.

3. Under **Manage**, select **Security**.

4. Then, in the **Security** blade, select **One-time bypass**. Click the **Add** button in the top menu:

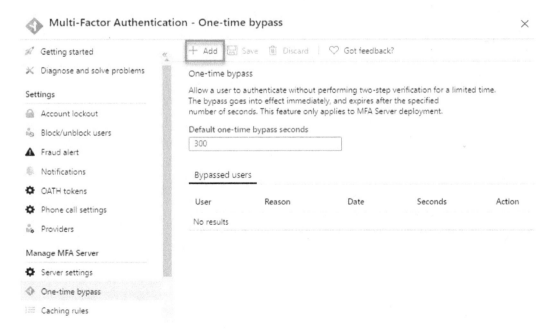

Figure 7.7 – One-time bypass

5. On the settings page, enter the username, including the full domain name, such as `username@domain.com`. Specify the number of seconds that the bypass should last and the reason for the bypass.

6. Click the **Add** button. The time limit will go into effect immediately.

We've now covered how to authenticate just once and bypass MFA using the one-time bypass feature.

Summary

In this chapter, we saw how to configure user accounts for MFA, and the different options available for tailoring the service for your organization. We examined the different capabilities depending on the AD licenses you have, and how MFA can be enabled for specific users, or set to not be required from specified locations.

We also looked at how to detect and be alerted of possible fraudulent actions, how to configure bypass options, how to configure trusted IPs, and how to configure verification methods.

In the next chapter, we'll move on to integrating Azure AD with an on-premises directory using Azure AD Connect, including how to configure password synchronization and single sign-on.

Questions

Answer the following questions to test your knowledge of the information in this chapter. You can find the answers in the *Assessments* section at the end of this book:

1. Can you allow trusted IP addresses to bypass MFA?

   a) Yes

   b) No

2. Can MFA be enabled using Conditional Access policies?

   a) Yes

   b) No

3. Are fraud alerts part of the MFA server (on-premises) offering?

   a) Yes

   b) No

Further reading

You can check out the following links for more information about the topics that are covered in this chapter:

- How it works—Azure MFA: `https://docs.microsoft.com/en-us/azure/active-directory/authentication/concept-mfa-howitworks`

- Deploying cloud-based Azure MFA: `https://docs.microsoft.com/en-us/azure/active-directory/authentication/howto-mfa-getstarted`

- Configuring Azure MFA settings: `https://docs.microsoft.com/en-us/azure/active-directory/authentication/howto-mfa-mfasettings`

8
Implementing and Managing Hybrid Identities

In the previous chapter, we covered how to configure user accounts for **multi-factor authentication (MFA)**, how to configure fraud alerts, bypass options, trusted IPs, and verification methods.

This chapter is the last part of the *Implement and Monitor Azure Infrastructure* objective. In this chapter, we are going to cover how to implement and manage hybrid identities. We are going to install and configure Azure AD Connect to synchronize the identities from your on-premises Active Directory to Azure AD. Then, you will learn how to manage Azure AD Connect.

In the last part of this chapter, we will dive into password synchronization and password writeback. You will learn how to enable password synchronization in Azure AD Connect and the Azure portal, and how to manage password synchronization. Finally, we will look at how to install and use Azure AD Connect Health.

The following topics will be covered in this chapter:

- Understanding Azure AD Connect
- Installing Azure AD Connect
- Managing Azure AD Connect
- Managing password synchronization and password writeback
- Using Azure AD Connect Health

Understanding Azure AD Connect

Azure AD Connect is a service that you can use to synchronize your on-premises Active Directory identities with Azure. This way, you can use the same identities for authentication on your on-premises environment as well as in the cloud and other **Software as a Service (SaaS)** applications.

The Azure AD Connect sync service consists of two parts: the **Azure AD Connect sync component**, which is a tool that is installed on a separate server inside your on-premises environment, and the **Azure AD Connect sync service**, which is part of Azure AD. The sync component can sync data from Active Directory and SQL Servers to Azure. There is also a third component, named the **Active Directory Federation Services (ADFS)** component, which can be used in a scenario where ADFS is involved. To monitor the on-premises identity infrastructure and the different Azure AD components, you can use a tool named **Azure AD Connect Health** – which we cover in more detail at the end of the chapter, in the *Using AD Connect Health* section. The following diagram illustrates the architecture of Azure AD Connect:

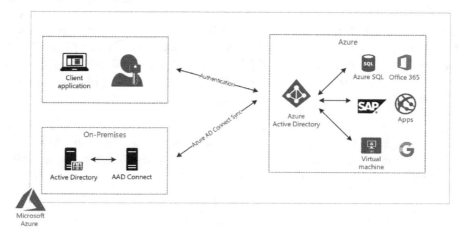

Figure 8.1 – Azure AD Connect architecture

Azure AD Connect offers support for your users to sign in with the same passwords to both on-premises and cloud resources. It provides three different authentication methods for this: the password hash synchronization method, the pass-through authentication method, and the Federated SSO method (in conjunction with ADFS).

Azure AD password hash synchronization

Most organizations only have a requirement to enable users to sign in to Office 365, SaaS applications, and other Azure AD-based resources. The password hash synchronization method is well suited to those scenarios.

Using this method, hashes of the user's password are synced between the on-premises Active Directory and Azure AD. When there are any changes to the user's password, the password is synced immediately, so users can always log in with the same credentials on-premises as well as in Azure.

This authentication method also provides Azure AD Seamless **Single Sign-On (SSO)**. This way, users are automatically signed in when they are using a domain-joined device on the corporate network. Users only have to enter their username when using Seamless SSO. To use Seamless SSO, you don't have to install additional software or components on the on-premises network. You can push this capability to your users using group policies.

Azure AD pass-through authentication

Azure AD pass-through authentication offers the same capability as Azure AD password hash synchronization. Users can log in to their Azure resources as well as on-premises resources using the same credentials. The difference is that the passwords don't sync with Azure AD using pass-through authentication. The passwords are validated using the on-premises Active Directory and are not stored in the Azure AD at all.

This method is suitable for organizations that have security and compliance restrictions and aren't allowed to send usernames and passwords outside the on-premises network. Pass-through authentication requires an agent to be installed on a domain-joined Windows server that resides inside the on-premises environment. This agent then listens for password validation requests and only makes an outbound connection from within your network. It also offers support for MFA and Azure AD Conditional Access policies.

Azure AD pass-through authentication offers Azure AD Seamless SSO as well.

In the next section, we are going to install Azure AD Connect and synchronize some on-premises users with Azure.

Installing Azure AD Connect

Azure AD Connect is installed on an on-premises server with Active Directory installed and configured on it. The first step is to download Azure AD Connect. After downloading, we can install it on a domain controller. We will be installing the AD Connect tool using Express Settings, which sets the most common options – such as password hash synchronization. We will then rerun the AD Connect tool to enable other options, such as password writeback.

> **Important note**
>
> For this demonstration, I have already deployed a Windows Server 2016 virtual machine in Azure and installed and configured Active Directory on it. Configuring Active Directory is beyond the scope of the exam and this book. Make sure that when you configure Active Directory Domain Services, the forest name matches one of the existing verified custom domains in Azure AD. Otherwise, you will receive a warning message when you install Azure AD Connect on your domain controller, which will state that SSO is not enabled for your users. For installing Active Directory on a Windows Server 2016 machine, you can refer to the following link: `https://blogs.technet.microsoft.com/canitpro/2017/02/22/step-by-step-setting-up-active-directory-in-windows-server-2016/`.

Therefore, perform the following steps:

1. Before downloading Azure AD Connect, add at least one user to your on-premises Active Directory.

2. Navigate to the Azure portal at `https://portal.azure.com`.

3. In the top bar, search for and select **Azure Active Directory**.

4. In the left-hand menu, click **AD Connect**.

5. You will see the link **Download AD Connect**.

6. Download and store it on a local drive on your *on-premise domain controller* and run it.

7. The installation wizard starts with the **Welcome** screen. Select the checkbox to agree with the license terms:

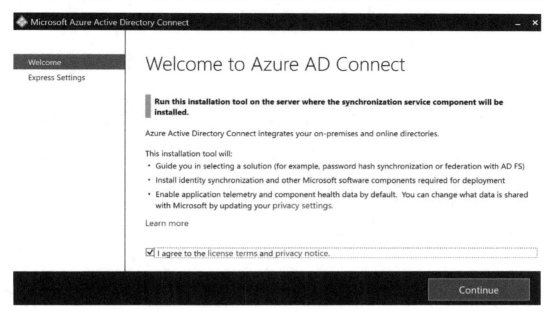

Figure 8.2 – Azure AD Connect Welcome screen

8. Select **Use express settings** on the next screen:

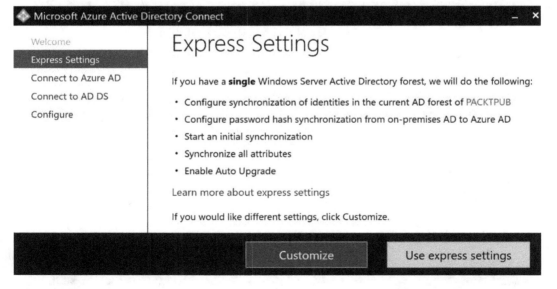

Figure 8.3 – Installing Azure AD Connect using express settings

9. On the next screen, provide the username and password of a global administrator account for your Azure AD and click **Next** (this account must be a school or organization account and cannot be a Microsoft account or any other type of account):

Figure 8.4 – Provide global administrator credentials

10. On the **Connect to AD DS** screen, enter the username and password of an enterprise administrator account, and click **Next** as follows:

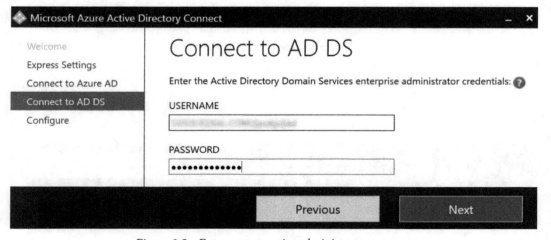

Figure 8.5 – Enter an enterprise administrator account

The last screen will give you an overview of what is going to be installed, as follows:

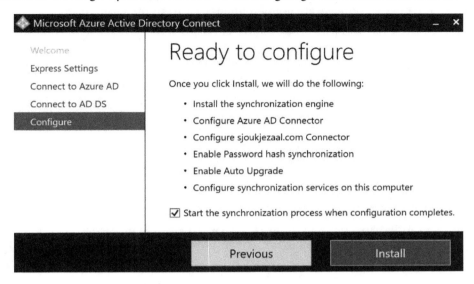

Figure 8.6 – Ready to configure

11. Click **Install**.

12. This will install Azure AD Connect on your domain controller. The synchronization process of user accounts with Azure AD will automatically be started after configuration.

13. After successful configuration, you will see the following outcome:

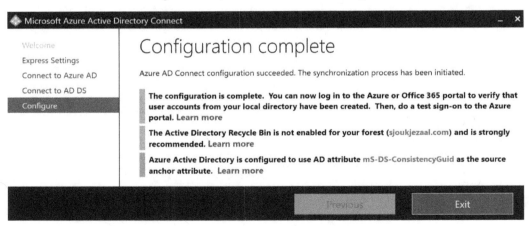

Figure 8.7 – Configuration complete

14. Click **Exit** to close the installer.

In this demonstration, we installed Azure AD Connect on an on-premises domain controller. In the next section, we are going to manage it from the Azure portal.

Managing Azure AD Connect

Azure AD Connect can be managed from the Azure portal after installation and configuration on the on-premises domain controller. To manage it, you have to perform the following steps:

1. Navigate to the Azure portal by opening `https://portal.azure.com`.

2. In the left-hand menu, select **Azure Active Directory**.

3. Under **Manage**, select **Azure AD Connect**. In the **Azure AD Connect** blade, as shown in the following screenshot, you can see that syncing is enabled, that the last sync was more than a day ago, and that **Password Hash Sync** is enabled:

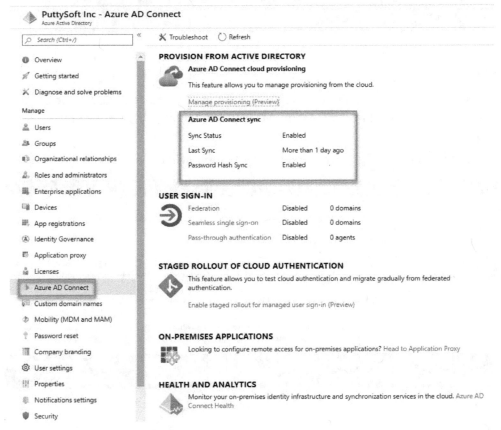

Figure 8.8 – Azure AD Connect settings

4. You can also set the three authentication methods under **USER SIGN-IN**. Here, you can set the authentication method to **Federation, Seamless single sign-on**, or **Pass-through authentication**. You can monitor the health of your on-premises infrastructure and synchronization services under **HEALTH AND ANALYTICS**.

5. To check whether the users are synced, you can go to the **User overview** blade. Here, you will find your synced users, as shown in the following screenshot:

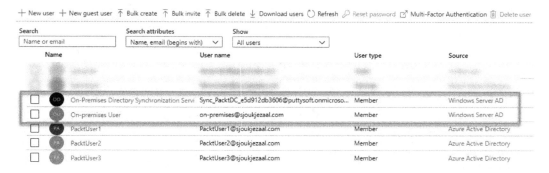

Figure 8.9 – Synced users

> **Important note**
>
> *Azure AD Connect sync* synchronizes changes in your on-premises directory using a scheduler. There are two scheduler processes: one for password synchronization and another for object/attribute sync and maintenance tasks. For more information on how to configure this or how to create a custom scheduler using PowerShell, you can refer to the following tutorial: `https://docs.microsoft.com/en-us/azure/active-directory/hybrid/how-to-connect-sync-feature-scheduler`.

In this demonstration, we managed Azure AD Connect from the Azure portal. In the next section, we are going to cover how to manage password writeback in more detail.

Managing password synchronization and password writeback

Password writeback is used for synchronizing password changes in Azure AD back to your on-premises Active Directory environment. This setting is enabled as part of Azure AD Connect, and it provides a secure mechanism to send password changes from Azure AD back to an on-premises Active Directory.

It provides the following features and capabilities:

- **Enforcement of on-premises Active Directory password policies**: When a user resets their password, the on-premises Active Directory policy is checked to ensure it meets the password requirements before it gets committed to the directory. It checks the password complexity, history, password filters, age, and other password restrictions that are defined in the on-premises Active Directory.

- **Zero-delay feedback**: Users are notified immediately after changing their password if their password doesn't meet the on-premises Active Directory policy requirements. This is a synchronous operation.

- **Supports password writeback when an administrator resets it from the Azure portal**: When an administrator resets the password in the Azure portal, the password is written back to the on-premises Active Directory (when a user is federated or password hash synchronized). This functionality doesn't work from the Office admin portal.

- **Doesn't require any inbound firewall rules**: Password writeback uses Azure Service Bus for communicating with the on-premises Active Directory, so there is no need to open the firewall. All communication is outbound and goes over port 443.

- **Supports password changes from the access panel and Office 365**: When federated or password hash synchronized users change their passwords, those passwords are written back to your on-premises Active Directory as well.

In the next demonstration, we are going to enable password writeback.

Managing password writeback

To enable password writeback, we need to make some changes to both the configuration of Azure AD Connect on the on-premises domain controller, and from the Azure portal.

Enabling password writeback in Azure AD Connect

To enable password writeback in Azure AD Connect, we have to take the following steps:

1. Log in to your on-premises domain controller using **Remote Desktop** (**RDP**) and start the Azure AD Connect wizard again.

2. On the **Welcome to Azure AD Connect** page, select **Configure** as follows:

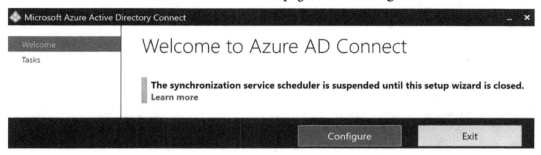

Figure 8.10 – Welcome screen

3. On the **Additional tasks** screen, select **Customize synchronization options**, and then select **Next** as follows:

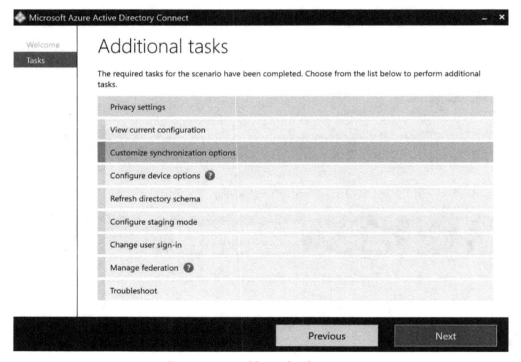

Figure 8.11 – Additional tasks screen

4.   Provide Azure AD global administrator credentials and then select **Next** as follows:

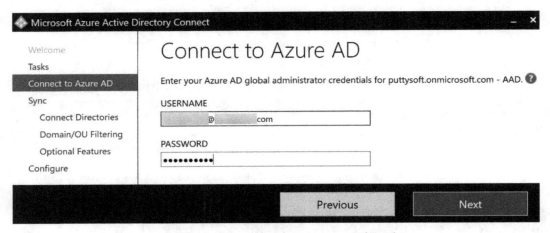

Figure 8.12 – Providing administrator credentials

5.   On the **Connect your directories** screen, select **Next** as follows:

Figure 8.13 – Connecting your directories

6. On the **Domain and OU filtering** screen, select **Next** again, as follows:

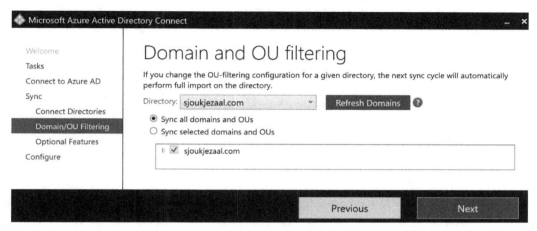

Figure 8.14 – Domain and OU filtering screen

7. On the **Optional features** screen, select the box next to **Password writeback** and then select **Next** as follows:

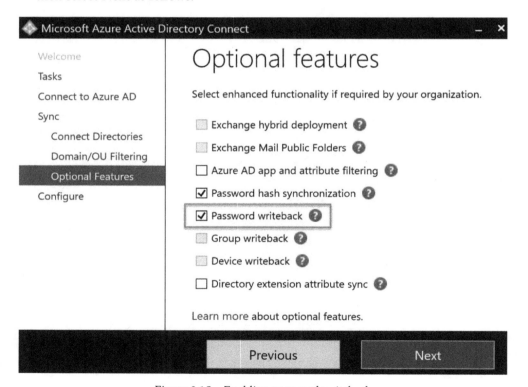

Figure 8.15 – Enabling password writeback

8. On the **Ready to configure** page, select **Configure**.

9. When the configuration is finished, select **Exit**.

We have now enabled password writeback on the domain controller. In the next section, we are going to enable it in the Azure portal as well.

Enabling password writeback in the Azure portal

To enable password writeback in the Azure portal, we have to perform the following steps:

1. Navigate to the Azure portal by opening `https://portal.azure.com`.

2. In the left-hand menu, select **Azure Active Directory**.

3. Under **Manage**, select **Password reset**.

4. In the **Password reset** blade, under **Manage**, select **On-premises integration**. Set the **Write back passwords to your on-premises directory?** option to **Yes** and set the **Allow users to unlock accounts without resetting their password?** option to **Yes** as follows:

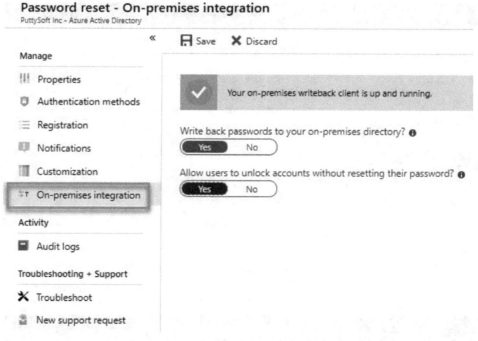

Figure 8.16 – Enabling password writeback

5. Click **Save**.

We have now completely configured password writeback in Azure AD Connect and the Azure portal. In the next section, we are going to cover how to manage password synchronization.

Password synchronization

Next, we are going to cover password synchronization. We installed Azure AD Connect using the **Express Settings** option. Password hash synchronization is automatically enabled if you use this option.

If you install Azure AD Connect using the custom settings, password hash synchronization is available on the **User sign-in** screen and you can enable it there, as shown in the following screenshot:

Figure 8.17 – Enabling password hash synchronization during installation

Now we have configured the AD Connect agent, the final step is to monitor the synchronization process, which we will cover in the next, final section.

Using Azure AD Connect Health

When you integrate your on-premises directories with Azure AD, it is important that the synchronization between your local directory and the Azure Directory is in a healthy state. Azure provides Azure AD Connect Health, which helps you do the following:

- Monitor and gain insights into AD FS servers, Azure AD Connect, and AD domain controllers.

- Monitor and gain insights into the synchronizations that occur between your on-premises AD DS and Azure AD.

- Monitor and gain insights into your on-premises identity infrastructure that is used to access Office 365 or other Azure AD applications.

There are three separate agents, one for each scenario;

- The standard **Azure AD Connect** agent automatically installs a health monitoring service to monitor the standard synchronization of users and passwords.

- The **Azure AD Connect Health Agent for AD DS** extends the monitoring capabilities to your on-premise Active Directory.

- The **Azure AD Connect Health Agent for AD FS** monitors and alerts the Federated Services synchronization process.

As with the main AD Connect agent, the download link for the other AD Connect Health agent can be accessed in the Azure portal:

1. Navigate to the Azure portal at `https://portal.azure.com`.

2. In the top bar, search for and select **Active Directory**.

3. In the left-hand menu, click **AD Connect**.

4. On the main page, click **Azure AD Connect Health** under **Health and Analytics**.

5. You will see the links to the AD Connect Health agents under **Get Tools**.

6. Click and download the link **Download Azure AD Connect Health Agent for AD DS**.

In the next example, we will install the Azure AD Connect agent for AD DS, therefore ensure you have downloaded the AD DS agent in the above *steps 1-6*:

1. Log on to your Windows domain controller.

2. Copy the Azure AD Connect Health Agent for AD DS to the local drive on your domain controller.

3. Run the `AdHealthADDSAgentSetup.exe`.

4. In the window that appears, click **Install**.

5. Once the installation has finished, click **Configure Now**.

6. You will be prompted to log on using an Azure AD – use an account with global administrator rights to your Azure AD directory.

7. The health metrics for the domain controller will now start being sent to Azure. Wait an hour for data to be sent through.

8. Navigate to the Azure portal at `https://portal.azure.com`.

9. In the top bar, search for and select **Active Directory**.

10. In the left-hand menu, click **AD Connect**.

11. On the main page, click **Azure AD Connect Health** under **Health and Analytics**.

12. On the left-hand menu, click **AD DS Services**.

13. Your domain will be listed. Click on it to see your Directory Services Health statistics as in the following example:

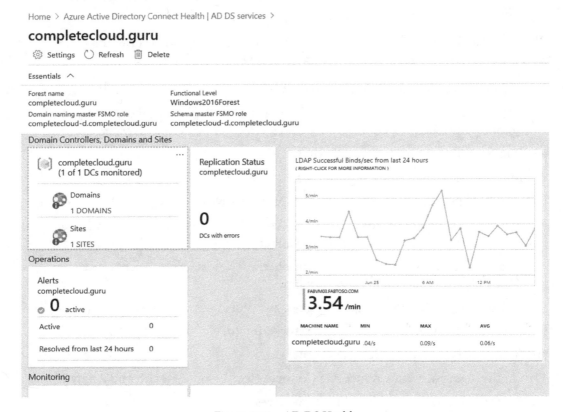

Figure 8.18 – AD DS Health

We have seen in this section how the Azure Active Directory Connect Health agents can ensure not just the health of the synchronization process but of your on-premise directory as well.

Summary

In this chapter, we covered the last part of the *Implement and Monitor Azure Infrastructure*. We covered how to implement and manage hybrid identities by covering Azure AD Connect, and you learned how to install and manage it after installation. We also covered how to enable password writeback and password hash synchronization.

In the next chapter, we will start with the *Implement Management and Security Solutions* objective by covering how to migrate servers to Azure.

Questions

Answer the following questions to test your knowledge of the information in this chapter. You can find the answers in the *Assessments* section at the end of this book:

1. If you use the **Express Settings** option when installing Azure AD Connect, password hash synchronization is disabled by default:

    a) Yes

    b) No

2. When you want to enable password synchronization, you only have to do so inside the Azure portal:

    a) Yes

    b) No

3. If the on-premises forest name doesn't match one of the Azure AD custom domain names, you cannot install Azure AD Connect:

    a) Yes

    b) No

Further reading

You can check out the following links for more information about the topics that were covered in this chapter:

*   What is hybrid identity with Azure Active Directory?: `https://docs.microsoft.com/en-us/azure/active-directory/hybrid/whatis-hybrid-identity`

*   What is federation with Azure AD?: `https://docs.microsoft.com/en-us/azure/active-directory/hybrid/whatis-fed`

*   What is password hash synchronization with Azure AD?: `https://docs.microsoft.com/en-us/azure/active-directory/hybrid/whatis-phs`

*   User sign-in with Azure Active Directory Pass-Through Authentication: `https://docs.microsoft.com/en-us/azure/active-directory/hybrid/how-to-connect-pta`

*   Azure AD Connect sync: Understand and customize synchronization: `https://docs.microsoft.com/en-us/azure/active-directory/hybrid/how-to-connect-sync-whatis`

- Azure AD Connect user sign-in options: `https://docs.microsoft.com/en-us/azure/active-directory/hybrid/plan-connect-user-signin`

- Tutorial: Enabling password writeback: `https://docs.microsoft.com/en-us/azure/active-directory/authentication/tutorial-enable-writeback`

- Implement password hash synchronization with Azure AD Connect sync: `https://docs.microsoft.com/en-us/azure/active-directory/hybrid/how-to-connect-password-hash-synchronization`

- AD Seamless Single Sign-On: `https://docs.microsoft.com/en-us/azure/active-directory/hybrid/how-to-connect-sso-faq`

Section 2: Implement Management and Security Solutions

In this section, you will learn how to migrate, implement, and secure workloads in Azure.

This section contains the following chapters:

9
Managing Workloads in Azure

In the previous chapter, we learned how to implement and manage hybrid identities, as well as how to install and configure Azure AD Connect and the Azure AD Connect Health agents. Then, we migrated some users.

This chapter is the first chapter of the *Implement Management and Security Solutions* objective. In this chapter, we will cover Azure Migrate and the different tooling that is integrated with Azure Migrate that's used to create assessments and migrate servers, databases, web applications, and data to Azure. We are also going to migrate some Hyper-V VMs to Azure using Azure Migrate.

Next, we will see how to ensure that VMs are kept up to date using Azure Update Management.

Finally, we will then look at backup and recovery options, including how to implement a disaster recovery plan using Azure Backup and Azure Site Recovery.

In this chapter, we will cover the following topics:

- Understanding Azure Migrate
- Selecting Azure Migrate tools
- Migrating on-premises severs to Azure
- Using Azure Update Management
- Protecting Virtual Machines with Azure Backup
- Implementing disaster recovery

Understanding Azure Migrate

You can use Azure Migrate to migrate your on-premises environment to Azure. It offers a centralized hub in the Azure portal, which can be used to track the discovery, assessment, and migration of your on-premises infrastructure, applications, and data. It offers the following features:

- **Unified Migration platform**: The migration journey can be started, executed, and tracked directly from the Azure portal.

- **Range of tools**: Assessments and migrations are offered from the hub in the Azure portal. This is integrated with other Azure services (such as **Azure Site Recovery**), other tools, and **independent software vendor** (**ISV**) offerings.

- **Different workloads**: Azure Migrate provides assessments and migration for the following:

  a) **Servers**: You can use the Azure Migrate Assessment tool, Azure Migrate Server Migration tool, and more, for assessing and migrating your servers to Azure VMs.

  b) **Databases**: You can use **Database Migration Assistant** (**DMA**) to assess databases and the **Database Migration Service** (**DMS**) for migrating on-premises databases to Azure SQL Database or Azure SQL Managed Instances.

  c) **Web applications**: You can use the Azure App Service Assistant to assess and migrate on-premises web applications to Azure App Service.

  d) **Virtual desktops**: You can use different ISV tools to assess and migrate on-premises **virtual desktop infrastructures** (**VDIs**) to Windows Virtual Desktop in Azure.

  e) **Data**: Azure Data Box products can be used to quickly and cost-effectively migrate large amounts of data to Azure.

In the next section, we are going to look at the different tools that Azure Migrate has to offer.

Selecting Azure Migrate tools

Azure Migrate offers a variety of tools that can be used to assess and migrate on-premises environments, databases, applications, and more. These will be covered in more detail in the upcoming sections.

Azure Migrate Server Assessment tool

The Azure Migrate Server Assessment tool is used to assess your on-premises infrastructure. It can assess Hyper-V VMs, VMware VMs, and physical servers. It can help you identify the following:

- **Azure readiness**: It can help you define whether your on-premises machines are ready for migration to Azure. It scans the OS of the VM to assess whether your on-premises machines are ready for migration to Azure. The VMs that are assessed can be added to the following categories:

  a) **Ready for Azure**: The machine is up to date and can be migrated as is to Azure without any changes. It will start in Azure with full Azure support.

  b) **Ready with Conditions**: Machines in this category might start in Azure but might not have full Azure support. This can be the case for machines that run an older version of Windows Server that isn't supported in Azure. It's recommended that you follow the remediation guidance that is suggested in the assessment to fix these issues.

  c) **Not ready for Azure**: Machines in this category will not start in Azure. For instance, you might have a machine with a disk of more than 64 TB attached to it. These disks cannot be hosted in Azure. For these machines, it is also recommended to follow the remediation guidance and fix the issues that appear before they are migrated. Note that for the machines in this category, right-sizing and cost estimation is not done.

  d) **Readiness unknown**: Azure Migrate wasn't able to determine the readiness of the machine because of insufficient metadata that was collected from the on-premises environment.

- **Azure sizing**: Based on historical data, such as the growth of the disks of the VM, Azure Migrate can estimate the size of Azure VMs after migration.

- **Azure cost estimation**: Azure Migrate can also estimate the costs for running your on-premises servers in Azure.

- **Dependency analysis**: When dependency analysis is enabled, Azure Migrate is capable of understanding cross-server dependencies, and it will also provide you with optimal ways to move dependent servers to Azure.

Before you can execute a server assessment, a lightweight appliance needs to be installed on the on-premises environment. Then, the appliance needs to be registered with the server assessment tool in Azure. This appliance does the following:

- It discovers the on-premises machines in your current environment. It connects to the server assessment and sends machine metadata and performance data to Azure Migrate.

- The appliance discovery process is completely agentless; it doesn't require anything to be installed on the discovered machines.

- Discovered machines can be placed into groups. Typically, you place machines into groups that you want to migrate to the same batch.

- For each group, you can create an assessment. You can figure out your migration strategy by analyzing the different assessments.

Azure Migrate Server Migration tool

The Azure Migrate Server Migration tool can be used to migrate the machines to Azure. It supports the migration of physical and virtualized servers using agent-based replication. A wide range of machines can be migrated to Azure, such as Hyper-V or VMware VMs, on-premises physical servers, VMs running in private clouds, VMs running in public clouds such as **Amazon Web Services** (**AWS**) or **Google Cloud Platform** (**GCP**), and VMs virtualized by platforms such as Xen and KVM.

You can migrate machines after assessing them, or without an assessment.

Database Migration Assistant

You can assess your on-premises SQL Server databases from Azure Migrate using Microsoft's **Database Migration Assistant** (**DMA**). Your on-premises databases can be assessed for migration to Azure SQL DB, Azure SQL managed instances, or Azure VMs running SQL Server. DMA allows you to do the following:

- Assess on-premises SQL Server instances for migrating to Azure SQL databases. The assessment workflow helps you to detect migration blocking issues, as well as partially supported or unsupported features.

- Discover issues that can affect an upgrade to an on-premises SQL Server. These are organized into different categories: breaking changes, behavior changes, and deprecated features.

- Assess on-premises **SQL Server Integration Services (SSIS)** packages that can be migrated to an Azure SQL Database or Azure SQL Database managed instance.

- Discover new features in the target SQL Server platform that the database can benefit from after an upgrade.

Database Migration Service

To perform the actual migration of the databases, Azure Migrate integrates with the Azure **Database Migration Service (DMS)**. This tool is used to migrate the on-premises databases to Azure VMs running SQL, Azure SQL DB, and Azure SQL managed instances.

Web App Migration Assistant

You can use the Azure App Service Migration Assistant directly from the Azure Migrate hub. This tool can help you assess and migrate your on-premises web apps to Azure.

It offers the following features:

- **Assess web apps online**: You can assess on-premises websites for migration to Azure App Services using the App Service Migration Assistant.

- **Migrate web apps**: Using the Azure App Service Migration Assistant, you can migrate .NET and PHP web apps to Azure.

The assessment process starts by providing a public endpoint that is scanned to provide a detailed list specifying the different technologies that are used. These technologies are then compared to other sites that are hosted on Azure App Service to generate an accurate assessment report for your site.

Offline data migration

Azure Migrate also supports offline data migration. You can use the different Azure Data Box products to move large amounts of data offline to Azure.

With the Azure Data Box solution, you can send terabytes of data into Azure quickly. You can order this Data Box solution from the Azure portal. By doing this, you will receive a Data Box storage device with storage capacity that can be used to transfer the data securely.

Azure offers the following three different types of storage devices:

- **Data Box**: This device, with 100 TB of capacity, uses standard **network-attached storage (NAS)** protocols and common copy tools. It features AES 256-bit encryption for safer transit.

- **Data Box Disk**: This device has a capacity of 8 TB with SSD storage (with packs of up to 5 for a total of 40 TB). It has a USB/SATA interface and has 128-bit encryption.

- **Data Box Heavy**: This is a self-contained device that is designed to lift 1 PB of data to the cloud.

You can copy this data from your servers to one of the devices and ship this back to Azure. Then, Microsoft will upload this data to the Azure data center from the device. This entire process is tracked in the Azure Migrate hub in the Azure portal to deliver insights regarding all the steps of the data migration process.

The Data Box solution is well-suited for scenarios where there is no or limited network connectivity and where data sizes larger than 40 TB need to be migrated. It is also an ideal solution for one-time migrations and initial bulk transfers, followed by incremental transfers over the network.

For incremental transfers over the network, Azure offers the following services:

- **Azure Data Box Gateway**: Data Box Gateway is a virtual device based on a virtual machine that's been provisioned in your virtualized environment or hypervisor. The virtual device resides in your on-premises environment and you write data to it using the **Network File System (NFS)** and **Server Message Block (SMB)** protocols. The device then transfers your data to Azure block blobs, page blobs, or Azure Files.

- **Azure Stack Edge Pro**: Azure Stack Edge Pro is a physical device supplied by Microsoft for secure data transfer. This device resides in your on-premises environment, and you can write data to it using the NFS and SMB protocols. Data Box Edge has all the gateway capabilities of Data Box Gateway. Additionally, Data Box is equipped with AI-enabled edge computing capabilities that help analyze, process, or filter data as it moves to Azure block blobs, page blobs, or Azure Files.

The following diagram shows an overview of the steps that are taken to store the data on the device and ship it back to Azure:

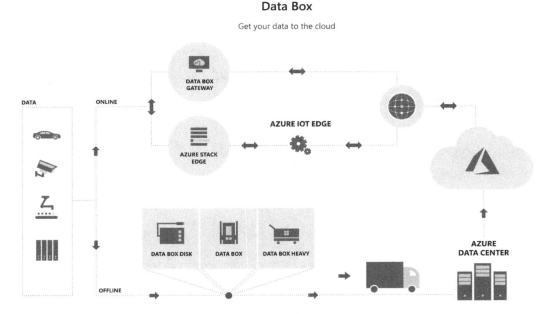

Figure 9.1 – Azure Data Box

In this section, we covered the different migration tooling that's integrated into Azure Migrate. In the next section, we are going to migrate on-premises machines to Azure.

Migrating on-premises servers to Azure

In the upcoming demo, we are going to use Azure Migrate to assess a Hyper-V environment and migrate VMs that are deployed inside the Hyper-V environment to Azure. This demo is going to be divided into different steps. The first step is to create an assessment for the on-premises environment. Due to this, we need an Azure Migrate project in the Azure portal.

Creating an Azure Migrate project

Before we can create an assessment in the Azure portal, we need to create an Azure Migrate project. Therefore, we have to perform the following steps:

1. Navigate to the Azure portal by opening `https://portal.azure.com`.

2. Click **Create a resource** and type `Azure Migrate` into the search bar.

3. Click **Create**.

4. In the **Overview** blade, click **Assess and migrate servers**.

5. The first step is to add a tool to Azure Migrate. Select **Create project**.

6. The **Create** blade will open. In the **Migrate project** tab, add the following settings:

    a) **Subscription**: Pick a subscription.

    b) **Resource group**: `PacktMigrateProjectResourceGroup`.

    c) **Migrate project**: `PacktMigrateProject`.

    d) **Geography: United States**.

7. Click **Create**.

8. After creation, you'll see the tools that were added to the project, as shown in the following screenshot:

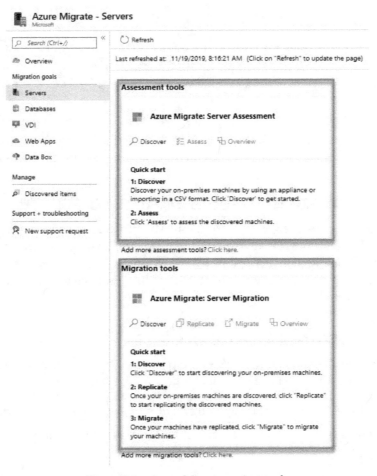

Figure 9.2 – Azure Migrate project tools

Now that we have set up the Azure Migrate project and added the necessary tools, we are going to download and install the appliance in the Hyper-V environment.

Downloading and installing the appliance

The next step is to download and install the appliance in the Hyper-V environment. Therefore, we need to perform the following steps:

1. In the Azure Migrate **Project Overview** blade, under **Migration goals**, select **Servers**.

2. Under **Assessment tools | Azure Migrate: Server Assessment**, select **Discover**.

3. In the **Discover machines** blade, under **Are your machines virtualized?**, select **Yes, with Hyper-V**.

4. You first need to create a project key, enter `PacktMigrate` next to **Name your appliance**, and then click **Generate Key**.

5. Once generated, click the **Download** button to download the appliance to your local disk:

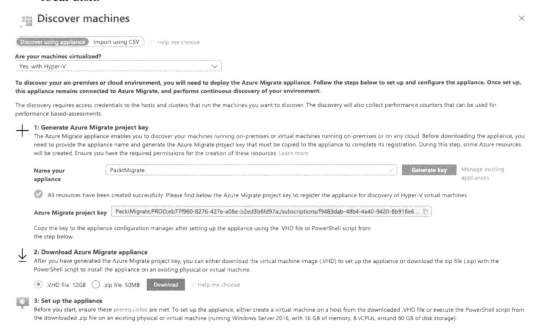

Figure 9.3 – Downloading the Azure Migrate appliance

6. Once the appliance has been downloaded, we can import it to Hyper-V. First, extract the `.zip` file.

7. Open Hyper-V Manager and, from the top menu, select **Action | Import Virtual Machine…**:

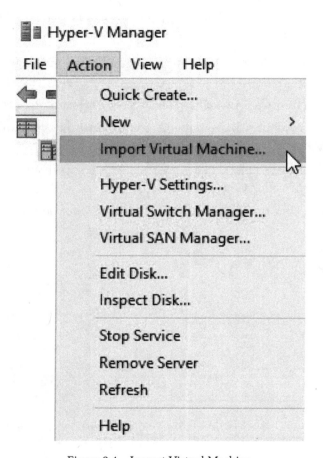

Figure 9.4 – Import Virtual Machine

8. Follow the steps in the wizard to import the VM into Hyper-V Manager.

9. In the **Import Virtual Machine Wizard | Before you begin** section, click **Next**.

10. In **Locate Folder**, navigate to the folder where you unzipped the appliance after downloading it.

11. In **Select Virtual Machine**, select the appliance VM.

12. From the **Choose Import Type** screen in the wizard, select **Copy the virtual machine (create a new unique ID)**:

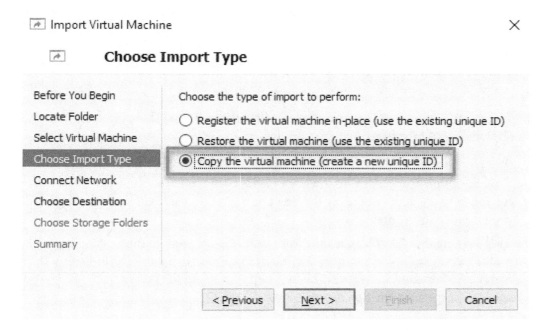

Figure 9.5 – Copying the VM

13. Click **Next**.

14. In **Choose Destination**, leave the default settings as is and click **Next**.

15. In **Storage Folders**, leave the default setting as well and click **Next**.

16. In **Choose Network**, specify the virtual switch that the VM will use. Internet connectivity is required in order to send the data to Azure from the appliance.

17. In the **Summary** screen, review your settings and click **Finish**.

18. Now that you've imported the appliance into Hyper-V, you can start the VM.

> **Important note**
>
> For this demonstration, I've created a couple of VMs inside the Hyper-V environment that can be used for the assessment and migration processes. You can create some VMs yourself as well.

Now we have downloaded the appliance and imported and started the VM. In the next section, we are going to configure the appliance so that it can communicate with Azure Migrate.

Configuring the appliance and starting continuous discovery

The appliance needs to be configured before it can send data about your on-premises machines to Azure. Therefore, we have to perform the following steps:

1. Go back to Hyper-V Manager and select **Virtual Machines**. Then, right-click the **appliance VM | Connect**.

2. Accept the license terms and provide the password for the appliance. Click **Finish**.

3. Now, the configuration of Windows Server has been finalized. This means that you can log in to the appliance using the `Administrator` username and the password you provided in the previous step.

4. After logging into the VM, click the Microsoft Migrate icon on the desktop again or wait for Internet Explorer to open automatically along with the wizard to set up the server as a configuration server. The first step is to set up the prerequisites. First, you need to accept the license terms:

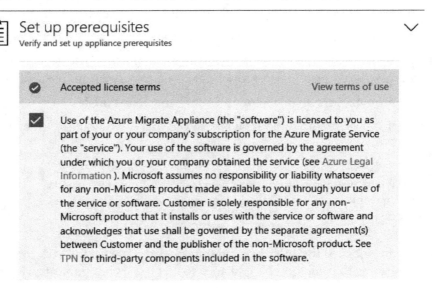

Figure 9.6 – Accepting the license terms

5. Check the box to be automatically redirected to the next step.

6. After that, your network connectivity will be checked, a check will be performed to ensure that the time of the VM is in sync with the internet time server, and the latest Azure Migrate updates will be installed. Once you've done this, you will see the following output:

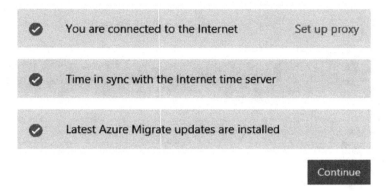

Figure 9.7 – Checking the prerequisites

7. Click **Continue**.

8. The next step is to register the appliance with Azure Migrate. Click the **Login** button and log in with your Azure credentials (make sure that you use an account that has permission to create applications in Azure AD since Azure Migrate will create an application in there):

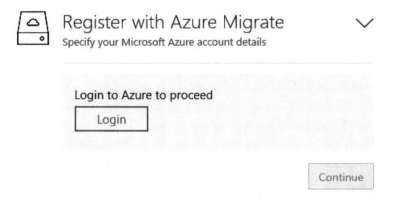

Figure 9.8 – Logging in with your Azure account

9. A new tab will open where you can log in using your credentials. By doing so, you'll be signed into Azure PowerShell at the appliance and can close the tab.

10. You will be taken back to the **Register with Azure Migrate** step in the configuration process. Now that you are signed into Azure Migrate, you can select the subscription where the Migrate project will be created. Then, select the project to migrate from the **Migrate project** dropdown and specify the appliance's name. You can obtain this value from the URL in Internet Explorer, as shown in the following screenshot:

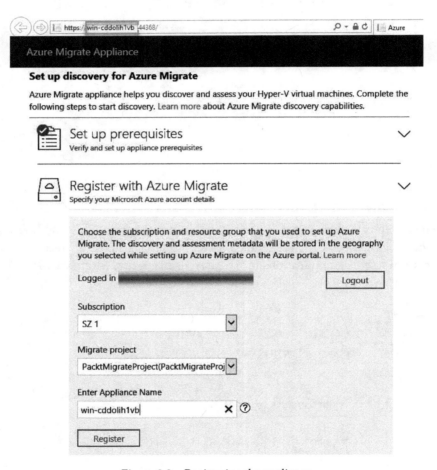

Figure 9.9 – Registering the appliance

11. Click **Register**.

12. When the appliance has been successfully registered, you can click **Continue**.

13. In the next step, we are going to configure continuous discovery. To do so, we need to provide the Hyper-V host's details. For this, use an account that has permissions on the VMs that need to be discovered and give it a friendly name:

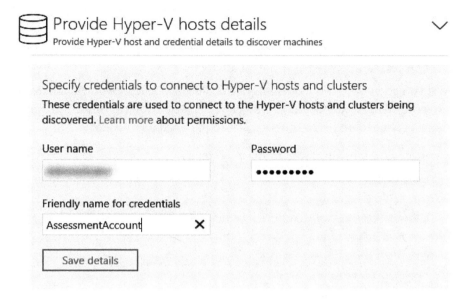

Figure 9.10 – Adding an assessment account

14. Click **Save details**.

15. Finally, you have to specify the list of Hyper-V hosts and clusters to discover. Click **Add** and specify the host IP address or FQDN. Make sure that the VMs that need to be assessed are turned on. If validation fails for a host, review the error by hovering over the icon in the **Status** column. Fix the issues and then validate it again, as shown in the following screenshot:

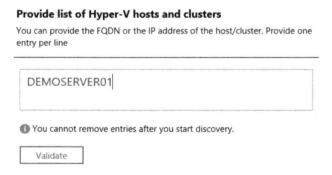

Figure 9.11 – Validating the host IP address

16. Click **Validate**.

17. After validation, click **Save and start discovery** to start the discovery process.

18. The appliance will start collecting and discovering the VMs. It will take around 15 minutes for the metadata of the discovered VMs to appear in the Azure portal. If you want to use performance data in your assessment, it is recommended that you create a performance-based assessment after 24 hours or more.

19. Now, if you go back to the Azure Migrate project in the Azure portal and select **Discovered items** under **Manage**, you can select the subscription where the project has been created, along with the project and the appliance's name.

The discovered VMs will be displayed in the overview:

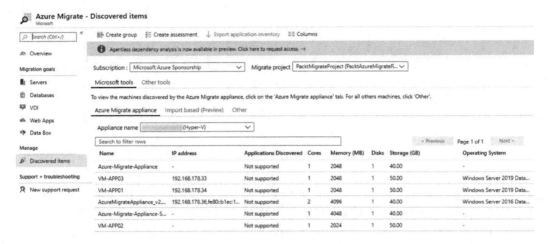

Figure 9.12 – Discovered VMs

In this section, we have configured the appliance and configured continuous discovery. In the next section, we are going to create an assessment from the Azure portal.

Creating and viewing an assessment

In this section, we are going to create an assessment in the Azure portal. To do so, perform the following steps:

1. Open the Azure Migrate project that we created earlier and, under **Manage**, select **Discovered items** again.

2. From the top menu, select **Create assessment**.

3. Add the following values:

a) **Discovery source**: Leave the default value as it is, which is **Machines discovered from Azure Migrate appliance.**

b) **Assessment name**: `PacktAssessment1`.

c) **Select or create a group**: Create a new one and call it `PacktVMGroup`.

d) **Appliance name**: Leave the default as it is.

e) Select the VMs that you want to add to the assessment:

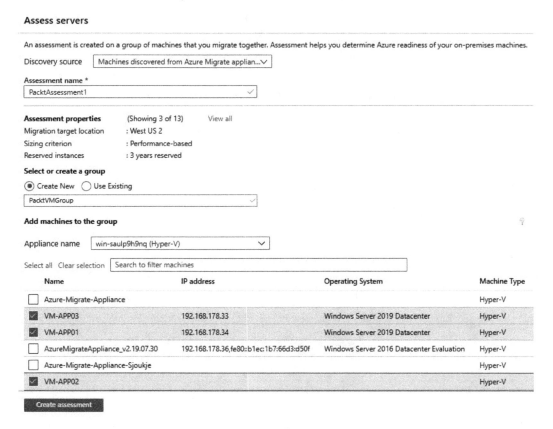

Figure 9.13 – Selecting VMs

4. Click **Create assessment.**

5. It will take some time before the assessment is ready.

6. Once the assessment has been created, you can view it by going to **Servers | Azure Migrate: Server Assessment | Assessments**:

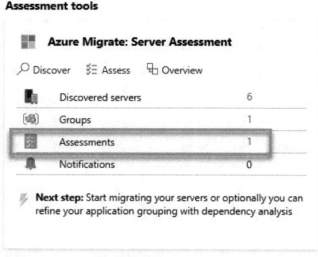

Figure 9.14 – Reviewing the assessment

7. Select the assessment from the list. You will get an overview of the assessed machines, as shown in the following screenshot:

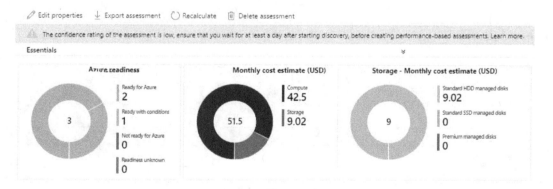

Figure 9.15 – Overview of the assessment

8. Click on **Export assessment** from the top menu to download the assessment as an Excel file.

In this section, we created and viewed an assessment. In the upcoming sections, we are going to migrate the VMs to Azure. We will start by preparing the Hyper-V host server.

Preparing the Hyper-V host

In this section, we are going to prepare the Hyper-V host so that it can migrate the VMs that we assessed previously to Azure. To do so, perform the following steps:

1. In the Azure Migrate project overview blade in the Azure portal, under **Migration goals**, select **Servers | Azure Migration | Server Migration | Discover**:

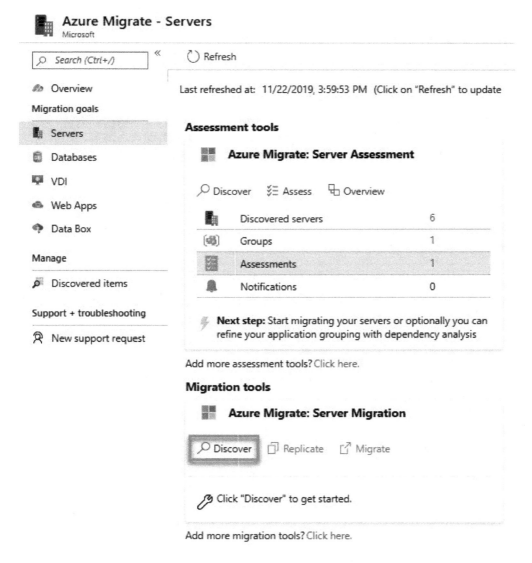

Figure 9.16 – Azure Migrate: Server Migration

2. A new blade will open where you can specify what type of machines you want to migrate. Ensure that you select the following values:

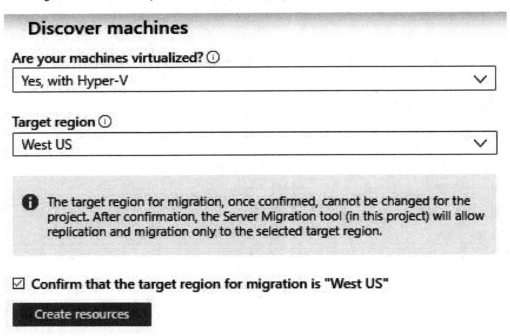

Figure 9.17 – Selecting the required machine types for migration

3. Click **Create resources**. This will start the creation of a deployment template, which creates an Azure Site Recovery vault in the background.

4. Now that the deployment template has been created, we can download the Hyper-V replication provider software installer. This is used to install the replication provider on the Hyper-V servers. Download the Hyper-V replication provider and the registration key file:

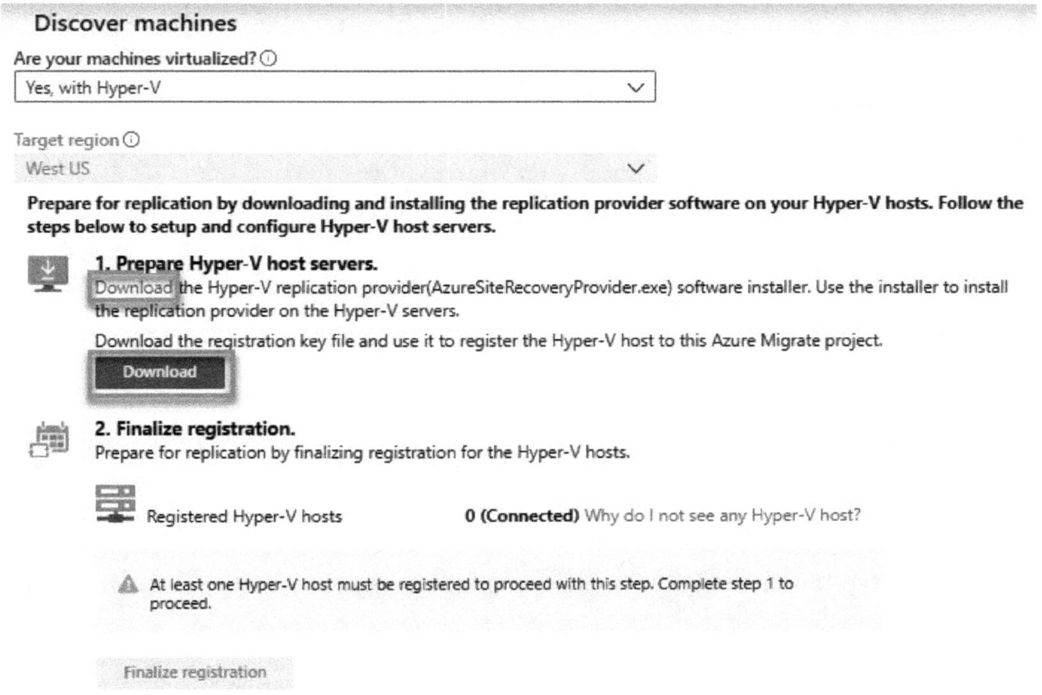

Figure 9.18 – Downloading the Hyper-V replication provider software

5. The registration key is needed to register the Hyper-V host with Azure Migrate Server Migration. Note that the key is valid for 5 days after it is generated.

6. Copy the provider setup file and registration key file for each Hyper-V host (or cluster node) running the VMs you want to replicate. Now, run the `AzureSiteRecoveryProvider.exe` file. In the first screen of the installer, you will be asked whether you want to use the Microsoft update. Click **Yes**.

7. In the next screen, you have to provide an installation directory. Leave the default as it is and click **Install**.

8. After installation, you need to register the server in the key vault that was automatically created when we created the Azure Migrate project. Then, click **Register**.

9. Now, you need to specify the vault settings. First, import the key file that we downloaded in the previous step. This will automatically fill in the **Subscription**, **Vault name**, and **Hyper-V site name** fields:

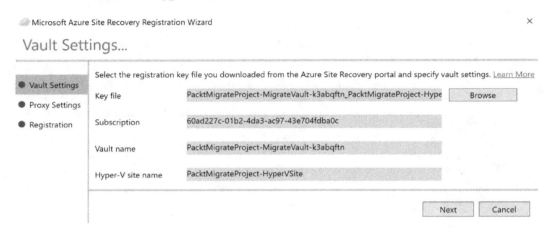

Figure 9.19 – Registering the key vault

10. Click **Next**.

11. In the next screen, you have to specify whether you want to connect to Azure Site Recovery directly or by using a proxy. Keep the default setting as it is and click **Next**:

Figure 9.20 – Connecting to Azure Site Recovery

12. The server will be registered, and all the settings configured. You can close the installer by clicking **Finish**.

13. Now, go back into the Azure portal, go to the **Discover machines** blade, and click the **Finalize registration** button:

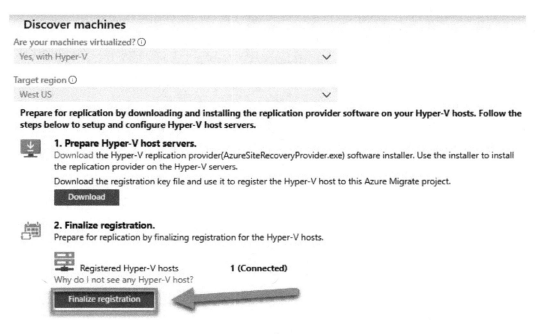

Figure 9.21 – Finalize registration

14. Now, the registration has been finalized and the VMs have been discovered. This can take up to 15 minutes.

Now that we have prepared the Hyper-V host for migration, we are going to replicate (migrate) the servers to Azure.

Replicating the Hyper-V VMs

In this section, we are going to replicate the VMs to Azure Site Recovery. To do so, perform the following steps:

> **Important note**
>
> For this part of the demonstration, you should have a resource group with a storage account, VNet, and subnet created in the same region as the target region of the discovery, which we selected in the *Prepare Hyper-V host discovery* step in the previous section. In this demonstration, this is the West US region.

1. Open the Azure Migrate project from the Azure portal. Then, under **Migration goals**, click **Servers | Azure Migrate: Server Migration** and then click **Replicate**:

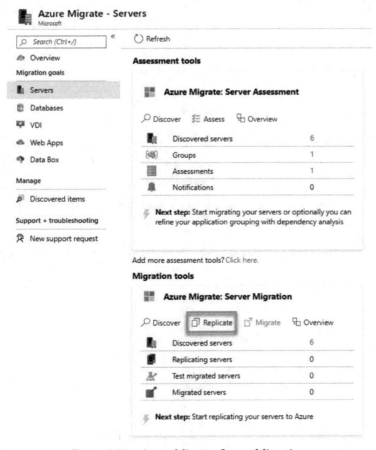

Figure 9.22 – Azure Migrate: Server Migration

2. In the **Replicate** blade, add the following values:

   Are your machines virtualized?: **Yes, with Hyper-V**.

3. Then, click **Next: Virtual Machines**. Add the following values to the **Virtual machines** blade:

   a) **Import migration settings from an Azure Migrate assessment?**: **Yes, apply migration settings from an Azure Migrate assessment**.

   b) Select the group that we created earlier, as well as the assessment.

   c) Select the VMs that you want to migrate:

Replicate

Source settings   **Virtual machines**   Target settings   Compute   Disks   Review + Start replication

Select the virtual machines to be migrated.

Import migration settings from an assessment? * ⓘ

Yes, apply migration settings from an Azure Migrate assessment	⌄

> ⓘ Only VMs that can be replicated through the specified appliance (selected in the source settings step) and having a Azure Migrate assessment are shown. We will automatically populate compute and disk information from the latest Azure Migrate assessment published to this project. You have the option of overriding the selections.

Select group * ⓘ

PacktVMGroup	⌄

Select assessment * ⓘ

PacktAssessment1	⌄

* Virtual machines ⓘ

Search to filter machines			< Previous Page 1 Next >

Name	Azure VM Readi...	Host Name
☑ VM-APP03	⚠	DEMOSERVER01
☑ VM-APP01	✓	DEMOSERVER01
☑ VM-APP02	⚠	DEMOSERVER01

Selected items : 3

« Previous: Source settings	Next: Target settings

Figure 9.23 – Selecting the assessment as input

4. Click **Next: Target settings**.

5. In **Target settings**, the target region that you'll migrate to will be automatically selected. Select the subscription, the resource group, the storage account, and the virtual network and subnet that the Azure VMs will reside in after migration. You can also select whether you want to use the Azure hybrid benefit here:

Replicate

Source settings	Virtual machines	**Target settings**	Compute	Disks	Review + Start replication

Select target properties for migration. Migrated machines will be created with the specified properties.

> ⚠️ The region (West US) selected for the migration is different from the region (West US 2) specified in the selected assessment (PacktAssessment1). Costs may vary based on the selected region.

Region * ⓘ	West US ⌄
Subscription * ⓘ	Microsoft Azure Sponsorship ⌄
└─ Resource group * ⓘ	AzureMigratedVMsResourceGroup ⌄
* Replication Storage Account ⓘ	azuremigratestorage1 (Standard) ⌄
Virtual Network * ⓘ	MigrateVNet ⌄
Subnet * ⓘ	default ⌄

Azure Hybrid Benefit

Apply Azure Hybrid Benefit and save up to 49% vs. pay-as-you-go virtual machine costs with an eligible Windows Server license.

Already have an eligible Windows Server License? * ⓘ

(Yes (**No**))

« Previous: Virtual machines	Next: Compute

Figure 9.24 – Target settings

6. Click **Next: Compute**.

7. In **Compute**, review the VM name, size, OS disk type, and availability set:

   a) **VM size**: If the assessment recommendations are used, the VM dropdown will contain the recommended size. Otherwise, Azure Migrate picks a size based on the closest match in the Azure subscription. You can also pick a size manually.

   b) **OS type**: Pick the OS type. You can choose between Windows or Linux.

   c) **OS disk**: Specify the OS (boot) disk for the VM.

d) **Availability set**: If the VM should be in an Azure availability set after migration, specify the set. The set must be in the target resource group you specify for the migration:

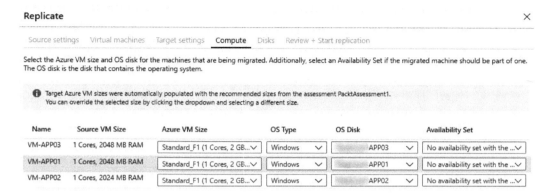

Figure 9.25 – Compute settings

8. Click **Next: Disks**.

9. In **Disks**, specify whether the VM disks should be replicated to Azure and select the disk type (standard SSD/HDD or premium-managed disks) in Azure:

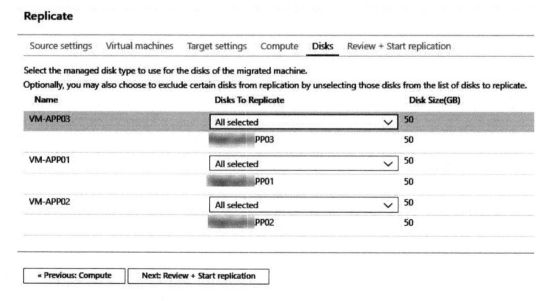

Figure 9.26 – Selecting the required disks

10. Then, click **Next: Review + start replication**.

11. Click **Replicate**.

    When the replication process has finished, you will see an overview of the replication VMs. Under **Migration goals | Servers | Azure Migrate: Server Migration | Replicating server**, you will see a list of all the replicated VMs and whether they have been replicated successfully. You can also view the health of the replication:

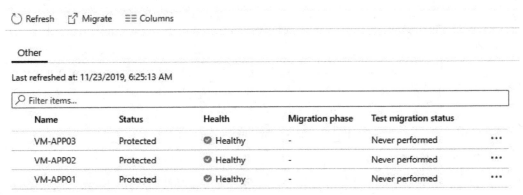

Figure 9.27 – Replication overview

> **Important note**
> It can take a while before machines are fully replicated depending on the size of the VM and your internet connection.

Now that we have replicated the VMs to Azure Site Recovery, we need to migrate them to Azure.

Replicating for the first time

If you are replicating VMs for the first time in the Azure Migrate project, the Azure Migration Server Migration tool will automatically provide the following resources:

- **Azure Service bus**: The Azure Migrate Server Migration tool uses the Azure Service Bus to send replication orchestration messages to the appliance.

- **Gateway storage account**: Server Migration uses the gateway storage account to store state information about the VMs that are being replicated.

- **Log storage account**: The Azure Migrate appliance uploads replication logs for VMs to a log storage account. Then, the replication information will be applied to the replicated managed disks.

- **Key vault**: The Azure Migrate appliance uses the key vault to manage connection strings for the service bus, as well as the access keys for the storage accounts that are used in replication.

Migrating Hyper-V VMs to Azure

For the actual migration, there are two steps: a test migration and a migration. To make sure everything's working as expected without impacting the on-premises machines, it is recommended to perform a test migration. During a test migration, the on-premises machines remain operational and can continue replicating. You can also use the replicated test Azure VM to validate the migration process, perform app testing, and address any issues prior to full migration. It is recommended to do this at least once for every machine before it gets migrated.

Running a test migration

To do a test migration, perform the following steps:

1. In the Azure Migrate project, under **Migration goals | Servers | Azure Migrate: Server Migration**, navigate to **Overview**:

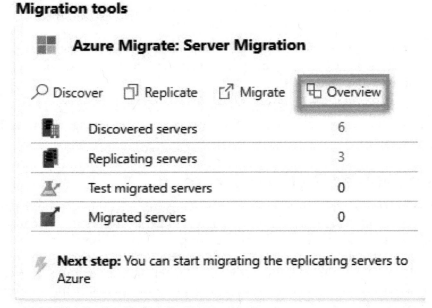

Figure 9.28 – Azure Migrate: Server Migration

2. From the **Overview** blade of **Azure Migrate: Server Migration**, click the **Test migration** button:

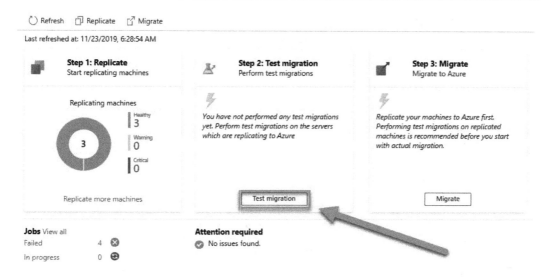

Figure 9.29 – Test migration

3. Select the VM that you want to migrate. Then, from the **Overview** page, select **Test migration** from the top menu:

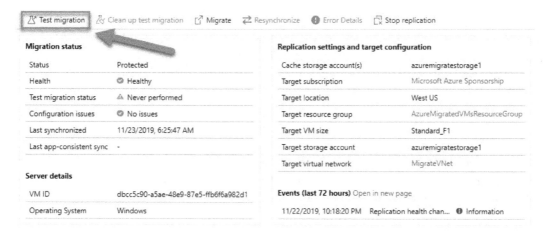

Figure 9.30 – Starting test migration for the VM

4. Select the virtual network that we created before we replicated the VMs and start the test migration by clicking the **Test migration** button.

5. After the migration has finished, you can view the migrated Azure VM by going to **Virtual machines** in the Azure portal. The machine name has a suffix of `-test`:

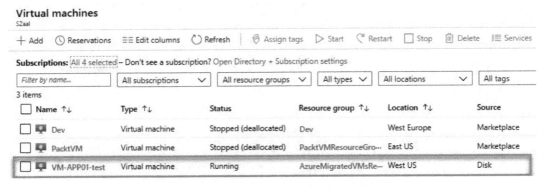

Figure 9.31 – Migrated VM after test migration

6. Clean up the test migration after finishing the test migration. You can also remove the test VM.

Now that we have executed a test migration, we can migrate the VM to Azure.

Migrating VMs to Azure

In this final part of this demonstration, we are going to migrate the VM. We did a successful test migration in the previous step, which means we can start migrating the VMs properly. The migration of VMs is also done by the **Azure Migrate: Server Migration** tool:

1. From the **Overview** blade of **Azure Migrate: Server Migration**, click the **Migrate** button.

2. In the **Migrate** blade, you need to specify whether you want to shut down the machines before migration to minimize data loss. Select **Yes**. Then, select the machines that you want to migrate:

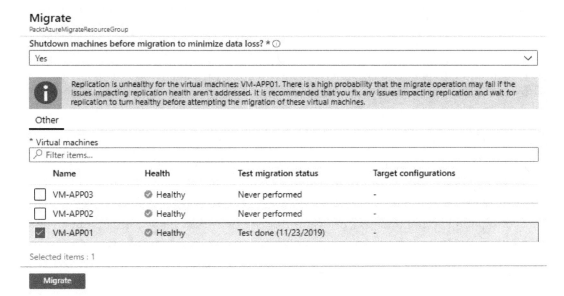

Figure 9.32 – Selecting the VMs to migrate

3. Click **Migrate**.

4. The migration process will be started.

5. When the migration process has completed, you can view the migrated Azure VM by going to **Virtual Machines** in the Azure portal:

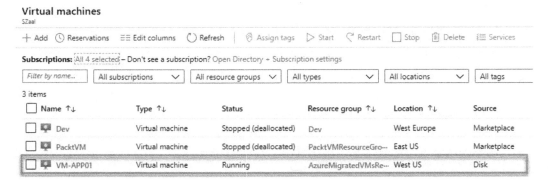

Figure 9.33 – Migrated VM

> **Important note**
> It can take a while before the VMs are fully migrated.

In this section, we migrated a VM to Azure. Next, we will look at how to keep our servers healthy and safe using Azure Update Management.

Using Azure Update Management

In the following walk-through, we will set up automated patching on a VM. This can be performed on either a VM you have migrated, or one you have created yourself.

Update Management can be enabled on the **Virtual Machine** blade, which will create an Azure Automation account for you, or by creating an Azure Automation account first and then adding VMs to it – we will be doing the latter.

A prerequisite of Update Management is also a Log Analytics workspace. If you have not created a workspace yet, or you deleted the workspace you created in the first chapter, go back to *Chapter 1, Implementing Cloud Infrastructure Monitoring*, the *Viewing alerts in Log Analytics* section, and create a new workspace.

> **Important note**
> It is best to keep your resources, such as VMs, automation accounts, and Log Analytics accounts, in the same region. So, when following these steps, and when creating the Log Analytics workspace, make sure you always use the same locations.
>
> However, pairing automation accounts with Log Analytics workspaces is *not* supported in all regions. For example, if you have a Log Analytics workspace in East US, the automation account *must* be created in *East US 2*.
>
> Refer to the following link for more details on region pairings: `https://docs.microsoft.com/en-gb/azure/automation/how-to/region-mappings`.
>
> For this example, we are assuming that all resources are in East US, which means that the automation account will be created in East US 2

Next, we will create the automation account as follows:

1. Navigate to the Azure portal by opening `https://portal.azure.com`.

2. Type `Automation` into the search bar and select **Automation Accounts**.

3. Click **Add**.

4. Enter the following details:

   a) **Name**: `PacktAutomation`.

   b) **Subscription**: Select your subscription.

   c) **Resource Group**: Select the resource group in which your Log Analytics workspace is created.

   d) **Location**: **East US 2**.

   e) **Create Azure Run As Account**: Leave as **Yes**.

5. Click **Create**.

The next step is to link the automation account with the Log Analytics workspace and then add your VMs:

1. Navigate to the Azure portal by opening `https://portal.azure.com`.

2. Type `Automation` into the search bar and select **Automation Accounts**.

3. Select the automation account you just created.

4. Click **Update Management**.

5. First you need to link your workspace to your automation account. Select your workspace from the drop-down list and then click **Enable**. Refer to the following screenshot for an example:

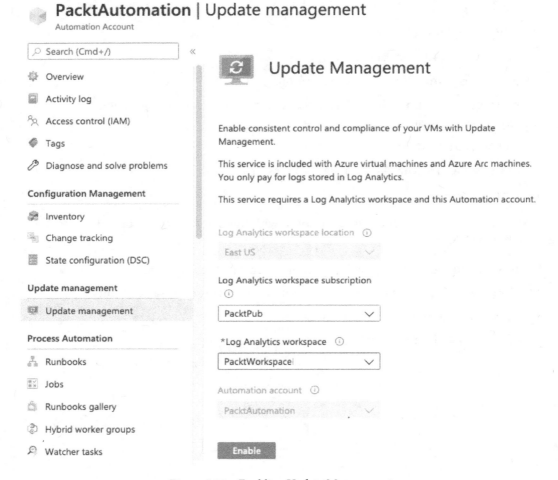

Figure 9.34 – Enabling Update Management

> **Tip**
>
> If your workspace is not available, or if you see an error, this is because your workspace or automation account is not in a support region. In the preceding example, the automation account is in East US 2, and the Log Analytics workspace is in East US.

It can take a few minutes for the Update Management feature to be enabled. You may need to navigate away from the **Update Management** blade and back again. Eventually you will see a screen like the following:

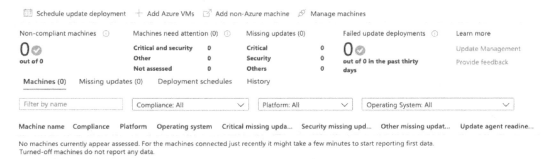

Figure 9.35 – Empty Update Management view

Once the feature is enabled, the next step is to onboard VMs to use the service by performing the following steps.

6. Still in the Update Management blade of your automation account, click **Add Azure VMs**.

7. All VMs in your subscription will be listed. Select the VMs you wish to enable updates for and then click **Enable**, as in the following example:

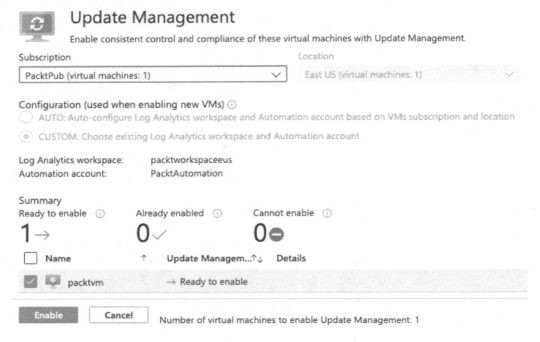

Figure 9.36 – Enabling Update Management

It can take 15 minutes to enable the necessary agents on each VM, and a few hours before they are scanned and the relevant patching information received. Once completed, you can use the **Update Management** view to see the compliance state of all onboarded VMs, as in the following example:

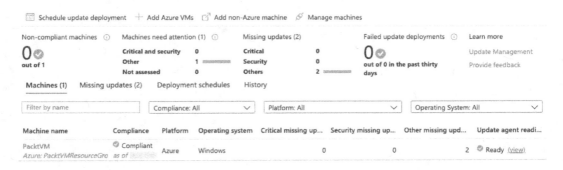

Figure 9.37 – Update Management compliance

Through the Update Management tool, you can view the status of all enabled VMs' patches.

The final step is to schedule the installation of patches:

1. Navigate to the Azure portal by opening `https://portal.azure.com`.

2. Type `Automation` into the search bar and select **Automation Accounts**.

3. Select the automation account you just created.

4. Click **Update Management**.

5. Click **Schedule deployment** and complete the details as follows:

   a) **Name**: `Patch Tuesday`

   b) **Operating System**: **Windows**

   c) **Maintenance Window (minutes)**: `120`

   d) **Reboot options: Reboot if required**

6. Under **Groups to update**, click **Click to Configure**.

7. Select your subscription and **Select All** under **Resources Groups**.

8. Click **Add**, followed by **OK**.

9. Then, click **Schedule Settings**.

10. Set the following details:

    a) **Start date**: The first **Tuesday** of the month

    b) **Recurrence: Recurring**

    c) **Recur Every**: `14 days`

11. Click **OK**.

12. **Click Create.**

Through the Update Management feature, you can control how your VMs are patched, and when and what updates to include or exclude. You can also set multiple schedules and group servers by resource group, location, or tag.

Next, we will examine another important aspect of managing VMs – Backup and Restore.

Protecting VMs with Azure Backup

Part of ensuring a reliable service is the ability to back up and restore VMs. Single VMs can still become unavailable, corruption may occur, or they may get deleted accidentally.

Azure Backup allows you to create regular or one-off virtual backups, and the backups are replicated to a paired region, with the option to use locally redundant or zone-redundant replication. These options mean your backups would not be available in another region, but it costs less and is therefore a cost-effective solution for non-critical workloads.

It isn't just Azure VMs you can protect with Azure Backup. You can also back up the following:

- On-premises VMs using the **Microsoft Azure Recovery Services (MARS)** agent
- Azure File shares
- SQL Server in Azure VMs
- SAP HANA databases in Azure VMs
- Azure Database for PostgreSQL servers

To enable backups on VMs, you need an Azure Recovery Services vault:

1. Navigate to the Azure portal by opening `https://portal.azure.com`.
2. Click **+ Create a resource**.
3. Search for and select **Backup and Site Recovery**.
4. Click **Create**.
5. Complete the following details:

    a) **Resource Group**: `PacktRecoveryVaultResourceGroup`

    b) **Vault name**: `PacktRecoveryVault`

    c) **Region**: **East US**

6. Click **Review and Create**, followed by **Create**.

Once your recovery vault is created, you can define backup policies that define when and how often backups occur:

1. Navigate to the Azure portal by opening `https://portal.azure.com`.

2. In the search bar, search for and select **Recovery Services vaults**.

3. Select your recovery vault.

4. From the left-hand menu, click **Backup policies**.

5. From here you can create new policies – note that two are created automatically:

   a) **Hourly Log Backup** – Hourly SQL Server backups

   b) **DefaultPolicy** – Standard nightly backup

6. From the left-hand menu, click **Backup**.

7. Leave the default options – **Azure** and **Virtual Machine**. Click **Backup**.

8. Next to **Policy**, select **DefaultPolicy**.

9. Under **Virtual Machines** click **Add**.

10. Select the VMs you wish to have included in the backups and click **OK**.

11. Click **Enable Backup**.

This will create a nightly backup for the VMs selected. You can view the status of your backups by selecting **Backup Items** underneath **Protected Items** on the left-hand menu of the **Recovery Services Vault** blade.

Sometimes, you may need to create a one-off backup of a single VM. This can be achieved through the **Virtual Machine** blade:

1. Navigate to the Azure portal by opening `https://portal.azure.com`.

2. In the search bar, search for and select **Virtual Machines**.

3. Select the VM you wish to back up.

4. From the left-hand menu, click **Backup** under **Operations**.

5. Click **Backup now**.

6. Click **OK**.

This will trigger a one-off backup. Wait for approximately 30 minutes for the backup to complete. From the **Backup** blade, click **View all jobs** to see the status of your backup, as in the following example:

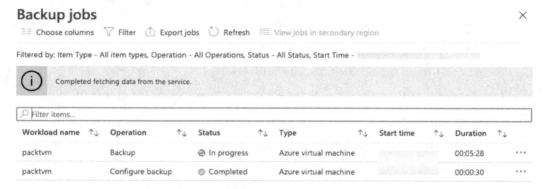

Figure 9.38 – Virtual Machine backup status

Once the backup is complete, navigate back to the **Backup** blade on the VM. You will now see the restore point from the backup. Click the ellipses to the right and you have two options – restore the entire VM or perform a file restore – as in the following example:

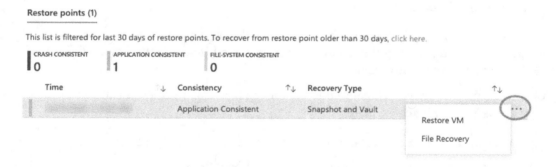

Figure 9.39 – VM restore Points

The file restore option prompts you to download a script that you will run on the VM. This will mount the backup as a drive that you can then use to copy individual files from.

To restore the entire VM, choose the **Restore VM** option. When restoring a VM, you then have two more options – either create a completely new VM by specifying **Resource Group, Virtual Network**, and the name of the new VM, or by choosing to replace the existing VM, in which case select a storage account where temporary files will be stored and then click **Restore**, as in the following example:

Restore Configuration

◯ Create new   ◉ Replace existing

    ⓘ The disk(s) from the selected restore point will replace the disk(s) in your exisiting VM. Learn more about In-Place Restore.

Restore Type ⓘ   Replace Disk(s)   ⌄

Staging Location * ⓘ   packtstorage (StandardLRS)   ⌄

                                                               Can't find your storage account ?

Restore

Figure 9.40 – Restoring a VM

By default, you can only restore a VM to the same region where it was backed up. If you wish to have the option to create cross-region restores, this must be explicitly enabled on your vault by following these steps:

1. Navigate to the Azure portal by opening `https://portal.azure.com`.
2. In the search bar, search for and select **Recovery Services vaults**.
3. Select your recovery vault.
4. From the left-hand menu, click **Properties**.
5. Under **Backup | Backup Configuration**, click **Update**.
6. Set **Cross Region Restore** to **Enable**.
7. Click **Save**.

Azure Backup enables you to protect your infrastructure. However, the timeframes between backups are limited. For services that don't store transient data, for example, web servers that don't store data, daily backup plans are sufficient.

However, in the event of a disaster scenario, consideration must be given to how quickly services can be restored, and at what point in time they need to be restored too. If a service needs a fast recovery time, or more granular restore points, then another feature can be used – **Azure Site Recovery**.

Implementing disaster recovery

Business-critical applications may have different backup and recovery requirements to servers that don't change often or store data. For applications that need to be recovered quickly, or that need to be backed up every few minutes or hours, Azure provides site recovery services.

Whereas Azure Backup takes a one-off backup at daily intervals, Azure Site Recovery continually replicates your VM's disks to another region. In the event that you need to failover, a new VM is created and attached to the replicated disk.

When determining a disaster recovery plan, two key points you need to consider are the following:

- **Recovery Time Object (RTO)** – The maximum amount of elapsed time between a failure and having your service back and running.

- **Recovery Point Objective (RPO)** – The maximum amount of data loss you are prepared to accept. If your data does not change very often, an RPO of 1 day may be sufficient, whereas for transactional data, an RPO of 5 minutes may be required.

The replication process of Azure Site Recovery ensures the lowest possible RTO. Because data is constantly being replicated to another region, when a failover occurs, the data is already where you need it – in other words, you don't need to wait for a restore to happen.

To support the lowest possible RPO, Site Recovery takes two different types of snapshots when replicating your VM:

- **Crash Consistent** – Taken every 5 minutes. Represents all the data on the disk at that moment in time.

- **App-consistent** – Crash-consistent snapshots of the disks, plus an in-memory snapshot of the VM every 60 minutes.

As well as protecting your Azure VMs across regions, you can also protect on-premises Hyper-V or VMware VMs, or even VMs in AWS.

Once your VMs are being replicated by your Site Recovery services vault, you can perform a full failover in the event of your primary region outage or run disaster recovery tests by performing test failovers or your production systems without disrupting the live service.

In the next exercise, we will create a new Recovery Services vault, protect a VM, and then perform a test failover.

> **Important note**
>
> We're going to use the same VM we used in the previous example, and for the purposes of this walk-through, it is based on a VM built in the East US region. The replicated VM will therefore be in West US. If your VM is built in a different region, you will replace West US with the paired region for your VM.

The recovery vault must also exist in the region where we want to replicate our VMs. Therefore, because we created our vault in East US, the same as our VM, we first need to create a second vault in the West US region:

1. Navigate to the Azure portal by opening `https://portal.azure.com`.

2. Click **+ Create a resource**.

3. Search for and select **Backup and Site Recovery**.

4. Click **Create**.

5. Complete the following details:

    a) **Resource Group**: `PacktDRVaultResourceGroup`

    b) **Vault name**: `PacktDRVault`

    c) **Region**: **West US**

6. Click **Review and Create**, followed by **Create**.

Once the new vault has been created, we can set up our VM replication:

1. Navigate to the Azure portal by opening `https://portal.azure.com`.

2. In the search bar, search for and select **Recovery Services vaults**.

3. Select your DR recovery vault.

4. From the left-hand menu, click **Site Recovery**.

5. Under **Azure Virtual Machines**, click **Enable Replication**.

6. Enter the following details:

    a) **Source Location**: **East US**

    b) **Source Subscription**: Your subscription

    c) **Source Resource Group**: The resource group containing your VM

    d) **Disaster Recovery between Availability Zones?**: **No**

7. Click **Next**.

8. Select your VM. Click **Next**.

9. By default, the target location will be the same as the vault, in this instance, West US. A target resource group, virtual network, and storage account will automatically be created in the target location. If you wish to edit any of the names or details, click **Customize**. For this example, we will accept the defaults. Click **Enable replication**.

You will be taken back to the Site Recovery view of the **Site Recovery Vault** blade and replication will be enabled on the VM. Once replication has been enabled, you can view the status in the **Recovery Services Vault** blade by clicking **Replicated Items** – all protected items are listed with their status.

From here, you can failover or perform test failovers of individual servers. Alternatively, in the **Recovery Plans** view, we can create recovery plans that allow you to define groups of servers. In the next walk-through, we will create a plan for our server and then perform a test failover:

1. Navigate to the Azure portal by opening `https://portal.azure.com`.

2. In the search bar, search for and select **Recovery Services vaults**.

3. Select your DR recovery vault.

4. From the left-hand menu, under **Manage**, click **Recovery Plans**.

5. Click **+ New Recovery plan**.

6. Fill in the following details:

    a) **Name**: `PacktRecoveryPlan`

    b) **Source**: **East US**

    c) **Target**: **West US**

    d) **Allow items with the deployment model**: **Resource Manager**

7. Click **Select items**.

8. Select your VM and click **OK**.

9. Click **Create**.

With our recovery plan created, we can now run a test failover:

1. Still in the **Recovery Plans** view, click the ellipses next to your recovery plan, as in the example in the following screenshot:

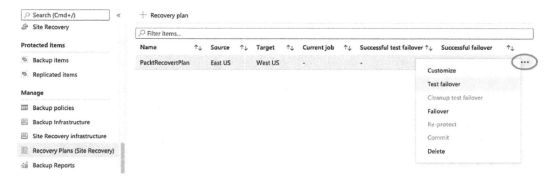

Figure 9.41 – Recovery plans

2. Click **Test failover**.

3. Select **Latest processes (low RTO)**.

4. Select the virtual network in the alternate region that has been created for you.

5. Click **OK**.

Once completed, a new VM has been created in the paired region. We can confirm this by going to the Azure Portal search bar, then typing and selecting the VM.

We will see our original VM plus a copy of it in the paired region. Note how the private IP address is the same. However, because this is a test failover, there is no public IP address, as in the example screenshot:

Name ↑↓	Type ↑↓	Status	Resource group ↑↓	Public IP address	Private IP address	Location ↑↓
PacktVM	Virtual machine	Running	packtvmresourcegroup	52.152.226.77	10.0.0.4	East US
PacktVM-test	Virtual machine	Running	PacktVMResourceGro···	-	10.0.0.4	West US

Figure 9.42 – Replicated VM

The final step in our test failover is to clean up our resources:

1. Navigate to the Azure portal by opening `https://portal.azure.com`.

2. In the search bar, search for and select **Recovery Services vaults**.

3. Select your DR recovery vault.

4. From the left-hand menu, under **Manage**, click **Recovery Plans**.

5. Click on the ellipses next to your recovery plan and click **Cleanup test failover**, as in the following screenshot:

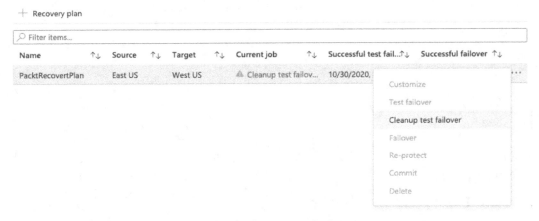

Figure 9.43 – Cleaning up the DR test

6. Enter some notes detailing your test, and then tick the **Testing is complete** checkbox

7. Click **OK**.

In this section, we can see that Azure Site Recovery provides a simple and easy mechanism to ensure that your workloads are protected against disaster scenarios, for both on-premises, Azure-hosted, or even AWS-hosted VMs.

Summary

In this chapter, we covered the first part of the *Implement Workloads and Security* objective. We learned how to use Azure Migrate so that we can migrate, as well as the different features and capabilities that it has to offer, including how to migrate VMs from a Hyper-V environment to Azure.

We have also covered how to keep your Azure VMs up to date with security patches, and the different options that are available for backing up and keeping your workloads protected.

In the next chapter, we will continue with this objective by covering how to protect your services against failures and spread workloads using load balancing technologies such as Azure Front Door, Traffic Manager, Application Gateway, Azure Firewall, and Network Security Groups.

Questions

Answer the following questions to test your knowledge of the information contained in this chapter. You can find the answers in the *Assessments* section at the end of this book:

1. Can you use Azure Migrate to migrate web applications?

   a) Yes

   b) No

2. Is Azure Migrate capable of visualizing cross-server dependencies so that groups of VMs can be migrated together?

   a) Yes

   b) No

3. Azure Migrate cannot be used to migrate physical machines.

   a) Yes

   b) No

4. Azure Site Recovery can provide protection for on-premises workloads as well as those hosted in Azure.

   a) Yes

   b) No

Further reading

You can check out the following links for more information about the topics covered in this chapter:

- Azure Migrate: `https://docs.microsoft.com/en-us/azure/migrate/migrate-services-overview`

- Migrating physical or virtualized servers to Azure: `https://docs.microsoft.com/en-us/azure/migrate/tutorial-migrate-physical-virtual-machines`

- Migrating VMware VMs to Azure (agentless): `https://docs.microsoft.com/en-us/azure/migrate/tutorial-migrate-vmware`

- Azure Site Recovery: `https://docs.microsoft.com/en-us/azure/site-recovery/`

- Azure Backup: `https://docs.microsoft.com/en-us/azure/backup/`

- Update Management: `https://docs.microsoft.com/en-us/azure/automation/update-management/manage-updates-for-vm`

10
Implementing Load Balancing and Networking Security

In the previous chapter, *Managing Workloads in Azure*, we looked at how to migrate servers and then manage security updates and protect them with Azure Backup and Azure Site Recovery.

In this chapter, we will look at other ways in which we can protect our services – preventing our applications from being overwhelmed and ensuring we control who has access to them. With this in mind, we will investigate the different ways in which we can distribute traffic among backend services using Azure Load Balancer, Traffic Manager, Application Gateway, and Azure Front Door, each of which provide different options depending on your requirements.

Next, we will look at how to secure network-level access to our platform by looking at **Network Security Groups (NSG)** and Azure Firewall.

Finally, we will examine a tool for specifically securing remote access to VMs.

This chapter will therefore cover the following topics:

- Understanding load balancing options
- Implementing Azure Load Balancer
- Implementing Azure Traffic Manager
- Understanding Azure Application Gateway
- Understanding Azure Front Door
- Choosing the right options
- Implementing network security and application security groups
- Understanding Azure Firewall
- Using Azure Bastion

Technical requirements

This chapter will use the Azure portal (`https://portal.azure.com`) for examples.

The source code for our examples can be downloaded from `https://github.com/ PacktPublishing/Microsoft-Azure-Architect-Technologies-Exam- Guide-AZ-303/tree/master/Ch10`.

Understanding load balancing options

Many PaaS options in Azure, such as Web Apps and Functions, automatically scale as demand increases (and within limits you set). For this to function, Azure places services such as these behind a load balancer in order to distribute loads between them and redirect traffic from unhealthy nodes to healthy ones.

There are times when either a load balancer is not included, such as with VMs, or when you want to provide additional functionality not provided by the standard load balancers, such as the ability to balance *between* regions. In these cases, we have the option to build and configure our own load balancers. There are a number of options you can choose from, each providing their own individual capabilities depending on your requirements.

> **Tip**
> Although we have stated that VMs require a load balancer if you wish to distribute loads between one or more VMs, when you create a VM scale set, Azure creates and manages a load balancer for you – there's no need to create one yourself.

Over the next few sections, we will be creating different load balancing technologies, and in order to test them, we will need three VMs running IIS. You will need to run through this script twice initially – to create two VMs in the same VNet and Availability Set, and then again to create a third in a different region:

1. Navigate to the Azure portal at `https://portal.azure.com`.

2. Click **Create Resource**.

3. Complete the details as follows:

    a) **Subscription**: Select your subscription.

    b) **Resource Group**: **Create New** on the first run, and use **Existing** on the second run – `PacktLbResourceGroup`.

    c) **Virtual machine name**: `PacktVM1/PacktVM2`.

    d) **Region**: **East US.**

    e) **Availability Options**: **Availability Set.**

    f) Availability Set: **Create New** on the first run, and use **Existing** on the second run.

    g) **Name**: `PacktAVS`.

       — **Fault domains: 2**

       — **Update domains: 5**

    h) **Image: Windows Server 2016 Datacenter.**

    i) **Azure Spot Instance: No.**

    j) **Size: Standard_B2ms.**

    k) **Username**: `PacktAdmin`.

    l) **Password**: `P@ssw0rd123!`.

    m) **Inbound Port Rules: Allow Selected Ports.**

    n) **Select inbound ports**: **HTTP (80)** and **RDP (3389)**.

4. Click **Next**, and then **Next** again to skip to **Networking**.

5. Complete the following:

    a) **Virtual Network**: **Create New** on the first run, and use **Existing** on the second run.

       — **Name**: `PacktLBVNET`

6. Leave the rest as their default settings.

7. Click **Review and create**, and then **Create**.

    Wait a few minutes for the first VM to be created. Once complete, repeat the preceding steps again, for **PacktVM2**. Keep everything else the same, ensuring you select the same VNet, subnets, and Availability Set you created in the first pass.

8. Finally, perform the preceding *again*, but this time with the following changes:

    a) **Virtual machine name**: `PacktVM3`

    b) **Region**: **West US**

    c) **Availability Options**: **None**

    d) **Virtual Network**: **Create New**

    e) **Name**: `PacktWestVNET`

The final steps in our preparation involve installing IIS on each VM and creating a default web page that displays the names of the VMs.

On each VM, perform the following tasks:

1. Connect to the VM using **Windows Remote Desktop**.

2. Once logged on, go to **Start Menu**, search for **PowerShell**, right-click the icon, and then choose **Run as Administrator**.

3. Enter the following commands:

```
Add-WindowsFeature Web-Server
Add-Content -Path "C:\inetpub\wwwroot\Default.htm" -Value
$($env:computername)
```

4. Disconnect from the VM.

5. Go back to the Azure portal at `https://portal.azure.com`.

6. In the search bar, type and then select **Virtual Machines**.

7. Select the VM to see its properties.

8. From the top right, next to **DNS Name**, click **Configure**.

9. Under the **DNS name** label, enter the name of the VM, for example, `PacktVM1`, `PacktVM2`, or `PacktVM2`.

10. Click **Save**.

11. Click the *Back* button in your browser to go back to the VM details blade. You may need to reload the page again in order to see the updated DNS entry.

12. In a new browser window, navigate to the VM's DNS name, for example, `http://packtvm1.eastus.cloudapp.net`, `http://packtvm2.eastus.cloudapp.net`, or `http://packtvm3.westus.cloudapp.net`.

13. You will then see a page showing the name of your VM.

> **Tip**
>
> We specifically only opened port 80 to keep configuration changes to a minimum. This is not safe for production workloads; in real environments, you should always use HTTPS.

You should now have two VMs set up, each running a basic website that displays the name of the VM when you navigate to it. We can start to deploy the different load balancing technologies, and we will start with the simplest – Azure Load Balancer.

Implementing Azure Load Balancer

Azure Load Balancer allows you to distribute traffic across VMs, allowing you to scale apps by distributing loads and offering high availability so that in the event a node becomes unhealthy, traffic is not sent to us.

Load balancers distribute traffic and manage the session persistence between nodes in one of two ways:

- The default is **Five-tuple Hash**. The tuple is composed of the source IP, source port, destination IP, destination port, and protocol type. Because the source port is included in the hash and the source port changes for each session, clients may use different VMs between sessions. This means that applications that need to maintain state for a client between requests will not work.

- The alternative is **Source IP Affinity**. This is also known as *session affinity* or *client IP affinity*. This mode uses a two-tuple hash (from the source IP address and destination IP address) or three-tuple hash (from the source IP address, destination IP address, and protocol type). This ensures that requests from a specific client are always sent to the same VM behind the load balancer, hence, applications that need to maintain state will still function.

> **Important note**
>
> To use an Azure load balancer, VMs must be built in a scale set, an Availability Set, or across Availability Zones. VMs built with Availability Sets have an SLA of 99.95%, whereas VMs built with Availability Zones have an SLA of 99.99%

Load balancers can be configured to be either internally (private) facing or external (public), and there are two SKUs for load balancers – *Basic* and *Standard*. The Basic tier is free, but only supports 300 instances, VMs in Availability Sets or scale sets, and HTTP and TCP protocols when configuring health probes. The Standard tier supports more advanced management features, such as zone-redundant frontends for inbound and outbound traffic, HTTPS probes, and you can have up to 1,000 instances. Finally, the Standard tier has an SLA of 99.99%, whereas the Basic tier offers no SLA.

To create an Azure load balancer, perform the following steps:

1. Navigate to the Azure portal at `https://portal.azure.com`.
2. Click + **Create Resource**.
3. Search for, and select, **Load Balancer**.
4. Click **Create**.
5. Complete the following details:

    a) **Subscription**: Your subscription

    b) **Resource group**: The resource group you created earlier – `PactkLBResourceGroup`

    c) **Name**: **PacktLB**

    d) **Region**: **East US**

    e) **Type**: **Public**

    f) **SKU**: **Basic**

    g) **Public IP address: Create New**

    h) **Public IP address name**: `PacktLB-IP`

    i) **IP address assignment: Dynamic**

    j) **Add a public IPv6 address: No**

6. Click **Review and Create**, and then **Create**.

    Wait for the load balancer to be created and then click **Go to resource**. To configure the load balancer, we must define a backend pool and load balancing rules. Load balancers can also have multiple frontend IP configurations, but we will use the default one that is created.

7. From the left-hand menu, click **Backend pools**.

8. Click + **Add**.

9. Set the name to `PacktWeb` and **VirtualNetwork** to the one we created for the VMs in East US – `PacktLBVNET`.

10. In **Associated to**, select **Virtual Machines**.

11. You are now presented with an option to add VMs. Click + **Add**.

12. Your VMs in *East US* will be listed (*not* `PacktVM3` in *West US*). Tick them both, and then click **Add**, as in the following screenshot:

Add virtual machines to backend pool

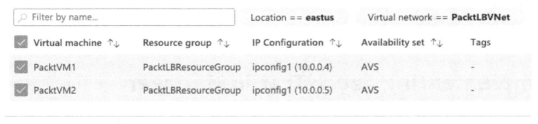

13. Click **Add** again. You will be taken back to the **Load Balancer** blade.

14. From the left-hand menu, click **Health probes**.

15. Click **Add**.

16. Enter the following details:

    a) **Name**: `PacktLBProbe`

    b) **Protocol**: **TCP**

Figure 10.1 – Adding VMs to the load balancer

      c) **Port**: **80**

      d) **Interval**: **5**

      e) **Unhealthy Threshold**: **2**

17. Click **OK** and wait for the health probe to be created.

18. From the left-hand menu, click **Load balancing rules**.

19. Click + **Add**.

20. Enter `PacktLbRule` for the name and leave everything else as its default setting.

21. Click **OK** and wait for the rule to be deployed.

Once deployed, whilst still in the **Load Balancer** blade, click **Overview** in the left-hand menu. In the top-right corner, you will see the public IP address of the load balancer. Copy this and then browse to it via HTTP.

You will see the name of one of your VMs. If you keep pressing the refresh button, you will occasionally be sent to the other server. We can also simulate a VM failing. Whichever VM you are currently seeing when browsing to your site, go into the VM blade for that VM and, from the **Overview** page, click **Stop**. Wait for 10 seconds, and then go back and refresh the browser again – this time you will be directed to the other VM.

Azure Load Balancer is a basic option for load balancing TCP traffic between two or more VMs. However, it only works within a particular region. If you wish to set up a load balancer between regions, for example, East US and West US, you need to use Azure Traffic Manager.

Implementing Azure Traffic Manager

When you wish to protect VMs or web apps across regions, for example, East US and West US, you cannot use an Azure load balancer. Instead we can use Azure Traffic Manager. Azure Traffic Manager is essentially a DNS router.

What this means is that, unlike a load balancer, which directs the flow of IP traffic from one address to another, Traffic Manager works by resolving a DNS entry, such as a web address, so a different backend IP address depending on the rules.

This enables us to direct users to the closest server available. Hence, traffic is distributed based on the user's location. If a particular region becomes unavailable, then all traffic will be directed to the healthy region.

With Azure Traffic Manager, we have a number of different options available for defining how to direct traffic as well as the two just mentioned. These are as follows:

- **Weighted**: Each endpoint is given a weight between 1 and 1,000. Endpoints are randomly assigned, but more traffic is sent to the higher-weighted endpoints.

- **Priority**: Defines a list of endpoints in priority order. All traffic goes to one particular point until that point degrades, at which point traffic gets routed to the next highest priority.

- **Performance**: Uses an internet latency table to send traffic to the fastest endpoint for the user.

- **Geographic**: Users are directed to endpoints based on their geographic location.

- **Multivalue**: Traffic Manager sends multiple healthy endpoints to the client. The client itself can then try each endpoint in turn and be responsible for determining which is the best to use.

- **Subnet**: A route based on a user's own subnet. This is useful for directing corporate users (in other words, those whereby you can pre-determine which network they are on, such as an office location).

In the next demonstration, we will set up Traffic Manager to direct traffic between our VMs in East US and West US:

1. Navigate to the Azure portal at `https://portal.azure.com`.
2. Click **+ Create Resource**.
3. Search for, and select, **Traffic Manager profile**.
4. Enter the following details:

    a) **Name**: `PacktTM`

    b) **Routing Method**: **Priority**

    c) **Subscription**: Your subscription

    d) **Resource Group**: `PacktLBResourceGroup`

5. Click **Create**.
6. Once created, click **Go to resource**.
7. In the **Traffic Manager** blade, from the left-hand menu, click **Endpoints**.
8. Click **+ Add**.

9. Enter the following details:

    a) **Type: Azure Endpoint**

    b) **Name:** `PacktVM1Endpoint`

    c) **Target resource type: Public IP address**

    d) **Public IP address:** `PacktVM1-ip` (or the public IP of your **PacktVM1**)

    e) **Priority: 1**

10. Click **Add**.

11. Repeat *steps 8 and 9*, this time with the following settings:

    a) **Type: Azure Endpoint**

    b) **Name:** `PacktVM2Endpoint`

    c) **Target resource type: Public IP address**

    d) **Public IP address:** `PacktVM3-ip` (or the public IP of your **PacktVM3**)

    e) **Priority: 100**

12. Click **Add**.

13. From the left-hand menu, click **Overview**.

14. Copy the DNS name and open this in a new browser window.

You will be directed to the **PacktVM1** server as this has the lowest priority. Go back to the Endpoints view, click **PacktVM1Endpoint**, change the priority to **1000**, and click **Save**. Then, wait for 5 minutes and refresh the browser looking at your server. The page should change to indicate you are now browsing to the **PacktVM3** server in West US.

If you wish to simulate a failed region, shut down the **PacktVM3** VM and all traffic will be directed back to the **PacktVM1** server.

The previous two technologies – Load Balancer and Traffic Manager – are relatively simple options that work only on the source and destination details.

Another alternative that can route more intelligently because it scans the actual contents of packets to determine which rule to follow is Application Gateway.

Understanding Azure Application Gateway

Azure Application Gateway is a web traffic load balancer that can be used to manage traffic to web applications. This web traffic load balancer operates at the application layer (Layer 7 in the OSI network reference stack).

It offers web load balancing, which is for HTTP(S) only. Traditional load balancers operate at the transport layer (Layer 4 in the OSI network reference stack), and route traffic—based on the source IP address and a port number—to a destination IP address and a port number. With Azure Application Gateway, traffic can be routed based on the incoming URL as well. For instance, if /pictures is part of the incoming URL, traffic can be routed to a particular set of servers that have been specifically configured for pictures. If /audio is part of the incoming URL, the traffic is routed to another set of servers, configured specifically for audio files. The following diagram shows the workflow of Azure Application Gateway:

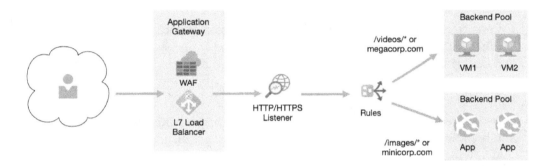

Figure 10.2 – Azure Application Gateway

Azure Application Gateway offers the following features and capabilities:

- **Web application firewall**: One of the features of the application gateway is its **web application firewall (WAF)**. It offers centralized protection of up to 40 web apps from common vulnerabilities and exploits. It is based on rules from the **Open Web Application Security Project (OWASP)** 3.1, 3.0, or 2.2.9. Common exploits include **cross-site scripting (XSS)** attacks and SQL injection attacks. With the WAF, you can centralize the prevention of such types of attacks, which makes security management a lot easier and gives a better assurance to the administrators than if this is handled in the application code. And, by patching a known vulnerability at a central location instead of in every application separately, administrators can react a lot faster to security threats.

- **URL Path-Based Routing**: This allows you to route traffic, based on URL paths, to different backend server pools.

- **Autoscaling**: Azure Application Gateway Standard_V2 offers autoscaling, whereby the number of application gateway or WAF deployments can scale based on incoming traffic. It also offers **Zone redundancy**, whereby the deployment can span multiple Availability Zones.

- **Static VIP**: This feature ensures that the VIP associated with the application gateway does not change following a restart. Additionally, it offers faster deployment and update times, and five times better SSL offload performance than the other pricing tier.

- **Secure Sockets Layer (SSL) termination**: Azure Application Gateway offers SSL termination at the gateway. After the gate, the traffic will be transported unencrypted to the backend servers. This will eliminate the need for costly encryption and decryption overheads. End-to-end SSL encryption is also supported for cases that require encrypted communication, such as when an application can only accept a secure connection, or for other security requirements.

- **Connection draining**: This feature will remove backend pool members during planned service updates. You can enable this setting at the backend **HTTP setting** and during rule creation. This setting can be applied to all the members of the backend pool. When this feature is enabled, Azure Application Gateway makes sure that all the deregistering instances in the pool do not receive any new requests.

- **Custom error pages**: You can create custom error pages using your customer layout and branding, instead of the default error pages that are displayed.

- **Multiple-site hosting**: With multiple-site hosting, more than one web app can be configured on the same application gateway. You can add up to 100 web apps to the application gateway, and each web app can be redirected to its own pool of backend servers.

- **Redirection**: Azure Application Gateway offers the ability to redirect traffic on the gateway itself. It offers a generic redirection mechanism that can be used for global redirection, whereby traffic is redirected from and to any port you define by using rules. An example of this could be an HTTP-to-HTTPS redirection. It also offers **path-based redirection**, where the HTTP-to-HTTPS traffic is only redirected to a specific site area and offers redirection to external sites.

- **Session affinity**: This feature is useful when you want to maintain a user session on the same server. By using gateway-managed cookies, the gateway can direct traffic from the same user session to the same server for processing. This is used in cases where session states are stored locally on the server for the user session.

- **WebSocket and HTTP/2 traffic**: The WebSocket and HTTP/2 protocols are natively supported by Azure Application Gateway. These protocols enable full-duplex communication between the client and the server over a long-running TCP connection, without the need for pooling. These protocols can use the same TCP connection for multiple requests and responses, which results in more efficient utilization of resources. These protocols work over the traditional HTTP ports 80 and 443.

- **Rewrite HTTP headers**: Azure Application Gateway can also rewrite the HTTP headers for incoming and outgoing traffic. This way, you can add, update, and remove HTTP headers while the request/response packets are moved between the client and the backend pools.

Azure Application Gateway comes in the following tiers:

- **Standard**: By selecting this tier, you are going to use Azure Application Gateway as a load balancer for your web apps.

- **Standard v2**: In addition to what is offered by the Standard tier, this tier offers autoscaling, zone redundancy, and support for static VIPs.

- **WAF**: By selecting this tier, you are going to create a WAF.

- **WAF v2**: In addition to the previous WAF tier, this tier offers autoscaling, zone redundancy, and support for static VIPs.

Azure Application Gateway comes in three different sizes. The following table shows an average performance throughput for each application gateway:

Page Response Size	Small	Medium	Large
6 KB	7.5 Mbps	13 Mbps	50 Mbps
100 KB	35 Mbps	100 Mbps	200 Mbps

As you can see, application gateways contain a lot of different configuration options. Now that we have a better understanding of them, we will now implement one through the portal.

Implementing the gateway

When setting up an app gateway, you need to configure a number of different components.

Frontend IP: This is the entry point for requests. This can be an internal IP, or a public IP, or both. However, there are two versions of the gateway service, V1 and V2. At present, only V1 supports purely private IPs.

Listeners: These are configured to accept requests to specific protocols, ports, hosts, or IPs. Listeners can be basic or multi-site. Basic listeners route traffic based on the URL, while multi-site listeners route traffic based on the hostname. Each listener is configured to route traffic to specified backend pools based on a routing rule. Listeners also handle SSL certificates for securing your application between the user and application gateway.

Routing rules: These bind listeners to backend pools. You specify rules to interpret the hostname and path elements of a request, and then direct the request to the appropriate backend pool. Routing rules also have an associated HTTP setting, which is used to define the protocol (HTTP or HTTPS), session stickiness, connection draining timeouts, and health probes, which are also used to help the load balancer part determine which services are available to direct traffic.

Backend pools: These define the collections of web servers that provide the ultimate service. Backend pools can be VMs, scale sets, Azure apps, or even on-premises servers.

To implement an application gateway, you first need a dedicated subnet. To do this, perform the following tasks:

1. Navigate to the Azure portal at `https://portal.azure.com`.

2. In the search bar, search for, and select, **Virtual Networks**.

3. Select the `PacktLBVnet` that your East US VMs were built in.

4. From the left-hand menu, click **Subnets**.

5. Click **+ Subnet**.

6. Next to **Name**, enter `ApplicationGateway`.

7. Leave everything else as the default settings and then click **OK**.

You should now see two subnets, the default subnet that contains your VMs, and the new `ApplicationGateway`, as in the following example:

Figure 10.3 – Creating the Application Gateway subnet

Next, we can create the application gateway by performing the following steps:

1. Navigate to the Azure portal at `https://portal.azure.com`.

2. Click **+ Create Resource**.

3. Search for, and select, **Application Gateway**.

4. Click **Create**.

5. Enter the following details:

    a) **Subscription**: Your subscription

    b) **Resource Group**: Select `PacktLBResourceGroup`

    c) **Application Gateway name**: `PacktAGW`

    d) **Region: East US**

    e) **Tier: Standard V2**

    f) **Enable autoscaling: No**

    g) **Scale Units:** 1

    h) **Availability Zone: None**

    i) **HTTP2: Disabled**

    j) **Virtual Network**: `PacktLBVNet`

    k) **Subnet**: `ApplicationGateway`

6. Click **Next: Frontends**.

7. Choose **Public** for **Frontend IP address**.

8. Next to **Public IP address**, click **Add new** and name it `PacktAGW-pip`.

9. Click **Next: Backends**.

10. Click **Add a backend pool**.

11. Name it `PacktAGWPool1`.

12. Under **Backend targets** select one of the two VMs you built in *East US* by setting **Target type** to **Virtual Machine** and then selecting the associated target's internal IP, as in the following example:

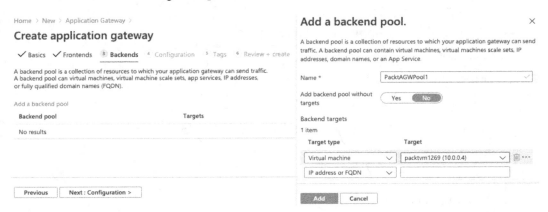

Figure 10.4 – Adding the backend targets

13. Click **Add**.

14. Add another backend pool by repeating *steps 10-13*, but this time name it `PacktAGWPool2`, and choose the other VM.

15. You should now have two backend pools, as in the following example:

Figure 10.5 – Adding two backend targets

16. Click **Next: Configuration**.

17. The final step is to create a rule to route traffic from the frontend to the backend pool. To do this, click **Add a routing rule**.

18. Name the rule `PacktAGWRule1`.

19. Under **Listener**, enter the following details:

    a) **Listener name**: `PacktAGWListener1`

    b) **Frontend IP: Public**

    c) **Protocol: HTTP**

    d) **Port**: `80`

20. Click **Backend targets** and complete the following details:

    a) **Target type: Backend pool**

    b) **Backend target**: `PacktAGWPool1`

21. Next to **Http settings**, click **Add new**.

22. Set **Http settings name** to `PacktAGWHttpSettings` and then click **Add**.

23. Click **Add** again.

24. We now need to create another routing rule for the second backend pool – repeat *steps 17-23*, this time with the following details:

    a) **Rule Name**: `PacktAGWRule2`

    b) **Listener**

    — **Listener name**: `PacktAGWListener2`

    — **Frontend IP: Public**

    — **Protocol: HTTP**

    — **Port**: `81`

    c) **Backend targets**

    — **Target type: Backend pool**

    — **Backend target**: `PacktAGWPool1`

    — **Http settings**: `PacktAGWHttpSettings` (the existing one you created in *step 22*)

25. Click **Add**. Your screen should look like this:

Figure 10.6 – Application Gateway configuration

26. This will add the rule. Then, click **Next: Tags**.

27. Click **Next: Review + create**.

28. Click **Create**.

Once built, click **Go to resource group** and then select your PacktAGW application gateway. On the **Overview** page, get the public IP of the application gateway, as in the following example:

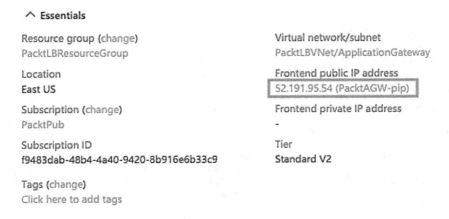

Figure 10.7 – Getting the application gateway public IP

We have configured our application gateway with two backends, one server in each, and we will route traffic from the same frontend IP to the different backends based on the port number – port 80 for one VM, and port 81 for the other.

To test this, open two browser windows and browse to the same IP in both windows. However, set one window to use port 81 by browsing to <IP address>:81, as in the following example:

PacktVM1

PacktVM2

Figure 10.8 – Browsing to the two different VMs via the application gateway

As already discussed, we have a number of options for configuring the rules to redirect traffic based on the port, hostname, or URL path to send users to one backend or the other.

Another important aspect of setting up the application gateway is ensuring that we only include backend services that are healthy, which we will investigate next.

Health probes

By default, the health of all the resources in the backend pool of Azure Application Gateway are monitored. When resources are considered unhealthy, they are automatically removed from the pool. They are monitored continuously and added back to the backend pool when they become available and respond to health probes again.

Azure Application Gateway offers default health probes and custom health probes. When a custom health probe is not created, Azure Application Gateway automatically configures a default health probe. The default health probe will monitor the resources in the backend pool by making an HTTP request to the IP addresses that have been configured for the backend pool. When HTTPS is configured, the probe will make an HTTPS request to test the health of the backend. The default health probe will test the resources in the backend pool every 30 seconds. A healthy response will have a status code between 200 and 399.

Custom health probes allow you to have more granular control over health monitoring. When using custom probes, you can configure the probe interval, the URL, and the path to test, and how many failed responses to accept, before marking the backend pool instance as unhealthy.

Monitoring

By using Azure Application Gateway, you can monitor resources, as follows:

- **Backend health**: The health of the individual servers in the backend pools can be monitored using the Azure portal, PowerShell, and the CLI. An aggregated health summary can be found in the performance diagnostic logs as well. The backend health report shows the output of the application gateway health probe to the backend instances. When probing is unsuccessful and the backend cannot receive traffic, it is considered unhealthy.

- **Metrics**: Metrics is a feature for certain Azure resources, whereby you can view performance counters in the portal. Azure Application Gateway provides seven metrics so that we can view performance counters: Current connections, Healthy host count, Unhealthy host count, Response status, Total requests, Failed requests, and Throughput.

- **Logs**: Azure provides different kinds of logs so that we can manage and troubleshoot application gateways. All the logs can be extracted from Azure Blob storage and viewed in different tools, such as Azure Monitor Logs, Excel, and Power BI. The following types of logs are supported:

  a) **Activity log**: All operations that are submitted to your Azure subscription are displayed in the Azure activity logs (formerly known as operational logs and audit logs). These logs can be viewed from the Azure portal, and they are collected by default.

  b) **Performance log**: This log reflects how the application gateway is performing. It is collected every 60 seconds, and it captures the performance information for each instance, including the throughput in bytes, the failed request count, the total requests served, and the healthy and unhealthy backend instance counts.

  c) **Firewall log**: In cases where the application gateway is configured with the WAF, this log can be viewed to display the requests that are logged through either detection or prevention mode.

  d) **Access log**: You can use this log to view application gateway access patterns and analyze important information. It is collected every 300 seconds. This information includes the caller's IP, response latency, requested URL, return code, and bytes in and out. This log contains one record per instance of the application gateway.

Logs can be stored using one of the following three options:

- **Storage account**: When logs need to be stored for a longer duration, storage accounts are the best solution. When they are stored in a storage account, they can be reviewed when needed.

- **Event Hubs**: Using Event Hubs, you can integrate the logs with other **security information and event management (SIEM)** tools to get alerts about your resources.

- **Azure Monitor Logs**: Azure Monitor Logs is the best solution for real-time monitoring of your application or for looking at trends.

Some basic monitoring can also be done from the **Overview** blade of the application gateway. There, you can find the sum total of requests, the sum total of failed requests, the sum throughput, and more.

Turning on the web application firewall

You can turn on the WAF after provisioning the application gateway. The WAF can be configured in the following modes:

- **Detection mode**: The application gateway WAF will monitor and log all threat alerts to a log file. You need to make sure that the WAF log is selected and turned on. The WAF will not block the incoming requests when it is configured in detection mode. This mode is useful for checking any potential impact a WAF could have on your application by monitoring what issues are logged.

- **Prevention mode**: In this mode, intrusions and attacks that have been detected by rules are actively blocked by the application gateway. The connection is terminated, and the attacker will receive a 403 unauthorized access exception.

Prevention mode continues to log such attacks in the WAF logs.

To enable the WAF, perform the following steps:

1. Open the application gateway resource again.

2. Under **Settings**, select **Web application firewall**. In the **WAF** blade, you have to switch the tier from **Standard V2** to **WAF V2**. Then, you can select **Detection** or **Prevention** for **Firewall mode**, and configure the required settings, as follows:

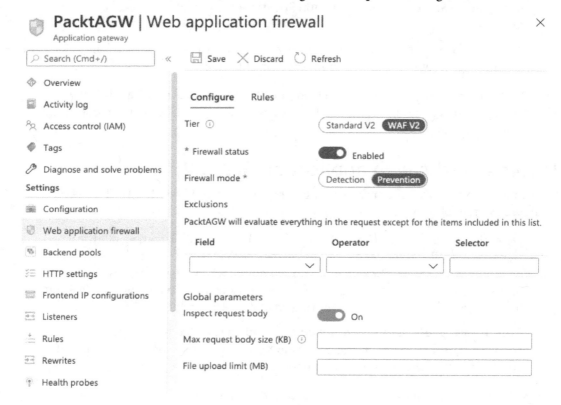

Figure 10.9 – Enabling the WAF

In this section, we covered the different ways of managing the Azure application load balancer. In the next section, we are going to look at Azure Front Door.

Understanding Azure Front Door

Azure Front Door offers a service that also works at the application layer (Layer 7). It is an **Application Delivery Network (ADN)** as a service, and it offers various load balancing capabilities for your applications.

Both Azure Front Door and Azure Application Gateway are Layer 7 (HTTP/HTTPS) load balancers. The difference between the two is that Front Door is a global service, whereas Application Gateway is a regional service. This means that Front Door can load balance between different scale units across multiple regions. Application Gateway is designed to load balance between different VMs/containers that are located inside the same scale unit.

Azure Front Door offers the following features and capabilities:

- **Accelerate application performance**: End users can quickly connect to the nearest Front Door **Point of Presence** (**POP**) using the split TCP-based anycast protocol. It then uses Microsoft's global network to connect the application to the backend.

- **Smart health probes**: Front Door increases application availability with smart health probes. These probes will monitor the backends for both availability and latency, and provide instant automatic failover when a backend goes down. This way, you can run planned maintenance operations on your applications without any downtime. Traffic is redirected to alternative backends during maintenance.

- **URL Path-Based Routing**: This allows you to route traffic to backend pools based on the URL paths of the request.

- **Multiple-site hosting**: This allows you to configure more than one web application on the same Front Door configuration. This allows a more efficient topology for deployments. Azure Front Door can be configured to route a single web application to its own backend pool or to route multiple web applications to the same backend pool.

- **Session affinity**: Azure Front Door offers managed cookies, which can be used to keep a user session on the same application backend. This feature is suitable in scenarios where the session state is saved locally on the backend for a user session.

- **Custom domains and certificate management**: If you want your own domain name to be visible in the Front Door URL, a custom domain is necessary. This can be useful for branding purposes. Also, HTTPS for custom domain names is supported and can be implemented by uploading your own SSL certificate or by implementing Front Door-managed certificates.

- **Secure Sockets Layer (SSL) termination**: Front Door offers SSL termination, which speeds up the decryption process and reduces the processing burden on backend servers. Front Door supports both HTTP and HTTPS connectivity between Front Door environments and your backends. Thus, you can also set up end-to-end SSL encryption, if this is required.

- **URL redirection**: To ensure that all communication between users and the application occurs over an encrypted path, web applications are expected to automatically redirect any HTTP traffic to HTTPS. Azure Front Door offers the functionality to redirect HTTP traffic to HTTPS. It also allows you to redirect traffic to a different hostname, redirect traffic to a different path, or redirect traffic to a new query string in the URL.

- **Application layer security**: The Front Door platform is protected by Azure DDoS Protection Basic. It also allows you to create rate-limiting rules to battle malicious bot traffic and configure custom WAF rules for access control. This can protect your HTTP/HTTPS workload from exploitation based on client IP addresses, HTTP parameters, and country codes.

- **URL rewrite**: You can configure an optional custom forwarding path to support URL rewrite in Front Door. This path can be used when the request is made from the frontend to the backend. You can configure host headers when forwarding this request.

- **Protocol support—IPv6 and HTTP/2 traffic**: Front Door natively offers end-to-end IPv6 connectivity and the HTTP/2 protocol. The HTTP/2 protocol enables full-duplex communication between application backends and a client over a long-running TCP connection.

We have seen in this section how Front Door, in fact, many of the load balancing options, can perform similar roles, so in the next section we will look at how to differentiate between them.

Choosing the right options

Each of the four load balancing options has its own use cases, and in fact, for some solutions, you may want to combine them. The following flowchart will help you to determine which to use based on your requirements:

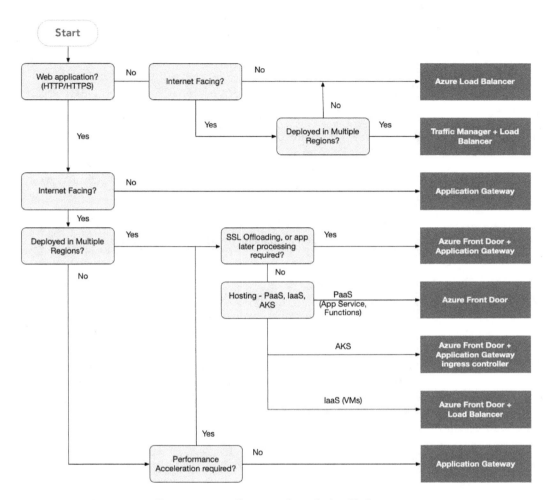

Figure 10.10 – Choosing the right load balancer

For the AZ-303 exam, you may be asked questions about not just how to implement the different options, but also when you should use one technology over another.

Load balancing is a key aspect of defining your communications strategy within a solution. Next, we will look at another equally important topic – network security, and we will start with network and application security groups.

Implementing network security and application security groups

Some Azure components, such as VMs and application gateways, must be connected to a subnet within a VNet. When you create subnets, you can optionally attach an **NSG**, which can be used to control what ports can route into it.

When we created our VMs, an NSG is created and attached to the subnet they are in, and when we chose the option to open RDP and HTTP, these ports were added to that NSG.

To see the details of the NSG that was created, in the Azure portal, in the search box, type and then select **Network Security Groups**.

You will see a list of groups; one should be called `PacktVM1-nsg` and be in the `PacktLBResourceGroup`. Click on the NSG, as in the following example:

Resource group (change)	: PacktLBResourceGroup		Custom security rules	: 2 inbound, 0 outbound
Location	: East US		Associated with	: 0 subnets, 1 network interfaces
Subscription (change)	: PacktPub			
Subscription ID	:			
Tags (change)	: Click here to add tags			

Inbound security rules

Priority	Name	Port	Protocol	Source	Destination	Action	
300	⚠ RDP	3389	TCP	Any	Any	⊘ Allow	⋯
310	HTTP	80	TCP	Any	Any	⊘ Allow	⋯
65000	AllowVnetInBound	Any	Any	VirtualNetwork	VirtualNetwork	⊘ Allow	⋯
65001	AllowAzureLoadBalancerInBound	Any	Any	AzureLoadBalancer	Any	⊘ Allow	⋯
65500	DenyAllInBound	Any	Any	Any	Any	⊗ Deny	⋯

Outbound security rules

Priority	Name	Port	Protocol	Source	Destination	Action	
65000	AllowVnetOutBound	Any	Any	VirtualNetwork	VirtualNetwork	⊘ Allow	⋯
65001	AllowInternetOutBound	Any	Any	Any	Internet	⊘ Allow	⋯
65500	DenyAllOutBound	Any	Any	Any	Any	⊗ Deny	⋯

Figure 10.11 – Example NSG ruleset

In the preceding screenshot, we can see five inbound rules and three outbound. The top two inbound rules highlighted in red were created when we created our VM when we specified allowing RDP (3389) and HTTP (80).

The three inbound and outbound rules highlighted in green are created by Azure and cannot be removed or altered. These define a baseline set of rules that must be applied in order for the platform to function correctly while blocking everything else. As the name suggests on these rules, `AllowVnetInBound` allows traffic to flow freely between all devices in that VNet, and the `AllowAzureLoadBalancerInBound` rule allows any traffic originating from an Azure load balancer. `DenyAllInBound` blocks everything else.

Each rule requires a set of options to be provided:

- **Name** and **Description** – These are for your reference. These have no bearing on the actual service; it just makes it easier to determine what it is or what it is for.

- **Source** and **Destination port** – The port is, of course, the network port that a particular service communicates on – for **RDP**, this is `3389`, for **HTTP** it is `80`, and for **HTTPS** it is `443`. Some services require port mapping, that is, the source may expect to communicate on one port, but the actual service is communicating on a different port.

- **Source** and **Destination location** – The source and destination locations define where traffic is coming *from* (the source) and where it is trying to go *to* (the destination). The most common option is an IP address or list of IP addresses, and these will typically be used to define external services.

  For Azure services, we can either choose **Virtual Network** – in other words, the destination is any service on the virtual network to which the NSG is attached, or a service tag, which is a range of IPs managed by Azure. Examples may include the following:

  a) **Internet**: Any address that doesn't originate from the Azure platform

  b) **AzureLoadBalancer**: An Azure load balancer

  c) **AzureActiveDirectory**: Communications from the Azure Active Directory service

  d) **AzureCloud.EastUS**: Any Azure service in East US

  As we can see from these examples, with the exception of the internet option, they are IP sets that belong to Azure services. Using service tags to allow traffic from Azure services is safer than manually entering the IP ranges (which Microsoft publish) as you don't need to worry about them changing.

- **Protocol** – **Any**, **TCP**, **UDP**, or **ICMP**. Services use different protocols, and some services require TCP and UDP. You should always define the least access. Therefore, if only TCP is required, only choose TCP. The ICMP protocol is used primarily for `Ping`.

- **Priority** – Firewall rules are applied one at a time in order, with the lowest number being applied last. Azure applies a **Deny all** rule to all NSGs with the lowest priority. Therefore, any rule with a higher priority will overrule this one. **Deny all** is a failsafe rule – this means that everything will be blocked by default unless you specifically create a rule to allow access.

Although `PacktVM1-nsg` was created for us when we created our VMs, we can, of course, create them individually through the Azure portal by clicking the + **Create Resource** button and choosing **Network Security Group**.

Once created, you assign a single NSG to multiple subnets; this makes it easier to control traffic across VNet and subnets in a consistent manner. However, be aware that each subnet can only have one rule associated, and the subnet and NSG must be in the same subscription.

A common practice is to create separate subnets for separate services – for example, one for websites, one for business processes, and one for databases – with separate NSG rulesets on each only allowing traffic between them. As we can see from the following example, this ensures that external services can only access the front tier, with the middle and backend tiers restricting access to the adjacent tier:

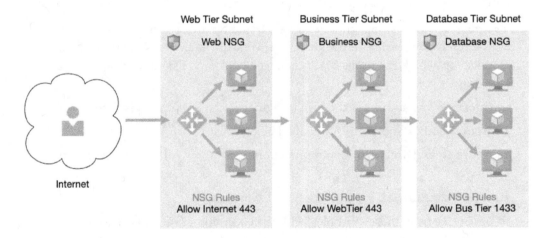

Figure 10.12 – Enabling security with subnet/NSG

When configuring the destination or source for NSG rules, another option is to use an **Application Security Group** (**ASG**).

An ASG is another way of grouping together resources instead of just allowing all traffic to all resources on your VNet. For example, you may want to define a single NSG that applies to all subnets. However, across those subnets, you may have a mixture of services, such as business servers and web servers.

You can define an ASG and attach your web servers to that ASG. In your NSG, you then set the HTTP inbound rule to use the ASG at the destination rather than the VNet. In this configuration, even though you have a common NSG, you can still uniquely allow access to specific server groups by carefully designing your subnets, NSGs, and ASGs. However, as your organization grows, managing NSGs and ASGs across subscriptions can become unwieldy. In this scenario, a better option is to create a central firewall that all your services route through.

Understanding Azure Firewall

Whereas individual NSGs and application security form part of your security strategy, building multiple layers, especially in enterprise systems, is a great way to secure your platform.

Azure Firewall is a cloud-based, fully managed network security appliance that would typically be placed at the edge of your network. This means that you would not typically have one firewall per solution, or even subscription – instead, you would have one per region and have all other devices, even those in different subscriptions, route through to it, as in the following example:

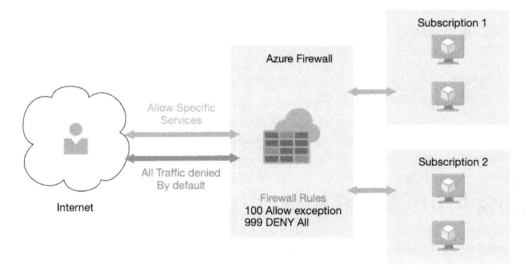

Figure 10.13 – Azure Firewall in the hub/spoke model

Azure Firewall offers some of the functionality that can be achieved from NSGs, including network traffic filtering based on the port and IP or service tags. Over and above these basic services, Azure Firewall also offers the following:

- **High Availability and Scalability** – As a managed offering, you don't need to worry about building multiple VMs with load balancers, or how much your peak traffic might be. Azure Firewall will automatically scale as required, is fully resilient, and supports Availability Zones.

- **FQDN Tags** and **FQDN Filters** – As well as IP addressing and service tags, Azure Firewall also allows you to define **Fully Qualified Domain Names** (**FQDNs**). FQDN tags are similar to service tags, but support a wider range of services, such as Windows Update.

- **Outgoing SNAT** and **Inbound DNAT** support – If you use public IP address ranges for private networks, Azure Firewall can perform **Secure Network Address Translation** (**SNAT**) on your outgoing requests. Incoming traffic can be translated using **Destination Network Address Translation** (**DNAT**).

- **Threat Intelligence** – Azure Firewall can automatically block incoming traffic originating from IP addresses known to be malicious. These addresses and domains come from Microsoft's own Threat Intelligence feed.

- **Multiple IPs** – Up to 250 IP addresses can be associated with your firewall, which helps with SNAT and DNAT.

- **Monitoring** – Azure Firewall is fully integrated with Azure Monitor for data analysis and alerting.

- **Forced Tunneling** – You can route all internet bound traffic to another device, such as an on-premises edge firewall.

Azure Firewall provides an additional and centralized security boundary for your systems, thereby ensuring an extra layer of safety.

In the final section, we will investigate how we can better secure a common operation for IaaS-based workloads – accessing VMs over **Remote Desktop Protocol** (**RDP**) or **Secure Shell** (**SSH**).

Using Azure Bastion

When working with VMs, it is common to connect to them using RDP or SSH, which, in turn, requires port 3389 (RDP) or 22 (SSH) to be opened on your VM.

If the VM is connected to an internal network, in other words, you need to use a VPN or an ExpressRoute to connect to your VM, this isn't a problem. However, connecting via RDP to a public IP on your VM is considered insecure, especially if you have to provide this access for all the VMs in your subscription.

One potential solution is to use a jump box, or bastion host in your subscription – a dedicated VM that has RDP open that can then be used to access other VMs using the internal IP addresses. However, this still means at least one VM is open and is susceptible to port scans and attacks.

Another more secure alternative is to use the Azure Bastion service. Azure Bastion is a portal-based solution, meaning you can access your VMs via the Azure portal over HTTPS. This enables you to protect the service using **Role-Based Access Control (RBAC)** and also negates the need for a public IP address or to open ports 3389 or 22 directly to your VMs from the public network.

To set up Azure Bastion, you first require a dedicated subnet. The subnet that contains the bastion must be called AzureBastionSubnet and must have a prefix of at least /27 – refer to the following example:

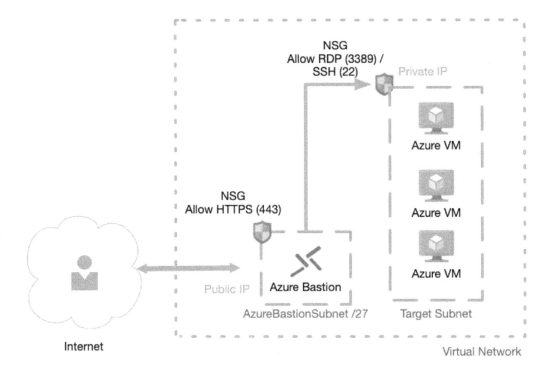

Figure 10.14 – Azure Bastion

If your bastion service is in a different VNet to your VMs, you need to peer (connect) the networks together.

Let's walk through an example of setting up a bastion service. We will first need to set up a new subnet to host the bastion:

1. Navigate to the Azure portal at `https://portal.azure.com`.

2. In the search bar, search for, and select, **Virtual Networks**.

3. Select the `PacktLBVnet` that your **East US** VMs were built in.

4. From the left-hand menu, click **Subnets**.

5. Click **+ Subnet**.

6. Next to **Name**, enter `AzureBastionSubnet`.

7. Leave everything else as the default settings and then click **OK**.

You should now see the new subnet.

With the subnet in place, we can now proceed with the building of the bastion service:

1. Navigate to the Azure portal at `https://portal.azure.com`.

2. Click **+ Create Resource**.

3. Search for, and select, **Bastion**, and then click **Create**.

4. Enter the following details:

   a) **Subscription**: Your subscription

   b) **Resource Group**: **Create New** – `PacktBastionResourceGroup`

   c) **Name**: `PacktBastion`

   d) **Region**: **East US**

   e) **Virtual network**: `PacktLBVnet`

   f) **Subnet**: `AzureBastionSubnet`

   g) **Public IP address**: **Create New**

   h) **Public IP address name**: `PacktBastion-pip`

5. Click **Review and create**.

6. Click **Create**.

Wait for the bastion service to be deployed. Once completed, we can connect to a VM using it:

1. In the Azure Search bar, search for, and select, **Virtual Machines**.

2. Select `PacktVM1` or another VM in the `PacktLBVnet` network.

3. From the left-hand menu, under **Operations**, click **Bastion**.

4. Enter your VM's username and password.

5. Click **Connect**.

6. A new browser window will open, and you will be logged in to your VM.

To disconnect from your session, log out from the VM as normal, and then choose the **Disconnect** option.

> **Tip**
> You may need to disable the pop-up blocker on your browser for the `portal.azure.com` website.

Azure Bastion provides a safer and easier way of connecting to your VMs in Azure through a web browser.

Summary

In this chapter, we have looked at different aspects of protecting workloads from security-based threats and how to respond to increasing loads.

We first looked at the load balancing options available and specifically, how to implement Azure Load Balancer, Azure Traffic Manager, and Azure Application Gateway. We also looked at Azure Front Door and how to choose between the different options.

We then went on to look at securing the network perimeter using NSGs, ASGs, and Azure Firewall for more advanced, centralized protection.

Finally, we saw how to build an Azure bastion to better protect and manage access to our VMs.

In the next chapter, we will continue with this objective by learning how to secure the Azure platform using **RBAC**, and how to simplify the management of our platform using management groups.

We will also investigate how to control internal access to the platform by enforcing governance and compliance in the form of Azure Policy and Azure Blueprints.

Questions

Answer the following questions to test your knowledge of the information contained in this chapter. You can find the answers in the *Assessments* section at the end of this book:

1. Azure Load Balancer can load balance traffic between different regions.

    a) Yes

    b) No

2. Azure Application Gateway can be used as a load balancer and as a web application firewall.

    a) Yes

    b) No

3. By using an NSG, you can block or deny access to a specific website by its URL.

    a) Yes

    b) No

4. When you set up an Azure Bastion service, you can only access VMs in the same VNet.

    a) Yes

    b) No

Further reading

You can check out the following links for more information about the topics covered in this chapter:

- Azure Load Balancer: `https://docs.microsoft.com/en-us/azure/load-balancer/load-balancer-overview`

- Azure Application Gateway documentation: `https://docs.microsoft.com/en-us/azure/application-gateway/`

- Network Security Groups: `https://docs.microsoft.com/en-us/azure/virtual-network/network-security-groups-overview`

- Application Security Groups: `https://docs.microsoft.com/en-us/azure/virtual-network/application-security-groups`

- Azure Firewall: `https://docs.microsoft.com/en-us/azure/firewall/overview`

- Azure Bastion: `https://docs.microsoft.com/en-gb/azure/bastion/`

11
Implementing Azure Governance Solutions

In the previous chapter, we covered how to protect our services from hacks and failures, and how to spread out workloads using different load balancing technologies in Azure, such as Application Gateway, Azure Firewall, Azure Load Balancer, Azure Front Door, and network security groups.

This chapter will cover the next part of the *Implement Management and Security Solutions* objective by covering management groups and **role-based access control (RBAC)**. You'll learn how to configure access to Azure resources by assigning RBAC roles from the Azure portal. You'll also learn how to configure management access by assigning global administrators to your Azure subscription and other resources. Then, you'll learn how to create custom roles that you can apply when custom permissions are needed by your users. Next, you will learn about Azure policies and how you can apply them to your Azure resources.

We will finish the chapter by looking at how to use Azure Blueprints for creating standardized platforms across subscriptions.

The following topics will be covered in this chapter:

- Understanding governance and compliance
- Understanding RBAC
- Configuring access to Azure resources by assigning roles
- Configuring management access to Azure
- Creating a custom role
- Azure Policy
- Implementing and assigning Azure policies
- Implementing and configuring Azure Blueprints
- Using hierarchical management

Technical requirements

The examples in this chapter use Azure PowerShell (`https://docs.microsoft.com/en-us/powershell/azure/`).

The source code for the sample application can be downloaded from `https://github.com/PacktPublishing/Microsoft-Azure-Architect-Technologies-Exam-Guide-AZ-303/tree/master/Ch11`.

Understanding governance and compliance

Many companies and organizations define rules that they expect their employees to abide by. Sometimes these rules align with a corporate mission, such as high levels of quality; some are for legal requirements, such as the need to protect customers' data.

When people and systems implement these rules, it is known as compliance, and the act of ensuring compliance is known as governance.

Sometimes, compliance and governance are manual tasks – a paper process to confirm the rules are being enacted. Azure provides you with tools to help you both codify and provide automatic governance to these compliance rules, thus automating much of it.

By placing automated systems within our organization, we speed up and ensure the highest compliance levels, thus saving costs while ensuring corporate principles are met.

We will begin by looking at one of the core governance toolsets in Azure, the ability to control what resources people can access.

Understanding RBAC

RBAC allows you to manage who has access to the different Azure resources inside of your tenant. You can also set what the users can do with different Azure resources.

It's good practice to assign permissions using the principle of least permissions; this involves giving users the exact permissions they need to do their jobs properly. Users, groups, and applications are added to roles in Azure, and those roles have certain permissions. You can use the built-in roles that Azure offers, or you can create custom roles in RBAC.

The roles in Azure can be added to a certain scope. This scope can be an Azure subscription, an Azure resource group, or a web application. Azure then uses access inheritance; that is, roles that are added to a parent resource give access to child resources automatically. For instance, a group that is added to an Azure subscription gets access to all the resource groups and underlying resources that are in that subscription as well. A user that is added to a **virtual machine** (**VM**) only gets access to that particular VM.

Let's start by looking at built-in roles.

Built-in roles

Azure offers various built-in roles that you can use to assign permissions to users, groups, and applications. RBAC offers the following three standard roles that you can assign to each Azure resource:

- **Owner**: Users in this role can manage everything and create new resources.
- **Contributor**: Users in this role can manage everything, just like users in the owner role, but they can't assign access to others.
- **Reader**: Users in this role can read everything, but they are not allowed to make any changes.

Aside from the standard roles, each Azure resource also has roles that are scoped to particular resources. For instance, you can assign users, groups, or applications to the SQL security manager so that they can manage all the security-related policies of the Azure SQL server. Alternatively, you can assign them to the VM contributor role, where they can manage the VMs, but not the **virtual network** (**VNet**) or storage accounts that are connected to a VM.

> **Important note**
>
> For an overview of all the built-in roles that Azure offers, refer to `https://docs.microsoft.com/en-us/azure/role-based-access-control/built-in-roles`.

While these built-in roles usually cover all possible use cases, they can never account for every requirement in an organization. To allow flexibility in role assignment, RBAC lets you make custom roles. We'll look at this feature in more detail in the next section.

Custom roles

You can also create custom roles in RBAC when none of the built-in roles suit your needs.

Custom roles can be assigned to the exact same resources as built-in roles and can only be created using PowerShell, the CLI, and the REST API. You can't create them in the Azure portal. In each Azure tenant, you can create up to 2,000 roles.

Custom roles are defined in JSON and, after deployment, they are stored inside the Azure **Active Directory** (**AD**) tenant. By storing them inside the Azure AD tenant, they can be used in all the different Azure subscriptions that are connected to the Azure AD tenant.

Configuring access to Azure resources by assigning roles

If a user in your organization needs permission to access Azure resources, you need to assign the user to the appropriate role in Azure. In this section, we are going to assign administrator access to a user for a VM. First, we need to run a script in Azure Cloud Shell to create the VM. Let's get started:

1. Navigate to the Azure portal by opening `https://portal.azure.com`.

2. Open Azure Cloud Shell.

3. First, we need to create a new resource group:

```
az group create --location eastus --name
PacktVMResourceGroup
```

4. Then, we need to create the VM:

```
az vm create '
--resource-group PacktVMResourceGroup '
--name VM1 '
--image win2016datacenter '
--admin-username packtuser '
--admin-password PacktPassword123
```

5. Now that we have the VM in place, we can configure access to the VM for the user. Open the `PacktVMResourceGroup` resource group and select **VM1** from the list.

    You will be redirected to the VM settings blade.

6. In the settings blade, select **Access control (IAM)** from the left menu and click on **+ Add | Add a role assignment** from the top menu:

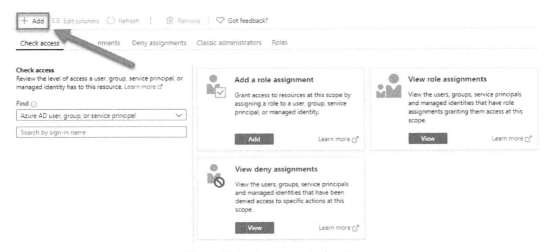

Figure 11.1 – Access control settings

7. In the **Add role assignment** blade, specify the following values:

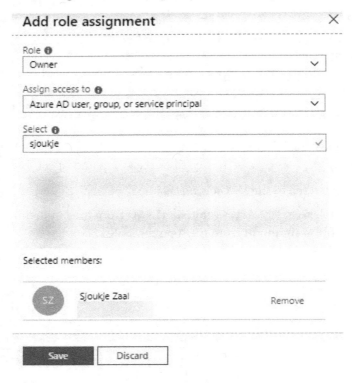

Figure 11.2 – The Add role assignment blade

8. Click on **Save**.

The user now has administrator permissions on the VM.

In this section, we assigned administrator access to a user for a VM. Now, we're going to learn how to configure management access to Azure.

Configuring management access to Azure

Management access to Azure can be configured at the subscription level. To do this, perform the following steps:

1. Navigate to the Azure portal by opening `https://portal.azure.com`.

2. Select **All services** from the left menu and type `subscriptions` into the search box. Then, select **Subscriptions**.

3. Select the subscription that you want to grant management access to from the list.

4. In the subscription settings blade, select **Access control (IAM)** and click on **Add |
Add co-administrator**, which can be found in the top menu:

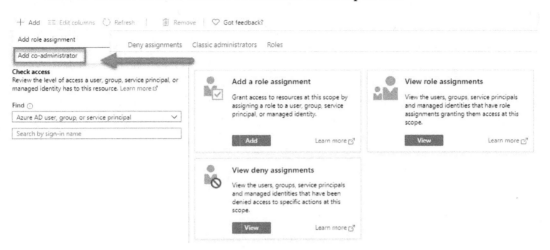

Figure 11.3 – The access control settings

5. In the **Add co-administrator** blade, search for and select a new user, as in the
following example:

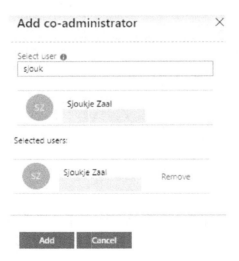

Figure 11.4 – Adding a co-administrator

6. Click on **Add**.

Now that we have configured management access to Azure, we are going to look at how to
create a custom role for our users.

Creating a custom role

In the following example, we will create a custom role that can only restart VMs in Azure. For this, you need to create a JSON file that will be deployed using PowerShell. We are going to be assigning that role to a user account inside the JSON file. Let's get started:

1. You can define the custom role by using the following JSON code. You should set Id to null since the custom role gets an ID assigned to it when it's created. We will add the custom role to two Azure subscriptions, as follows (replace the subscriptions in the AssignableScopes part with your subscription IDs):

```
{
  "Name": "Packt Custom Role",
  "Id": null,
  "IsCustom": true,
  "Description": "Allows for read access to Azure Storage,
Network and Compute resources and access to support",
  "Actions": [
  "Microsoft.Compute/*/read", "Microsoft.Storage/*/read",
  "Microsoft.Network/*/read",
  "Microsoft.Resources/subscriptions/resourceGroups/read",
  "Microsoft.Support/*"
  ],
  "NotActions": [
  ],
  "AssignableScopes": [
  "/subscriptions/********-****-****-****-************",  "/
subscriptions/********-****-****-****-************"
  ]
}
```

2. Save the JSON file in a folder named CustomRoles on the C: drive of your computer. Then, run the following PowerShell script to create the role. First, log in to your Azure account, as follows:

```
Connect-AzAccount
```

3. If necessary, select the right subscription:

```
Select-AzSubscription -SubscriptionId "********-****-
****-****-************"
```

Then, create the custom role in Azure by importing the JSON file into PowerShell:

```
New-AzRoleDefinition -InputFile "C:\CustomRoles\
PacktCustomRole.json"
```

In this section, we created a custom role that can only restart VMs in Azure. Now, we're going to take a look at how we can create policies using Azure Policy.

Azure Policy

With Azure Policy, you can create policies that enforce rules over your Azure resources. This way, resources stay compliant with service-level agreements and corporate standards. With Azure Policy, you can evaluate all the different Azure resources for non-compliance. For example, you can create a policy to allow only a certain size of VM in your Azure environment. When the policy is created, Azure will check all the new and existing VMs to see whether they apply to this policy.

Azure Policy differs from RBAC because Azure Policy focuses on controlling how resource properties can be configured, while RBAC focuses on user actions at different scopes. For example, a user can be granted the owner role in a resource group, which will give the user full rights to that resource group, but a policy might still prevent them creating a specific type of resource or setting a particular property.

Azure offers built-in policies and custom policies. Some examples of these built-in policies are as follows:

- **Allowed VM SKUs**: This policy specifies a set of VM sizes and types that can be deployed in Azure.

- **Allowed locations**: This policy restricts the available locations where resources can be deployed.

- **Not allowed resource types**: This policy prevents certain resource types from being deployed.

- **Allowed resource types**: This policy defines a list of resource types that you can deploy. Resource types that are not on the list can't be deployed inside the Azure environment.

- **Allowed storage account SKUs**: This policy specifies a set of storage account SKUs that can be deployed.

If the built-in policies don't match your requirements, you can create a custom policy instead. Custom policies are created in JSON and look similar to the following example.

The first part of the code sets the different properties:

```
{
"properties": {
"displayName": "Deny storage accounts not using only HTTPS",
"description": "Deny storage accounts not using only
HTTPS. Checks the supportsHttpsTrafficOnly property on
StorageAccounts.", "mode": "all",
"parameters": {
"effectType": {
"type": "string",
"defaultValue": "Deny",
"allowedValues": [
"Deny",
"Disabled"
],
"metadata": {
"displayName": "Effect",
"description": "Enable or disable the execution of the policy"
}
}
}
```

In the following part of the code, we are looking at the policy rule:

```
"policyRule": {
"if": {
"allOf": [
{
"field": "type",
"equals": "Microsoft.Storage/storageAccounts"
},
{
"field":
```

```
"Microsoft.Storage/storageAccounts/supportsHttpsTrafficOnly",
"notEquals": "true"
        }
    ]
},
"then": {
"effect": "[parameters('effectType')]"
        }
    }
  }
}
```

Policies are assigned at the management group level, the subscription level, or the resource group level.

Implementing and assigning Azure policies

To implement Azure policies, you have to assign them. In this section, we are going to assign an **Allowed location** policy to an Azure resource group. To do so, follow these steps:

1. Navigate to the Azure portal by opening `https://portal.azure.com`.

2. Open the `PacktVMResourceGroup` resource group.

3. Then, under **Settings**, select **Policies**.

4. Click on the **Getting started** menu item. You will see a page that looks similar to the following:

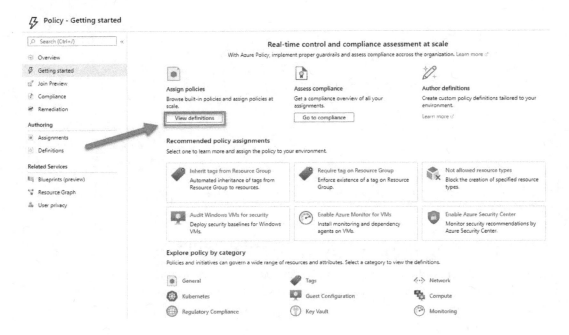

Figure 11.5 – Getting started with Azure policies

5. The first step is to view and select the policy definition. To do so, select the **View definitions** button, as shown in the preceding screenshot.

6. You will be taken to the available built-in and custom policies inside your subscription. On the right-hand side, type `Locations` into the search bar:

Figure 11.6 – Searching for a locations policy

7. Then, select the **Allowed locations** policy; you will be redirected to a blade where you can view the policy definition in JSON and assign the policy:

Figure 11.7 – Policy definition

8. Click on **Assign** in the top menu.

9.   To assign the policy, you have to fill in the following values:

A) In the **Basics** tab, apply the following values. For **Scope**, select a subscription and, optionally, a resource group. I've selected the `PacktVMResourceGroup` resource group for this demonstration:

Figure 11.8 – Add the resource group

B) In the **Parameters** tab, apply the following value. For **Allowed locations**, only select **East US**, as shown in the following screenshot:

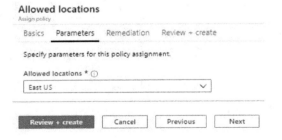

Figure 11.9 – Setting the allowed locations

Click **Review + create** and then **Create**.

10. Now, when we add a new resource to the resource group (such as a new VM) and set the location to **West Europe**, we will notice a validation error at the top left of the screen. When you click on it, you will see the following details on the right-hand side of the screen:

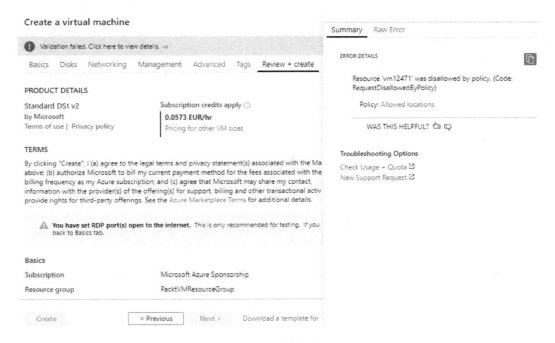

Figure 11.10 – Validation error

In this section, we learned how to assign a policy in Azure. Next, we will look at another method of controlling access to resources, and standardizing what resources are deployed within subscriptions, using Azure Blueprints.

Implementing and configuring Azure Blueprints

When new subscriptions are created within an Azure tenant, there will often be a set of components that always need to be in place. For example, every new subscription may need a VNet with pre-defined network security group rules set.

Azure Blueprints allows us to define and deploy resource groups, resources, role assignments, and policy assignments automatically as a subscription is created. A key feature of Blueprints is that the connection between what is defined and what is deployed is kept. In other words, through Blueprints, you can check for and correct any configuration drift.

Blueprints must first be defined, then versioned and published, and finally, assigned to a subscription. Whenever you modify a blueprint, it must be re-versioned and published before it can be used.

Creating a blueprint definition

Follow these steps to create a blueprint definition:

1. Navigate to the Azure portal at `https://portal.azure.com`.

2. In the top search bar, type `Blueprints` and select **Blueprints** under **Services**.

3. On the left-hand menu, select **Blueprint definitions**.

4. Click **+ Create blueprint**.

5. A list of sample and standard blueprints is listed, or you can create one from scratch. To make things easier, select **Basic Networking (VNET)**.

6. Give your blueprint a name, such as `CoreSystems`, and set **Definition location** to your subscription.

7. Click **Next: Artifacts**.

   You will be presented with a list of resources that will be deployed along with this blueprint. If you click on one of the artifacts, you will then be shown an ARM template that will be used, as in the following example:

Figure 11.11 – Example blueprint

8. If you clicked the artifact in *Step 7*, click **Cancel** to close the **Edit artifact** pane.

9. Click **Save Draft**.

Now, we have defined a blueprint; it will be in the draft state. Before we can assign it, we must publish the blueprint. When publishing a blueprint, you must set a version – this enables you to progressively update your blueprints as your platform matures. Versioning also allows you to keep a history of your changes.

Publishing and assigning a blueprint

Follow these steps for publishing and assigning a blueprint:

1. Still in the **Blueprints** blade, click **Blueprint definitions** on the left-hand menu.

2. Click your recently created blueprint.

3. Click **Publish blueprint**.

4. Enter a version number, for example, 1.0.

5. Click **Publish**.

6. Click **Assign Blueprint**.

7. Select a subscription to assign the blueprint to.

8. Select the location, for example, **East US**.

    Note the version number is **1.0**.

9. Scroll down the page to see a list of blueprint parameters that must be set – see the following screenshot for an example:

    —**Resource name prefix**: BP

    —**Resource Group: Location**: East US

    —**Address space for vnet**: `10.0.0.0/16`

    —**Address space for subnet**: `10.0.0.0/24`:

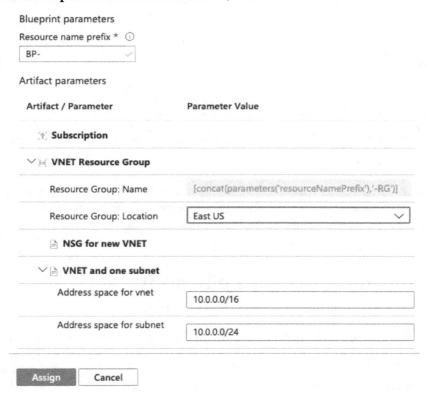

Figure 11.12 – Assigning blueprint parameters

10. Click **Assign**.

The new resource group and VNet will now be deployed to the subscription you selected; however, the process can take a few minutes.

Azure Blueprints allows you to create standardized deployments for your organization to help with governance and compliance.

In *Step 6* of the steps in the *Creating a blueprint definition* section, we created the blueprint within the subscription. In reality, when working with multiple subscriptions across large organizations, creating blueprints at this level might not make sense. In the next section, we will look at a better way of grouping subscriptions and resources using **management groups**.

Using hierarchical management

So far, we have been assigning roles, policies, and blueprints at either the resource or subscription level. In larger organizations, managing user access at the subscription level could become tedious and difficult to maintain.

Consider a multi-national company, with offices worldwide and individual departments at each location, such as HR, IT, and sales. Each department may wish to have multiple subscriptions – each one hosting a particular solution. In these cases, you may still want to maintain central control over user access.

Continually assigning rights at the subscription level would not scale very well – especially when employees join and leave the company or perhaps change roles.

Azure offers a feature called **management groups**, which can be set up to mirror your company's structure – either geographical or departmental or combining the two.

Roles, policies, and blueprints can then be assigned at each level, and those assignments flow down to the subscriptions and resources within them.

All structures start with a **root management group**, and you can add your management groups underneath. You can create *10,000* groups in total; however, you can only create *six levels* (excluding the root).

The following diagram shows an example structure broken down into region, environment (production and non-production), and finally, department. Different roles are applied at various levels, with the combined result taking effect at the subscription, resource group, or resource level:

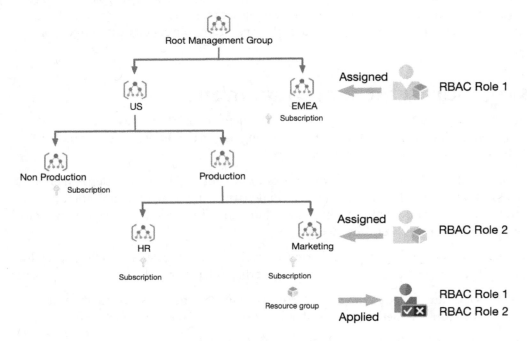

Figure 11.13 – Example management group structure

To set up a management group structure, perform the following steps:

1. Navigate to the Azure portal by opening `https://portal.azure.com`.

2. Search for and select **Management Groups**.

3. If you have not used **Management Groups** before, you will be presented with a **Start using Management groups** button – click it.

4. Enter a management group ID and display name, such as `Packt-US`.

5. Click **Save**.

6. You will now see your new management group and your subscription – note how they are both below **Tenant Root Group**, as in the following example:

Figure 11.14 – Example management groups

7. If you wish to move your subscription underneath your management group, click the ellipses to the right of your subscription and click **Move**. Select the management group and click **Save**.

8. To assign roles to the management group, click on the management group from the list, and then click **details**, as in the following screenshot:

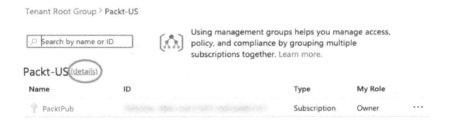

Figure 11.15 – Management group details

9. In the details view, click **Access control (IAM)** to add roles, as you did in the earlier section, *Configuring access to Azure resources by assigning roles*.

In this section, we created a single management group and moved a subscription within it. This allows easier control over user access, policies, and blueprints by allowing you to create management group structures that meet your organizational needs.

Summary

In this chapter, we finished the *Implement Management and Security Solutions* objective by covering how to configure access to Azure resources by assigning roles, configuring management access, creating a custom role, and assigning RBAC roles, as well as seeing how to implement and set Azure policies.

We then examined how Azure Blueprints can create standardized deployments for our subscriptions, and finally, we created a management group to organize resources better.

In the next chapter, we will start the *Implement Solutions for Apps* objective, beginning with Azure Web Apps, Azure Logic Apps, and Azure Functions.

Questions

Answer the following questions to test your knowledge of the information contained in this chapter. You can find the answers in the *Assessments* section at the end of this book:

1. With Azure Policy, can you assign permissions to users, giving them access to your Azure resources?

   a) Yes

   b) No

2. Suppose that you want to check whether all the VMs inside your Azure subscription use managed disks. Can you use Azure Policy for this?

   a) Yes

   b) No

3. Are custom policies created in XML?

   a) Yes

   b) No

4. Can you create up to 500 management groups in a three-layer structure?

   a) Yes

   b) No

Further reading

Check out the following links to find out more about the topics that were covered in this chapter:

- What is RBAC for Azure?: `https://docs.microsoft.com/en-us/azure/role-based-access-control/overview`

- Troubleshooting RBAC for Azure: `https://docs.microsoft.com/en-us/azure/role-based-access-control/troubleshooting`

- Overview of the Azure Policy service: `https://docs.microsoft.com/en-us/azure/governance/policy/overview`

- Creating a custom policy definition: `https://docs.microsoft.com/en-us/azure/governance/policy/tutorials/create-custom-policy-definition`

- Azure management groups: `https://docs.microsoft.com/en-us/azure/governance/management-groups/`

- Azure Blueprints: `https://docs.microsoft.com/en-us/azure/governance/blueprints/overview`

Section 3: Implement Solutions for Apps

In this section, we look at building solutions with **Platform as a Service (PaaS)** and serverless components, how they differ from traditional options, and how to enable authentication on them.

This section contains the following chapters:

- *Chapter 12, Creating Web Apps Using PaaS and Serverless*
- *Chapter 13, Designing and Developing Apps for Containers*
- *Chapter 14, Implementing Authentication*

12
Creating Web Apps Using PaaS and Serverless

In the previous chapter, we covered the last part of the *Implementing Management and Security Solutions* objective by learning how to manage user access through the use of **Role-Based Access Controls (RBAC)**, as well as how to implement governance controls with Azure Policy, Blueprints, and management groups.

This chapter introduces the *Implementing Solutions for Apps* objective. In this chapter, we are going to learn how to create Web Apps using different **Platform-as-a-Service (PaaS)** options. We are going to cover App Services and App Service plans, including alternative deployment mechanisms and how to make apps dynamically scale on demand. Next, we are going to create an App Service background task using WebJobs, before moving on to building microservice-based solutions with Azure Functions.

Finally, we will learn how Logic Apps help us rapidly build fully integrated applications using a drag and drop interface.

The following topics will be covered in this chapter:

- Understanding App Service
- Understanding App Service plans
- Using Deployment slots
- Setting up Automatic Scaling
- Understanding WebJobs
- Understanding diagnostics logging
- Using Azure Functions
- Building Azure Logic Apps

Let's get started!

Technical requirements

The examples in this chapter use Azure PowerShell (`https://docs.microsoft.com/en-us/powershell/azure/`), the Azure CLI (`https://docs.microsoft.com/en-us/cli/azure/install-azure-cli`), and Visual Studio 2019 (`https://visualstudio.microsoft.com/vs/`).

The source code for our sample application can be downloaded from `https://github.com/PacktPublishing/Microsoft-Azure-Architect-Technologies-Exam-Guide-AZ-303/tree/master/Ch12`.

Understanding App Service

App Service in Azure is part of Azure's PaaS and serverless solution, and you can use its services to host web apps, API apps, mobile apps, and logic apps. You can also host them inside App Service plans. Basically, this means that your apps are running on virtual machines that are hosted and maintained by Azure.

Azure App Service offers the following capabilities:

- **Multiple languages and frameworks**: Azure App Service supports ASP.NET, ASP.NET Core, Java, Ruby, Node.js, PHP, and Python. You can also run PowerShell and other scripts, which can be executed in App Services as background services.

- **DevOps optimization**: You can set up **continuous integration and continuous deployment (CI/CD)** with Azure DevOps, GitHub, BitBucket, Docker Hub, and Azure Container Registry. You can use the test and staging environments to deploy your apps, and you can manage these apps using the portal, PowerShell, or the CLI.

- **Global scale with high availability**: You can scale up (vertically) or out (horizontally) manually or automatically. This will be covered in the *Scaling out* and *Scaling up* sections later in this chapter.

- **Security and compliance**: App Service is ISO, SOC, and PCI compliant. You can authenticate users using Azure Active Directory or with social media logins, such as Google, Facebook, Twitter, and Microsoft accounts. You can also create IP address restrictions.

- **Visual Studio integration**: Visual Studio provides tools that can be used to create, deploy, and debug apps in App Service easily.

- **API and mobile features**: For RESTful API scenarios, Azure App Service provides turnkey **cross-origin resource sharing** (**CORS**) support. It also simplifies mobile app scenarios by enabling push notifications, offline data synchronization, authentication, and more.

- **Serverless code**: You can run scripts and code snippets on-demand without the need to provision or manage infrastructure, and you only have to pay for the resources that you are using.

Understanding App Service plans

App Service apps are hosted in App Service plans. You can configure all of the required settings, such as the compute resources, which region you want to deploy your apps to, and the costs, inside an App Service plan. You can choose from free plans, which are the most suitable for developing applications where you share all of the resources with other customers, and paid plans, where you can set the available CPU, whether you wish to host your apps on Linux or Windows VMs, and more.

Azure offers the following service plan options:

- **Dev/Test**: Free and shared are both offered for this service plan option. Your app runs in a shared environment on the same VM as other apps. This environment can also include apps from other Azure customers and users. Each app has a CPU quota and there is no ability to scale up or out. The free App Service plan can host up to 10 apps, and the shared plan can host up to 100 apps. These App Service plans are most suited for development and test apps or apps with less traffic.

There is no SLA support for these two plans. The shared service plan offers the ability to add custom domains. The service plan is shown here:

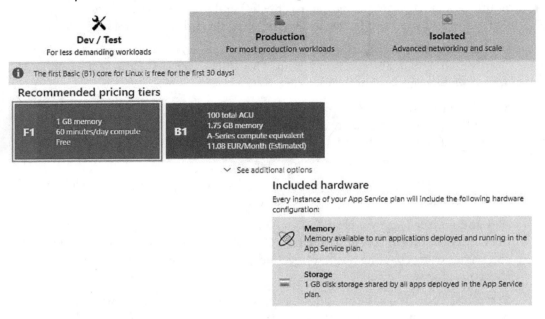

Figure 12.1 – Dev/Test App Service plans

- **Production**:

  A) **Basic**: The Basic tier is the first tier and is where you can choose between different pricing ranges. It offers three tiers, and the available cores and RAM doubles for every tier. Apps run on dedicated Linux or Windows VMs, and the compute resources are only shared between apps that are deployed inside the same App Service plan. All the apps inside the same App Service plan reside in an isolated environment that supports SSL and custom domains. The Basic tier can host an unlimited number of apps with a maximum of three instances and offers scaling to three instances, but you need to do this manually. This tier is most suitable for development and test environments and applications with less traffic.

  B) **Standard**: The Standard tier also has three tiers to choose from. It offers custom domains and SSL support, can also host an unlimited number of apps, offers autoscaling for up to 10 instances, and offers five deployment slots that can be used for testing, staging, and production apps. It also provides daily backups and Azure Traffic Manager.

C) **Premium**: Premium offers three types of tiers: Premium, Premium V2, and Premium V3. They both offer all of the features as the Standard tier, but the Premium tiers offer extra scaling instances and deployment slots. The Premium V2 tier runs on Dv2-series VMs, which have faster processors and SSD drives. The V3 tier has increased memory-to-core ratios, thus providing double the amount of RAM for the same number of cores. This effectively increases the performance of your application. This tier can host an unlimited number of apps and offers autoscaling for up to 20 instances. The dedicated compute plans in Azure portal are shown here:

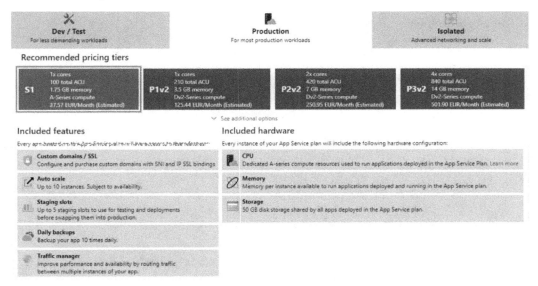

Figure 12.2 – Production App Service plans

- **Isolated**: The isolated tier offers full isolation for your applications by providing a private environment with dedicated VMs and virtual networks. This tier can host an unlimited number of apps, and you can scale up to 100 instances. These can be 100 instances in one App Service plan, or 100 different App Service plans. To create a private environment, App Service uses an **App Service Environment** (**ASE**). All the apps run on Dv2-series virtual machines, so this tier offers high-performance capabilities. The isolated App Service plan is most suitable for apps that need complete isolation because of high-security demands, for instance, but want to use all of the capabilities that Azure Web Apps offers, such as autoscaling and deployment slots. Inline creation of ASEs is not allowed in App Service creation, so you have to create an ASE separately.

Now that we have some basic information about App Service and the different App Service plans that Azure has to offer, we are going to create an Azure App Service Web App.

Creating an Azure App Service Web App

In this demonstration, we are going to create an Azure App Service Web App. We are going to deploy a basic .NET Core web API using the CLI. You can perform these steps either on your local computer, if you have the Azure CLI installed, or through the Azure Cloud Shell. For this demonstration, we will use the latter.

Go through the following steps:

1. Navigate to the Azure portal by going to `https://portal.azure.com/`.

2. Select the **Cloud Shell** button from the top-right menu bar in Azure portal.

3. The following commands will create an empty directory, and then a .NET WebAPI application inside it:

    ```
    mkdir webapi
    cd webapi
    dotnet new webapi
    ```

4. Next, we will set *git* with some user details, initialize this as a *git* repository, and then perform a commit:

    ```
    git config user.email "myemail@mydomain.com"
    git config user.name "brett"
    git init
    git add .
    git commit -m "first commit"
    ```

5. We need a deployment user we can use to push our code. We can do this with the following command (note that the username must be unique across Azure and cannot contain an @ symbol):

    ```
    az webapp deployment user set --user-name auniqueusername
    ```

6. You will be prompted for a password.

7. Next, we will create a resource group that will hold our Web App:

    ```
    az group create --location eastus --name PacktWebAppRSG
    ```

8. Create an App Service plan in the free tier:

```
az appservice plan create --name packtpubwebapi
--resource-group PacktWebAppRSG --sku FREE
```

9. Now, create a Web App:

```
az webapp create --name packtpubwebapi --resource-group
PacktWebAppRSG --plan packtpubwebapi
```

10. Now, set the local deployment mechanism to the local repository:

```
az webapp deployment source config-local-git --name
packtpubwebapi --resource-group PacktWebAppRSG
```

11. A deployment URL will be returned, along with your deployment username and publishing path – copy this value. Now, add this URL as the deployment target, as shown in the following example (replace the URL shown here with your own):

```
git remote add azure https://yourdeploymentusername@
packtpubwebapi.scm.azurewebsites.net/packtpubwebapi.git
```

12. Deploy the code:

```
git push azure master
```

13. Wait for a message stating that the deployment is complete. Then, open a browser to your application's URL and add /api/Values; for example:

```
https://packtpubwebapi.azurewebsites.net/api/Values
```

You will see the following output in your browser:

Figure 12.3 – Output API

With that, we have created and deployed the WebAPI inside an Azure Web App. In the next section, we are going to create documentation for the API.

Creating documentation for the API

It can be challenging for developers to understand the different methods of a web API when it is consumed. This problem can be solved by using Swagger, an open source software framework tool that offers developers support for automated documentation, code generation, and test case generation. This tool generates useful documentation and help pages for web APIs. It also provides benefits such as client SDK generation, API discoverability, and interactive documentation.

Swagger is a language-agnostic specification for describing REST APIs. It allows both users and computers to understand the capabilities of a service without any direct access to the implementation (source code, network access, documentation, and so on). This reduces the amount of time that's needed to accurately document a service, and it minimizes the amount of work that is needed to connect disassociated services to it.

Swagger uses a core specification document called `swagger.json`. It is generated by the Swagger toolchain (or third-party implementations of it) based on the service at hand. It describes how to access the API with HTTP(S) and what its capabilities are.

In this demonstration, we are going to create documentation for the API using Swagger. This can be created directly in Visual Studio using a NuGet package. Go through the following steps:

1. Clone or download the `PacktPubToDoAPI` sample application from the GitHub page that was referenced in the *Technical requirements* section at the beginning of this chapter.

2. Open the solution in Visual Studio.

3. In Visual Studio, from the top menu, go to **Tools | NuGet Package Manager**.

4. Open the **Browse** tab and search for `Swashbuckle.AspNetCore`. Select the package and install it for your application.

5. Wait for the package to be installed. Then, open `Startup.cs` and import the following namespace:

```
using Microsoft.OpenApi.Models;
```

6. Add the following line of code to the `ConfigureServices` method in `Startup.cs`:

```
services.AddSwaggerGen(c =>
{
c.SwaggerDoc("v1", new OpenApiInfo { Title = "To Do API",
Version = "v1" });
});
```

7. Add the following lines of code to the `configure` method in `Startup.cs`:

```
// Enable middleware to serve generated Swagger as a JSON
endpoint.
app.UseSwagger();

// Enable middleware to serve swagger-ui (HTML, JS, CSS,
etc.),
// specifying the Swagger JSON endpoint.
app.UseSwaggerUI(c =>
{
c.SwaggerEndpoint("/swagger/v1/swagger.json", "My To Do
API v1");
});
```

8. Now, run your application. When the API is displayed in the browser, replace the URL that's shown with the following:

```
https://localhost:44377/swagger/index.html
```

9.   The Swagger page will be displayed. It will look as follows:

Figure 12.4 – The Swagger page for the API

10. You can drill down into the different methods by clicking on them, as shown in the following screenshot:

Figure 12.5 – Different API methods

11. Now, you can publish this API directly from Visual Studio to the Web App that we created in the previous demonstration. Therefore, in the **Solution Explorer** window of Visual Studio, right-click on the project's name and select **Publish...**:

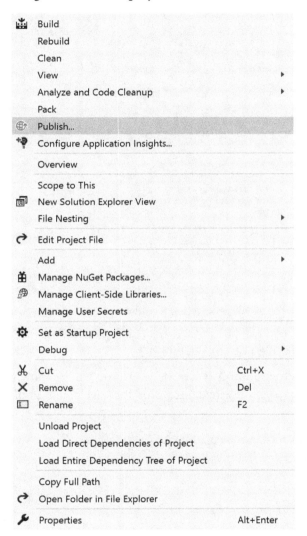

Figure 12.6 – Publishing the API

12. On the first screen of the wizard, select **Azure** and click **Next**.

13. On the next screen, select **App Service (Windows)** and then **Next**.

14. On the next screen, select the subscription where the API will be deployed, select **PacktPubToDoAPI**, and click **Next**:

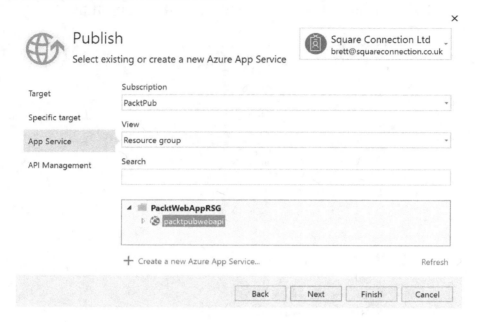

Figure 12.7 – Selecting the subscription and API

15. On the final step, you will be asked to create a façade for your APIs. At the bottom, click the **Skip this step** box and click **Finish**.

16. You will be taken to the **Publish** screen, where you will be shown the details of where your app will be published, including the URL. Click **Publish**:

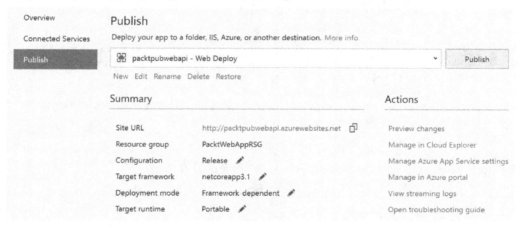

Figure 12.8 – Final publishing step

The API will be published and the browser will be opened to display the root URL of the Web App:

1. To display the API, append the following to the URL:

```
https://<our-azure-webapp-url/api/ToDo
```

2. To display the Swagger page, append the following to the root URL of the web API:

```
https://<your-azure-webapp-url/swagger/index.html
```

Now, we can navigate to Azure portal by going to `https://portal.azure.com/`.

Go to the **App Service** overview page. Then, under **API**, select **API definition**. Here, we can add the definition to App Service. Add the Swagger URL to the textbox and click **Save**.

This will update the API definition.

When publishing updates to production environments, there is always a danger that your changes may not deploy correctly. So, in the next section, we will look at an alternative way of deploying our apps that helps us easily roll back changes.

Using deployment slots

In the previous exercise, we deployed our API to the existing web app, completely overwriting the original code. Sometimes, it is better to perform a test deployment first to ensure any updates are working correctly.

We can achieve this in Azure using something called deployment slots. A deployment slot is essentially a side-by-side deployment of your Web App. It isn't the same as a test or development environment, which would have its own database and support infrastructure – it is a live deployment, except it doesn't replace the existing Web App.

Instead, it deploys to a different slot and gets a unique URL, but it still uses the same backend services, such as databases. We can then test the deployment to ensure everything is correct, and if we are happy, we can swap the slots – making the new version the live version.

Let's see an example of this in action. First, we need to change the pricing tier of the Web App as deployment slots are not available on the free tier. Follow these steps:

1. Go to Azure portal by going to `https://portal.azure.com`.

2. In the search bar, search for and select **App Services**.

3. Select the `packtpubwebpi` app we created at the start of this chapter

4. From the left-hand menu, click **Scale up (App Service plan)**.

5. Change the pricing **Tier** type from **Dev/Test** to **Production** and then select **S1**.

6. Click **Apply**.

7. From the left-hand menu, click **Deployment slots**.

8. You will see that a single slot exists called production, as shown in the following screenshot. At the top of the page, click **+Add slot**:

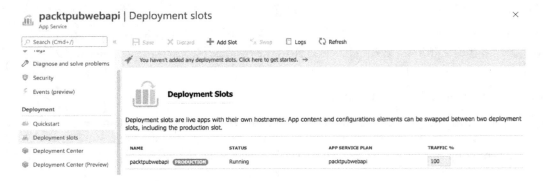

Figure 12.9 – Adding a new slot

9. Under **Name**, enter `Test`.

10. Click **Add**. Once completed, click **Close**.

We will now update our code and redeploy to the staging slot. Follow these steps:

1. Go to your `PacktToDoAPI` project in Visual Studio, right-click it, and choose **Publish**, as we did in the previous section.

2. You will be taken straight to the **Publish** page. Here, we want to create a new publishing profile, so click **New** underneath the existing profile:

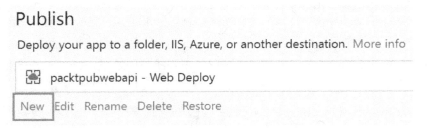

Figure 12.10 – Creating a new publishing profile

3. You will be taken through the same steps you were taken through previously. As you did previously, on the first page, select **Azure**.

4. On the next screen, select **App Service (Windows)**, then **Next**.

5. On the next screen, navigate to your app again, except this time click the arrow next to `packtpubwebapi`. A new option called **Deployment Slots** will appear.

6. Click the arrow next to **Deployment Slots** and your test deployment slot will appear. Click it, and then click **Finish**:

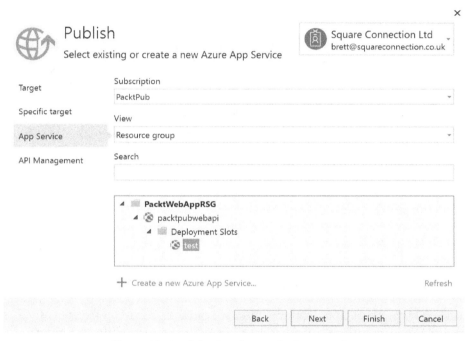

Figure 12.11 – Selecting the test deployment slot

7. Similar to what happened previously, you will be taken to the publish screen. Here, click **Publish**.

This time, when your app has been deployed, note that the URL it opens is slightly different – it is the same as your original URL, but with `-test` on the end; for example, `https://packtpubwebapi.azurewebsites.net/` has now become `https://packtpubwebapi-test.azurewebsites.net/`.

We can now test the new deployment on the test URL, even though our original deployment is still running on the original URL. Once we are ready to promote the test to production, we simply perform a swap of those slots.

To do this, go back to the **deployment slots** view in the Azure portal. At the top of the page, simply click the **Swap** button.

You will be asked to confirm which is the source slot and which is the destination. Because we only have two, they will be set up with test as the source and production as the destination. However, you can have multiple slots.

Then, click the **Swap** button. The actual swap process can take a couple of minutes, and once complete, everything that was in test will now be in production.

Using deployment slots is not only safer, but it is also quicker and produces less downtime for your apps.

In the next section, we will look at another way of minimizing disruption and downtime for our apps by using one of the key features of Azure PaaS components – scalability.

Setting up automatic scaling

The ability to scale applications is an important capability when you're fully leveraging the power of cloud technologies.

As a fully managed service, Web Apps are easy to scale – either manually or automatically. Auto-scaling is only available on S1 price tiers and above, so we will keep the Web App we published in the previous section at the S1 tier.

When we changed the pricing tier from free to S1, we performed a scale-up process. Auto-scaling uses scaling out. Scaling up involves adding more RAM and CPU to the individual instance of our Web App. Scaling out creates more instances of our Web Apps, and in front of them, Azure uses a load balancer to split traffic between those instances.

We can scale our apps based on different rules. We can scale based on a metric, such as CPU or RAM usage, or we can scale on a schedule – for example, scale up during core hours and scale back when we know the app won't be used as much.

In the following example, we will create two scale rules based on CPU metrics – one to scale out, and another to scale back in. Let's get started:

1. Go to Azure portal by going to `https://portal.azure.com`.
2. In the search bar, search for and select **App Services**.
3. Select the `packtpubwebpi` app we created at the start of this chapter.
4. From the left-hand menu, click **Scale out (App Service plan)**.

5. On the scale out view, click **Custom autoscale**.

6. Under **Autoscale setting name**, enter `PacktWebAutoScale`.

7. Keep **scale mode** set to **Scale based on a metric**.

8. Click **+Add a rule**.

9. Keep the default **metric** as-is, which is based on **CPU Percentage**, with a **Time aggregation** of **average**. We have a number of metrics available, including Data In/Out Memory Usage and TCP statistics.

10. With our metric chosen, we must decide on a threshold – accept the defaults, with **Operator** set to **Greater than** and the **Metric** threshold for triggering the action set to **70%**.

11. Next, you must set the **duration** – the default is **10 minutes**. This defines how long the threshold must be breached for before the scale out action can be triggered.

12. Finally, we can configure the number of additional instances to spin up when the threshold is breached. Again, we will leave the default as **1**. The **Cool down** setting defines how long we should wait after a single scale event before attempting to trigger another.

13. Click **Add**.

14. Before we finish, we can set an upper and lower limit on the number of instances, as well as the default count. This is useful for keeping control of costs.

15. We can now add another scale condition by clicking **+ Add a scale condition** at the bottom. This is important because we need to ensure that our instances scale back when traffic eases down.

16. Repeat steps *8-12* but this time, set the following values:

    a) **Metric name: CPU Percentage**

    b) **Operator: Less than**

    c) **Metric threshold to trigger scale condition**: `30`

    d) **Action | Operation: Decrease count by**

    e) **Action | Instance count**: `1`

17. Click **Add**.

18. You should see two scale settings, similar to what's shown in the following screenshot:

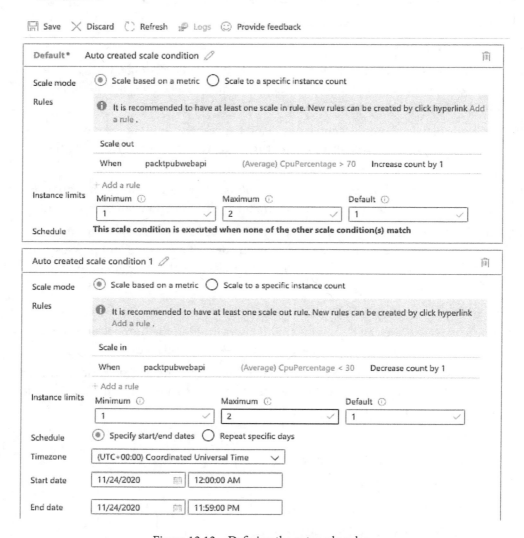

Figure 12.12 – Defining the autoscale rules

19. Click **Save**.

Using a combination of scale out and scale in events, you can fine-tune your app service so that it's always responsive while optimizing costs.

Before we leave Web Apps, we will look at an App Service component that allows you to create background jobs that run under the same Service plan.

Understanding WebJobs

Another feature of Azure App Service is **WebJobs**. With WebJobs, you can run scripts or programs as background processes on Azure App Service Web Apps, APIs, and mobile apps without any additional costs. Some scenarios that would be suitable for WebJobs are long-running tasks, such as sending emails and file maintenance, including aggregating or cleaning up log files, queue processing, RSS aggregation and image processing, and other CPU-intensive work. You can upload and run executable files such as the following:

- `.ps1` (using PowerShell)
- `.cmd`, `.bat`, and `.exe` (using Windows CMD)
- `.sh` (using Bash)
- `.py` (using Python)
- `.php` (using PHP)
- `.js` (using Node.js)
- `.jar` (using Java)

There are two different types of WebJobs:

- **Continuous**: This starts immediately after creating the WebJob. The work inside the WebJob is run inside an endless loop to keep the job from ending. By default, continuous WebJobs run on all the instances that the Web App runs; however, you can configure the WebJob so that it runs on a single instance as well. Continuous WebJobs support remote debugging.

- **Triggered**: This starts on a schedule, but when triggered manually, it runs on a single instance that is selected by Azure for load balancing. A triggered WebJob doesn't support remote debugging.

> **Important note**
> At the time of writing, WebJobs aren't supported for App Service on Linux.

> **Tip**
> Web Apps can time out after 20 minutes of inactivity. The timer can only be reset when a request is made to the Web App. The Web App's configuration can be viewed by making requests to the advanced tools site or in Azure portal. If your app runs continuous or scheduled (timer triggered) WebJobs, enable **Always On** to ensure that the WebJobs run reliably. The **Always On** feature is only available in the Basic, Standard, and Premium plans.

In the next demonstration, we are going to create a background task using WebJobs.

Creating an App Service background task using WebJobs

In this demonstration, we are going to create a continuous WebJob that executes a console application. This application will listen for queued messages when it starts up. To create the storage account and the queue, you can run the `CreateStorageAccountQueue` PowerShell script that's been added to this book's GitHub repository. To find out where to download the script, take a look at the *Technical requirements* section at the beginning of this chapter.

Once you have created the storage account, open Visual Studio 2019 and go through the following steps:

1. Create a new console application (.NET Core) project and name it `HelloWorldWebJob`.

2. First, we need to add two NuGet packages. Open **NuGet Package Manager** and install the following packages:

   ```
   Microsoft.Azure.WebJobs
   Microsoft.Azure.WebJobs.Extensions
   ```

3. In `Program.cs`, add the following `using` statement:

   ```
   using Microsoft.Extensions.Hosting;
   ```

4. Open `Program.cs` and replace the `Main` method with the following code:

   ```
   static void Main(string[] args)
   {
   var builder = new HostBuilder(); builder.
   ConfigureWebJobs(b =>
   {
   b.AddAzureStorageCoreServices();

   });
   var host = builder.Build(); using (host)
   {
   host.Run();
   }
   ```

```
}
```

5. The next step is to set up console logging, which uses the ASP.NET Core logging framework. Therefore, we need to install the following NuGet packages:

```
Microsoft.Extensions.Logging Microsoft.Extensions.
Logging.Console
```

6. In `Program.cs`, add the following `using` statement:

```
using Microsoft.Extensions.Logging;
```

7. Call the `ConfigureLogging` method on `HostBuilder`. The `AddConsole` method adds console logging to the configuration:

```
builder.ConfigureLogging((context, b) =>
{
b.AddConsole();
});
```

8. In version 3.x, the `Storage` binding extension needs to be installed explicitly. Import the following NuGet package:

```
Microsoft.Azure.WebJobs.Extensions.Storage
```

9. Update the `ConfigureWebJobs` extension method so that it looks as follows:

```
builder.ConfigureWebJobs(b =>
{
b.AddAzureStorageCoreServices(); b.AddAzureStorage();
});
```

10. Add a new class to the project and name it `Functions.cs`. Replace the code inside it with the following:

```
using Microsoft.Azure.WebJobs;
using Microsoft.Extensions.Logging;


namespace HelloWorldWebJob
{

public class Functions
```

```
{
public static void
ProcessQueueMessage([QueueTrigger("webjob")] string
message, ILogger logger)
{
logger.LogInformation(message);
}
}
}
```

11. Next, add a JavaScript JSON configuration file to the project and name it `appsettings.json`. Select the `appsettings.json` file in **Solution Explorer** and in the **Properties** window, set **Copy to Output Directory** to **Copy if newer**.

12. Switch to Azure portal and go to the **Overview** page of the storage account that we created in PowerShell. In the left navigation, under **Settings**, select **Access keys**. Here, copy the first connection string:

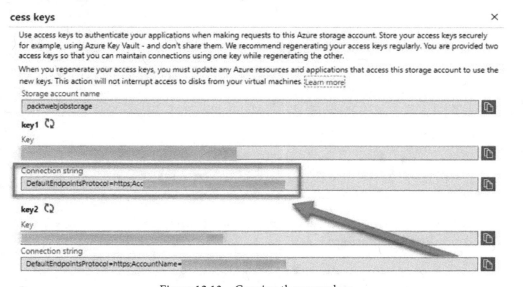

Figure 12.13 – Copying the access key

13. Add the following code to the `appsettings.json` file. Replace the value shown with the storage account's **Connection string**:

```
{
  "AzureWebJobsStorage": "{storage connection string}"
}
```

14. Switch back to Azure portal and open the **Overview** page of the storage account. Select **Storage Explorer** in the left menu and go to **Queues | webjob**. In the top menu, select **+ Add Message**:

Figure 12.14 – Storage Explorer

15. In the dialog, enter `Hello World` and then select **OK**. Now, run the project in Visual Studio and wait until the message is found. This can take up to 2 minutes due to the queue polling exponential backoff in the WebJobs SDK. The output in the console application will look as follows:

```
 C:\Program Files\dotnet\dotnet.exe
          "MaxPollingInterval": "00:01:00",
          "MaxDequeueCount": 5,
          "VisibilityTimeout": "00:00:00"
        }
info: Microsoft.Azure.WebJobs.Hosting.OptionsLoggingService[0]
      SingletonOptions
      {
        "LockPeriod": "00:00:15",
        "ListenerLockPeriod": "00:01:00",
        "LockAcquisitionTimeout": "10675199.02:48:05.4775807",
        "LockAcquisitionPollingInterval": "00:00:05",
        "ListenerLockRecoveryPollingInterval": "00:01:00"
      }
info: Microsoft.Azure.WebJobs.Hosting.OptionsLoggingService[0]
      BlobsOptions
      {
        "CentralizedPoisonQueue": false
      }
info: Microsoft.Azure.WebJobs.Hosting.JobHostService[0]
      Starting JobHost
info: Host.Startup[0]
      Found the following functions:
      HelloWorldWebJob.Functions.ProcessQueueMessage

info: Host.Startup[0]
      Job host started
Application started. Press Ctrl+C to shut down.
Hosting environment: Production
Content root path: C:\Users\SjoukjeZaal\Repos\Microsoft-Azure-Architect-Technologies-
Chapter14\WebJobs\HelloWorldWebJob\HelloWorldWebJob\bin\Debug\netcoreapp2.2\

info: Function.ProcessQueueMessage[0]
      Executing 'Functions.ProcessQueueMessage' (Reason='New queue message detected on
info: Function.ProcessQueueMessage[0]
      Trigger Details: MessageId: 3b4169e0-6804-4d88-b6d4-8e36f908ef62, DequeueCount: 1,
info: Function.ProcessQueueMessage.User[0]
      Hello World
info: Function.ProcessQueueMessage[0]
      Executed 'Functions.ProcessQueueMessage' (Succeeded, Id=992c6351-7062-4561-bf7f-7b54b92591ac)
```

Figure 12.15 – Output in the console application

16. To create an executable for the project, we need to publish it. Right-click on the project name in the **Solution Explorer** window and select **Publish**. In the wizard, select **Folder,** click the arrow next to **Publish**, and select **Create profile**:

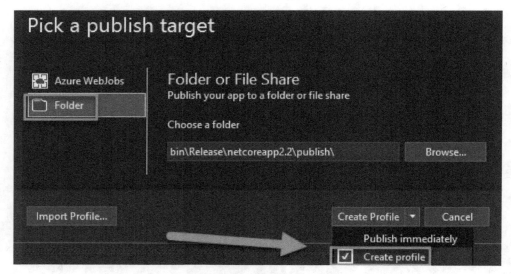

Figure 12.16 – Publishing the WebJob

17. On the next screen of the wizard, select the **Edit** button:

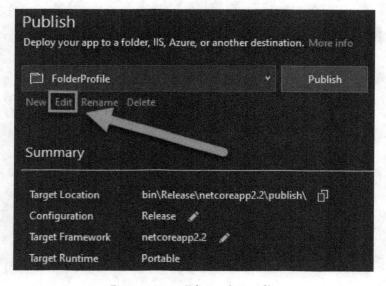

Figure 12.17 – Editing the profile

18. On the next screen, set **Target Runtime** to **win-x64** and click **Save**:

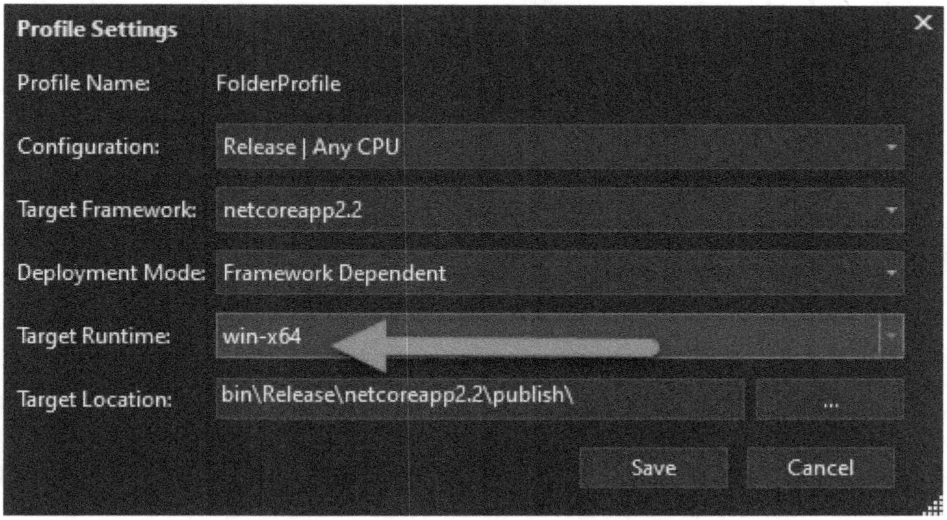

Figure 12.18 – Changing the target runtime

19. Click **Publish**.

20. The WebJob executable will now be published in the target location folder.

Now that we have created a console application that listens to the queue trigger, we can upload the executable to Azure App Service and create a new WebJob.

Deploying the WebJob to Azure App Service

Now that we have created the WebJob, we can deploy it to Azure App Service. Go through the following steps:

1. Navigate to Azure portal by going to `https://portal.azure.com/`.

2. Navigate to the **Overview** page of the `PacktPubToDoAPI` Web App that we created in the previous demonstration.

3. Under **Settings**, select **WebJobs**. Then, from the top menu, click **+ Add**.

4. On the next screen, add the following values:

    a) **Name**: `HelloWorldWebJob`.

    b) **File Upload**: Select the `HellWorldWebJob` executable; you can find it in the `bin` folder of the Visual Studio solution.

c) **Type: Triggered**.

d) **Triggers: Manual**:

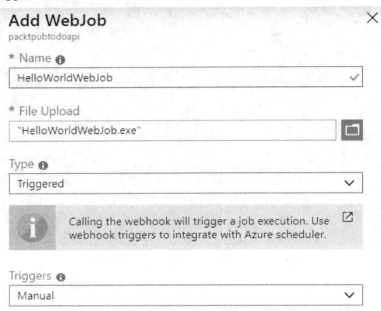

Figure 12.19 – WebJob settings

5.   Then, click **OK**. The WebJob will be created.

That concludes this demo. In the next and last section of this chapter, we are going to enable diagnostic logs for the `PacktPubToDoAPI` Web App.

Understanding diagnostic logging

Azure provides built-in diagnostics to assist with debugging an App Service app. App Service offers diagnostic logging for both the web server and the web application. These logs are separated into web server diagnostics and application diagnostics.

Web server diagnostics

Web server diagnostics offers the following types of logs that can be enabled and disabled:

- **Web server logging**: This type of log provides information using the W3C extended log file format for logging information about HTTP transactions. This is useful for determining the overall site metrics, such as how many requests are coming from a specific IP address and the number of requests that are handled.

- **Detailed error logging**: This type of log provides detailed information for any request that results in an HTTP status code of 400 or greater. This log contains information that can help you investigate why the server returned the error code. For each error in the app's filesystem, one HTML file is generated.

- **Failed request tracing**: This provides detailed information about failed requests. This includes the IIS components that were used to process the request and the time each component took. This is useful for improving the site's performance and isolating a specific HTTP error. One folder is generated for each error in the app's filesystem.

For web server logging, you need to select a storage account or filesystem. When selecting a storage account, you need to have a storage account in place with a blob container. You can also create a new one. If you store the logs on a filesystem, they can be downloaded as a ZIP file or accessed by FTP.

Application diagnostics

With application diagnostics, you can capture information that is produced by the app itself. To capture this information, the developer uses the `System.Diagnostics.Trace` class in the application code to log the information to the application diagnostics log. You can retrieve these logs at runtime for troubleshooting.

When you publish the application to App Service, deployment information is automatically logged with diagnostic logging, without it needing to be configured. This can give you information about why a deployment failed.

In the next section, we are going to enable diagnostic logging in Azure portal.

Enabling diagnostic logging

In this section, we are going to enable diagnostic logging for the `PacktPubToDoAPI` Web App. Let's go through the following steps:

1. Navigate to Azure portal by going to `https://portal.azure.com/`.

2. Go to the **Overview** page of the `PacktPubToDoAPI` Web App.

3. From the left menu, under **Monitoring**, select **App Service logs**. Here, you can enable the different types of logs:

Figure 12.20 – Enabling diagnostic logging

4. Enable diagnostic logging and click **Save** in the top menu.

In this section, we learned how to enable and use diagnostics to help troubleshoot issues. In the next section, we will examine another type of App Service called Azure Functions.

Using Azure Functions

Azure offers a variety of services that we can use to build applications. One of the possibilities it offers is the ability to create serverless applications. Serverless computing is the abstraction of servers, infrastructure, and operating systems. Serverless is consumption-based, which means that you don't need to anticipate any capacity. This is different from using PaaS services. With PaaS services, you still pay for the reserved compute.

Azure Functions is a serverless compute service that's used to run small pieces of code in the cloud. You can simply write the code you need in order to execute a certain task, without the need to create a whole application or manage your own infrastructure.

Azure Functions can be created from the Azure portal and from Visual Studio and can be created in a variety of programming languages. At the time of writing this book, the following languages are supported:

Language	1.x	2.x	3.x
C#	GA (.NET Framework 4.7)	GA (.NET Core 2.2)	GA (.NET Core 3.1)
JavaScript	GA (Node 6)	GA (Node 8 and 10)	GA (Node 11 and 12)
F#	GA (.NET Framework 4.7)	GA (.NET Core 2.2)	GA (.NET Core 3.1)
Java	N/A	GA (Java 8)	GA (Java 8) Preview (Java 11)
PowerShell	N/A	GA (PowerShell Core 6)	GA (PowerShell 7 & Core 6)
Python	N/A	GA (Python 3.6 and 3.7)	GA (Python 3.6, 3.7 and 3.8)
TypeScript	N/A	GA (Supported through transpiling to JavaScript)	GA (Supported through transpiling to JavaScript)

Functions can be created using ARM templates as well. They can be deployed on Windows or Linux and by using continuous deployment. At the time of writing this book, Linux is still in preview.

With Azure Functions, you can build solutions that process data, integrate various systems, and work with the **Internet of Things** (**IoT**), simple APIs, and microservices applications.

You can create a small task, such as image or order processing, and call this task from other applications or execute it based on a certain schedule. For that, Azure Move following line in here

Functions provides triggers such as `HTTPTrigger`, `TimerTrigger`, `CosmosDBTrigger`, `BlobTrigger`, `QueueTrigger`, `EventGridTrigger`, `EventHubTrigger`, `ServiceBusQueueTrigger`, and `ServiceBusTopicTrigger`.

Azure Functions uses an Azure storage account to store any code files and configuration bindings. It uses the standard version of Azure Storage, which provides blob, table, and queue storage for storing files and triggers. However, you can use the same App Service plans for the functions that you use for Web Apps and APIs. Azure Functions can also be deployed in **ASEs**. When using these App Service plans or an ASE to host your functions, the function will become non-serverless. This happens due to the compute power being pre-allocated and therefore prepaid.

Creating an Azure Function

To create an Azure Function in Azure portal, perform the following steps:

1. Navigate to Azure portal by going to `https://portal.azure.com`.

2. Click on **Create a resource** and type `Function` into the search bar. Select **Function App** and click **Create** to create a new function.

3. A new blade will open, where you have to specify the properties for the Function app. Add the following properties:

    a) **Subscription**: Select a subscription here.

    b) **Resource group**: Create a new resource group named `PacktFunctionApp`.

    c) **Function App Name**: Name it `PacktFunctionApp`.

    d) **Publish**: Code.

    e) **Runtime Stack**: Select **.NET Core**.

    f) **Location: East US**.

4. Click **Review + create** and then **Create** to create the function app.

5. Once the app has been created, open the resource in Azure portal. This will navigate you to the settings of the function app. To create a new function, click **+**, which can be found next to **Functions** in the top-left menu, and select **In-portal** to build your function directly into the Azure portal:

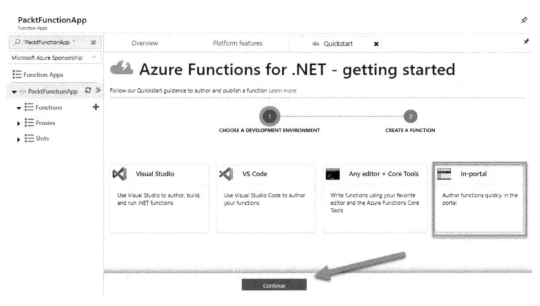

Figure 12.21 – Creating a new function in the Azure portal

6. Next, we need to select a trigger template. For this demonstration, we're going to select the **Webhook + API** trigger template. After that, click **Create**:

Figure 12.22 – Selecting the Webhook + API trigger template

7. The sample for your Azure Function app will be generated. Since the exam doesn't cover coding the Azure Function app, we are going to keep the default code for this demonstration.

8. By using the **Webhook + API** trigger, this function can be called manually or from other applications by using the function URL. To obtain this URL, click the **Get function URL** link at the top of the screen. You can use this URL to start executing the function:

Figure 12.23 – Obtaining the function URL

9. By default, the function URL consists of an application security key. You can change this authentication level in the settings if you want to:

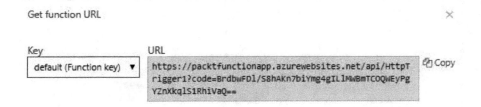

Figure 12.24 – Displaying the function URL

10. You can also test the function from the Azure portal. Click **Test** from the right-hand menu. A test value will already be provided for you. After that, click on **Run** from the lower right of the screen to test the function:

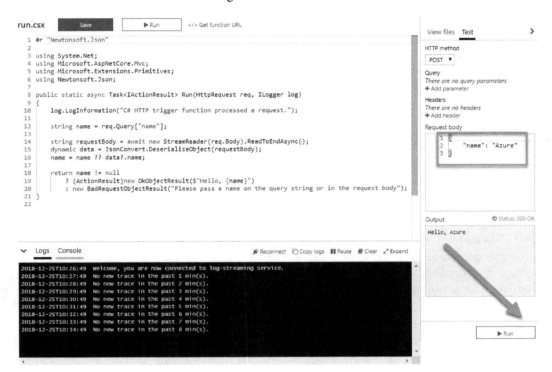

Figure 12.25 – Testing the function

> **Tip**
> The Function app is provided as an ARM template on GitHub. You can download it and deploy it inside your environment. It is recommended that you get familiar with the properties inside the template because there is a possibility that this will be part of the exam.

With that, we have learned how to easily create Azure Functions, which are ideal for microservices. Azure Functions are code-based and can become complex; however, many businesses need to be able to quickly create simple workflows. Microsoft provides a codeless service for creating such tasks known as Logic Apps.

Building Azure Logic Apps

Logic Apps is another serverless service provided by Azure. It differs from Azure Functions because Logic Apps is used for creating workflows. It manages durable processes, while Azure Functions is usually used for a short-lived chain element in a broader chain. Functions are used to execute short pieces of code that handle a single function in that broader chain. Azure Functions is completely code-based, while Logic Apps can be created using a visual designer. Both Logic Apps and Azure Functions can be used to integrate services and applications.

Logic Apps can also be used to automate business workflows. You can create a logic app in the Azure portal and developers can create them from Visual Studio as well. For both, the visual designer is used to moderate this process. Logic Apps offers connectors that can integrate a number of cloud services and on-premises resources and applications. For Logic Apps, there is also a connector available for calling an Azure Function app from a logic app.

These connectors can connect to different Azure services, third-party cloud services, different data stores and databases, and **line-of-business** (**LOB**) applications. Azure Logic Apps provides a number of pre-built connectors that you can leverage inside your workflow. Besides that, developers can also create their own connectors using Visual Studio. Besides using the Visual Editor, you can create and make adjustments to the Workflow Definition Language schema manually. This schema is created using JSON and can be created from scratch using Visual Studio, though it can be adjusted inside the Azure portal. Some capabilities can only be added to the schema directly and cannot be made from the Visual Editor. Examples of this include date and time formatting and string concatenation. Logic app definition files can be added to ARM templates and deployed using PowerShell, the CLI, or REST APIs.

> **Important note**
>
> You can refer to the following article for an overview of all the available connectors for Azure Logic Apps: `https://docs.microsoft.com/en-us/ azure/connectors/apis-list`.

Deploying the Logic App ARM template

In this demonstration, we are going to deploy the Logic App ARM template, which can be downloaded from this book's GitHub repository. The link to this was provided in the *Technical requirements* section at the beginning of this chapter. The logic app reads out RSS information on the CNN website and sends out an email from an Office 365 email account stating the RSS item's details. To deploy the Logic App ARM template, perform these steps:

1. Download the ARM template from this book's GitHub repository. If you want to look at the properties of the template, you can use Visual Studio Code or Notepad. The download link for Visual Studio Code was provided in the *Technical requirements* section as well. Create a new folder on your C drive and name it `MyTemplates`. Save the ARM templates in this directory.

2. We are going to deploy this template using PowerShell. Open PowerShell and add the following code (replace the subscription name):

   a) First, we need to log into our Azure account:

   ```
   Connect-AzAccount
   ```

   b) If necessary, select the right subscription:

   ```
   Select-AzSubscription -SubscriptionId "********-****-
   ****-****-************"
   ```

   c) Create a resource group:

   ```
   New-AzResourceGroup -Name PacktLogicAppResourceGroup -
   Location EastUS
   ```

   d) Deploy the template inside your Azure subscription, as follows:

   ```
   New-AzResourceGroupDeployment `
   -Name PacktDeployment `
   -ResourceGroupName PacktLogicAppResourceGroup `
   -TemplateFile c:\MyTemplates\template.json
   ```

3. After deployment, you need to go to the logic app's settings and configure an Outlook account that can be used to send the email. By default, the account details will be empty. To configure an account, open `PacktLogicApp` inside **Logic App Designer** and open the **Connections** step. There, click **Invalid connection** and connect to an Office 365 account, as shown in the following screenshot:

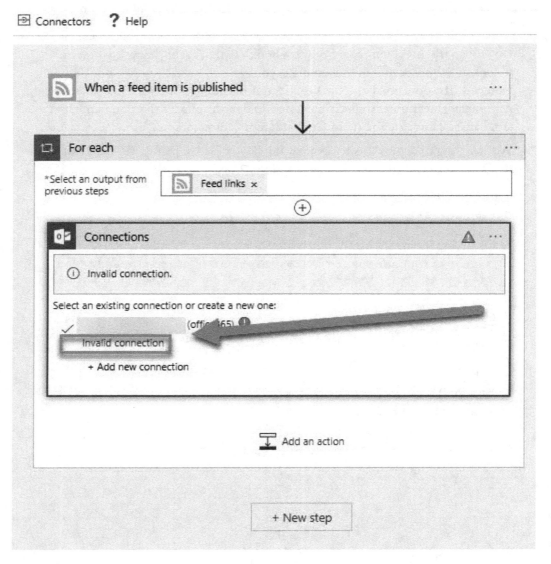

Figure 12.26 – Adding an email account to the logic app

> **Tip**
>
> You should become familiar with the different parts of the ARM template as they could be part of the exam questions.

Managing a Logic Apps resource

Now that the logic app has been deployed and is in place, we can learn how to manage it. You can use a variety of logging and monitoring capabilities to do so, all of which will be covered in this section.

Monitoring, logging, and alerts

After deploying and executing your logic app, you can look at the run and trigger history, performance, and status tabs in Azure portal. You can also set up diagnostic logging for real-time monitoring and more debugging capabilities for your logic app. You can also enable alerts to get notifications about problems and failures.

Viewing runs and trigger history

To view the runs and trigger history of your logic app, perform the following steps:

1. Navigate to Azure portal by going to `https://portal.azure.com`.

2. Click on **All Resources** in the left menu and select the resource group for the logic app.

3. A new blade will open, showing an overview of all the resources that are part of the logic app resource group.

4. Select the logic app itself. A new blade will open, showing an overview page of the logic app.

5. In the section that follows, under **Runs history**, you will see the runs of the logic app. To get details about a certain run, select one of the runs from the list.

6. The **Logic App Designer** window will open. Here, you can select each step to find out more about their input and output:

Figure 12.27 – Logic app detailed monitoring

7. To find out more about the triggered event, go back to the **Overview** pane. Then, under **Trigger History**, select the trigger event. Now, you can view various details, such as input and output information:

Figure 12.28 – Logic app trigger history

Now that we have finished setting up our logic app, we need to be able to monitor its health. We can do this using alerts.

Setting up alerts

In the case of problems and failures, you can set up alerts for Logic Apps. To set this up, perform the following steps:

1. On the Logic Apps, **Overview** pane in Azure portal, from the left-hand side menu under **Monitoring**, select **Alerts**.

2. Click **New alert rule** at the top of the screen. In the pane that opens, you can add a condition. Here, you can select a signal that the alerts subscribe to. Once you've selected a condition, you can add a threshold and select the period for which you wish the metric to be monitored:

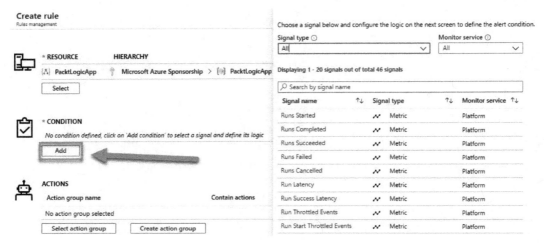

Figure 12.29 – Setting a condition for your alert

3. After that, you can create a new action group or select an existing one. Action groups allow you to trigger one or more actions so that you can notify others about an alert. You can select the following action types:

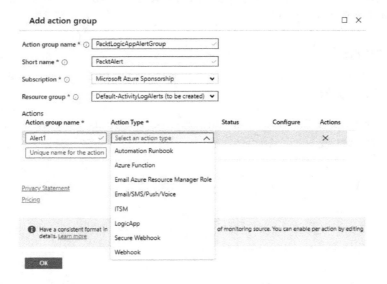

Figure 12.30 – Setting an action group for your alert

4. Finally, you need to specify an alert title, a description, and a severity level. Then, you need to enable the rule.

As you can see, Logic Apps offers an intuitive yet powerful alternative to code-based solutions, and allows you to quickly build simple workflows.

Summary

In this chapter, we covered the first part of the *Implementing Solutions for Apps* objective. We covered App Service and App Service plans, created an Azure App Service Web App, learned how to safely deploy updates without effecting live services, and how to automatically scale resources, depending on a number of factors. We also looked at creating an App Service background task using WebJobs, as well as enabled diagnostic logging for the Web App.

We then looked at Azure Functions and how they can be used to deploy microservice-based apps. Finally, we learned how Logic Apps offers a simple no-code alternative for business-driven workflows.

In the next chapter, we will continue with this objective by learning how to design and develop apps that run in containers.

Questions

Answer the following questions to test your knowledge of the information contained in this chapter. You can find the answers in the *Assessments* section at the end of this book:

1. When logging information using web server logs, developers need to add additional code to their web applications.

    a) True

    b) False

2. WebJobs are also supported on Linux.

    a) True

    b) False

3. You want to create a small coded task, such as processing an image, which can be called from other applications. Is Logic Apps the most suitable option for this?

    a) Yes

    b) No

Further reading

Check out the following links to find out more about the topics that were covered in this chapter:

- App Service on Linux documentation: `https://docs.microsoft.com/en-us/azure/app-service/containers/`

- ASP.NET Core Web API help pages with Swagger/OpenAPI: `https://docs.microsoft.com/en-us/aspnet/core/tutorials/web-api-help-pages-using-swagger`

- Running background tasks with WebJobs in Azure App Service: `https://docs.microsoft.com/en-us/azure/app-service/webjobs-create`

- Enabling diagnostic logging for apps in Azure App Service: `https://docs.microsoft.com/en-us/azure/app-service/troubleshoot-diagnostic-logs`

- Azure Functions documentation: `https://docs.microsoft.com/en-us/azure/azure-functions/`

- Monitoring statuses, setting up diagnostics logging, and turning on alerts for Azure Logic Apps: `https://docs.microsoft.com/en-us/azure/logic-apps/ logic-apps-monitor-your-logic-apps#find-events`

13
Designing and Developing Apps for Containers

In the previous chapter, we covered the first part of the *Implement Solutions for Apps* objective by learning how to create web apps using **Platform as a Service** (**PaaS**). Then, we covered Azure App Service and App Service plans. We also looked at using Azure Functions for creating smaller, microservices-based services and Logic Apps for building business workflows using a GUI.

In this chapter, we are going to learn how to design and develop apps that run in containers by implementing an application that runs on an Azure container instance, as well as creating a container image using a Dockerfile and publishing it to **Azure Container Registry** (**ACR**).

We are going to start by creating a container using **Azure Container Instances** (**ACI**), and then again through Azure Web Apps for Containers.

Finally, we are going to create an **Azure Kubernetes Service** (**AKS**) instance, deploy a simple web application to it, and view the generated application logs.

The following topics will be covered in this chapter:

- Understanding ACI
- Implementing an application that runs on ACI
- Creating a container image using a Dockerfile
- Publishing an image to ACR
- Understanding Web Apps for Containers
- Creating an App Service web app for containers
- Understanding AKS
- Creating an AKS cluster

Technical requirements

The examples in this chapter use Azure PowerShell (`https://docs.microsoft.com/en-us/powershell/azure/`), Visual Studio 2019 (`https://visualstudio.microsoft.com/vs/`), and Docker Desktop for Windows (`https://hub.docker.com/editions/community/docker-ce-desktop-windows`).

We will also be using a tool called `kubectl` – instructions for this depend on your OS and can be found here: `https://kubernetes.io/docs/tasks/tools/install-kubectl/`.

The source code for our sample application can be downloaded from `https://github.com/PacktPublishing/Microsoft-Azure-Architect-Technologies-Exam-Guide-AZ-303/tree/master/Ch13`.

Understanding ACI

With ACI, you can run your workloads in containers. It allows you to run both Linux and Windows containers. You can deploy containers that consist of applications, databases, caches, queues, and more. Everything that runs on a VM can also run inside ACI, without the need for us to manage the infrastructure. This is all handled for you so that you can focus on designing and building your applications.

ACI is a suitable solution for any scenario that can run in isolated containers, including simple applications, build jobs, and task automation. For scenarios where you need full container orchestration, including automatic scaling, service discovery across multiple containers, and coordinated application upgrades, AKS is recommended.

> **Important note**
> We will cover AKS in more depth later in this chapter.

Container image deployments can be automated and simplified using **continuous integration/continuous deployment (CI/CD)** capabilities with Docker Hub, ACR, and Visual Studio Team Services.

Containers in Azure use Docker, which is based on open standards. This means it can run on all major Linux distributions (and Windows Server 2019). Docker containers are lightweight and share the machine's system kernel on top of your OS. When your application is deployed inside a Docker container, the app is isolated from other applications or processes that are running on the same OS. You can compare this to creating different VMs to host different types of workloads or applications, but without the overhead of the virtualization and OS itself. Docker containers also share the same OS and infrastructure, whereas VMs need to have their own OS installed inside their own infrastructure.

With containers, you share the underlying resources of the Docker host and you build a Docker image that includes everything you need to run the application. You can start with a basic image and then add everything you need. Docker containers are also extremely portable. You can deploy a Docker container, including all its settings, such as configuration settings and a specific runtime, framework, and tooling, on a VM with Docker installed. Then, you can easily move that same container to Azure App Service on Linux, and the application will still run as expected. This solves the it-works-on-my-machine problem that (almost) all developers face. This makes Docker not a virtualization technology, but an application-delivery technology. The following figure shows an example of the key differences between VMs and containers:

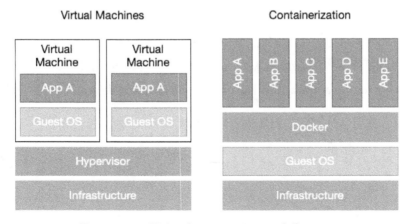

Figure 13.1 – VM and containerization differences

Docker containers are very suitable for building applications using the microservices architecture, where parts of an application are loosely coupled and divided into separate services that all collaborate with each other. Each service can then be deployed into a separate container and written in its own programming language using its own configuration settings. A service can consist of a web app, a web API, or a mobile backend. You can easily deploy multiple copies of a single application. The only thing to be aware of is that they all share the same OS. If your application needs to run on a different OS, you still have to use a VM.

In the next section, we are going to implement an application that runs on ACI. Then, we will create a container image using a Dockerfile and publish the image to ACR.

Implementing an application that runs on ACI

In the first part of this demonstration, we are going to create a simple application that can be packaged inside a Docker container. To create a Dockerfile for this application, you need to install Docker Desktop. For the installation URL, refer to the *Technical requirements* section, which can be found at the beginning of this chapter. To create the app, follow these steps:

1. Open Visual Studio and add a new project.

2. In the new project wizard, select **ASP.NET Core Web Application** and click **Next**.

3. Add the following values to create a new project:

    a) **Project name**: PacktACIApp.

    b) **Location**: Pick a location.

    c) **Solution name**: PacktACIApp.

4. Click **Create**.

5. On the next screen, select **Web Application** and click **Create**.

6. We are going to use the default web application, so no code changes are required.

7. Press *F5* to run the application. The output will look as follows:

Figure 13.2 – Web app output

In this section, we created an application that can be packaged inside a Docker container. We will continue with this in the next section.

Creating a container image using a Dockerfile

Now that we have our web API in place, we can create our Docker container image. We can generate this from Visual Studio directly. To do so, follow these steps:

1. Right-click the project file in Visual Studio Solution Explorer.

2. Then, select **Add | Docker Support…**:

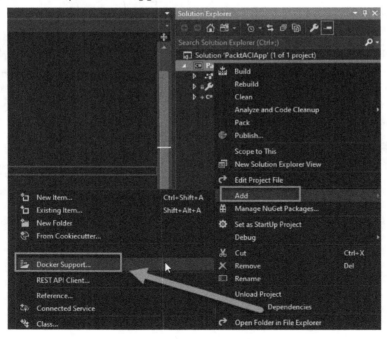

Figure 13.3 – Creating a Dockerfile

3. A wizard will open where you can generate a Dockerfile. Here, select **Linux**.

4. Visual Studio creates a **Dockerfile** for you. This defines how to create the container image for your web API project.

5. Your Dockerfile will now look as follows:

```
FROM mcr.microsoft.com/dotnet/core/aspnet:3.1-buster-slim
AS base
WORKDIR /app
EXPOSE 80
EXPOSE 443


FROM mcr.microsoft.com/dotnet/core/sdk:3.1-buster AS
build
WORKDIR /src
COPY ["PacktACIApp.csproj", ""]
RUN dotnet restore "./PacktACIApp.csproj"
COPY . .
WORKDIR "/src/."
RUN dotnet build "PacktACIApp.csproj" -c Release -o /app/
build


FROM build AS publish
RUN dotnet publish "PacktACIApp.csproj" -c Release -o /
app/publish


FROM base AS final
WORKDIR /app
COPY --from=publish /app/publish .
ENTRYPOINT ["dotnet", "PacktACIApp.dll"]
```

6. Now, when you run your project from Visual Studio, it will run inside a Docker container. Make sure that Docker is running and click the **Docker** button on the toolbar:

Figure 13.4 – Running the Dockerfile from Visual Studio

7. The browser will open once more, just like in the previous step, except now, your app will run in a container.

> **Tip**
> If you get a **Docker Volume sharing is not enabled** error when you first run your project, then open PowerShell as an administrator and run the following line of code: `docker run --rm -v c:/Users:/data alpine ls /data`. Restart Visual Studio and run the project again.

Now that we have created our Dockerfile, the next step is to deploy the image to ACR.

Publishing an image to ACR

In this section, we are going to publish the Docker image to ACR. We can do this directly from Visual Studio as well. Let's get started:

1. Right-click on the project file in Visual Studio Solution Explorer.
2. Click **Publish**.
3. In the wizard that opens, select **Docker Container Registry**, then **Azure Container Registry**.
4. Click **+ Create a new Azure Container Registry**.

5. On the next screen, add the following settings and click **Create**:

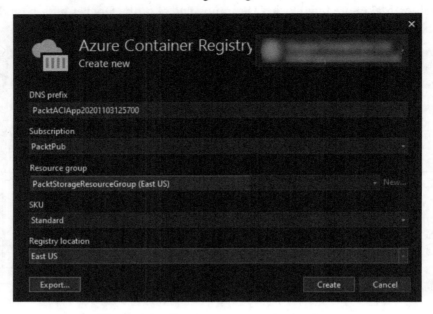

Figure 13.5 – Creating a new container registry

6. The **Create Registry** window will close. Click **Finish**.

7. You will be taken to the **Publish** page, as in the following example – click **Publish**:

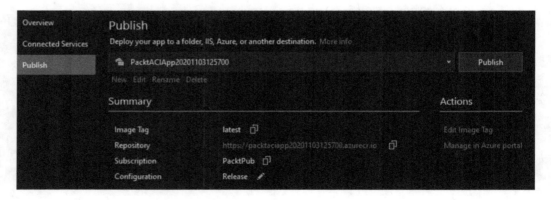

Figure 13.6 – Publishing the image

8. Now, navigate to the deployed container registry in the Azure portal by searching for `Container Registries` in the search bar. Under **Services**, click **Repositories**. You will see that the `packtaciapp` repository has been published. When you click on it, you will see the tag as well.

9. Click **latest** and you will see details of your image. Next to the Docker pull command, you will see the full path to the image – copy the command, but without the `docker pull` part:

Repository	: packtaciapp	Digest	
Tag	: latest	Manifest creation date	
Tag creation date	: 11/3/2020, 2:19 PM GMT	Platform	
Tag last updated date	: 11/3/2020, 2:19 PM GMT		

Docker pull command   docker pull packtaciapp20201103125700.azurecr.io/packtaciapp:latest

Figure 13.7 – Container image link

10. Under **Settings**, click **Access Keys**. Enable the **Admin user** switch, then copy the username and the password:

Figure 13.8 – Copying the username and password

We will need this password in the next step of the demonstration.

In the next section, we are going to create a container instance and push the new Docker image from ACR to ACI.

Pushing the Docker image from ACR to ACI

In this part of the demonstration, we are going to use the Azure CLI to deploy and push the Docker image from ACR to an Azure container image. To do so, follow these steps:

1. Navigate to the Azure portal by opening `https://portal.azure.com/`.

2. Open Azure Cloud Shell.

3. First, we need to create a new resource group:

    ```
    az group create --name PacktACIResourceGroup --location
    eastus
    ```

4. Run the following command to create a Windows Azure container instance and push the Docker image from ACR. `--image` will use the link you copied in *step 9* of the *Publishing an image to ACR* section. This will result in the following code:

    ```
    az container create \
    --name packtaciapp \
    --resource-group PacktACIResourceGroup \
    --os-type linux \
    --image packtaciapp20190917053041.azurecr.io/
    packtaciapp:latest \
    --ip-address public
    ```

5. When executing the preceding code, you will be prompted to provide a username and password. Here, paste in the username and password that you copied from the ACR settings in the previous section.

6. Once the container instance has been deployed, navigate to it in the Azure portal – type and select **Container Instances** in the search bar.

7. The deployed image will be listed. Click on it and from the overview blade, copy the IP address:

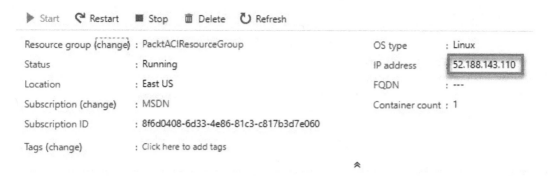

Figure 13.9 – Deployed container instance

8. Paste the IP address into a browser. The application will be displayed:

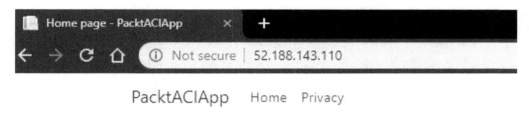

Figure 13.10 – An app running in ACI

In this section, we pushed the Docker image from ACR to an Azure container image. In the next section, we will see how we can use container images in Azure Web Apps.

Understanding Web App for Containers

Web App for Containers is part of the Azure App Service on Linux offering. You can pull web app container images from Docker Hub or a private Azure container registry, and Web App for Containers will deploy the containerized app with all the dependencies to Azure. The platform automatically takes care of OS patching, capacity provisioning, and load balancing.

In the next demo, we are going to create an App Service web app for containers.

Creating an App Service web app for containers

In this demonstration, we are going to create a web app for containers. This can be created from the Azure portal. Let's go through the following steps:

1. Navigate to the Azure portal by opening `https://portal.azure.com/`.
2. Click **Create a resource**, type `Web App for Containers` into the search bar, and create a new one.
3. Add the following values, as shown in the screenshot that follows:

    a) **Subscription**: Select a subscription.

    b) **Resource Group**: **PacktContainerResourceGroup**.

    c) **Name**: `PacktContainerApp`.

    d) **Publish**: **Docker Container**.

    e) **Operating System**: **Linux**.

    f) **Region**: **East US**.

    g) **App Service Plan/Location**: A new App Service plan will automatically be created. If you want to pick an existing one or change the default settings, you can click on it and change the settings:

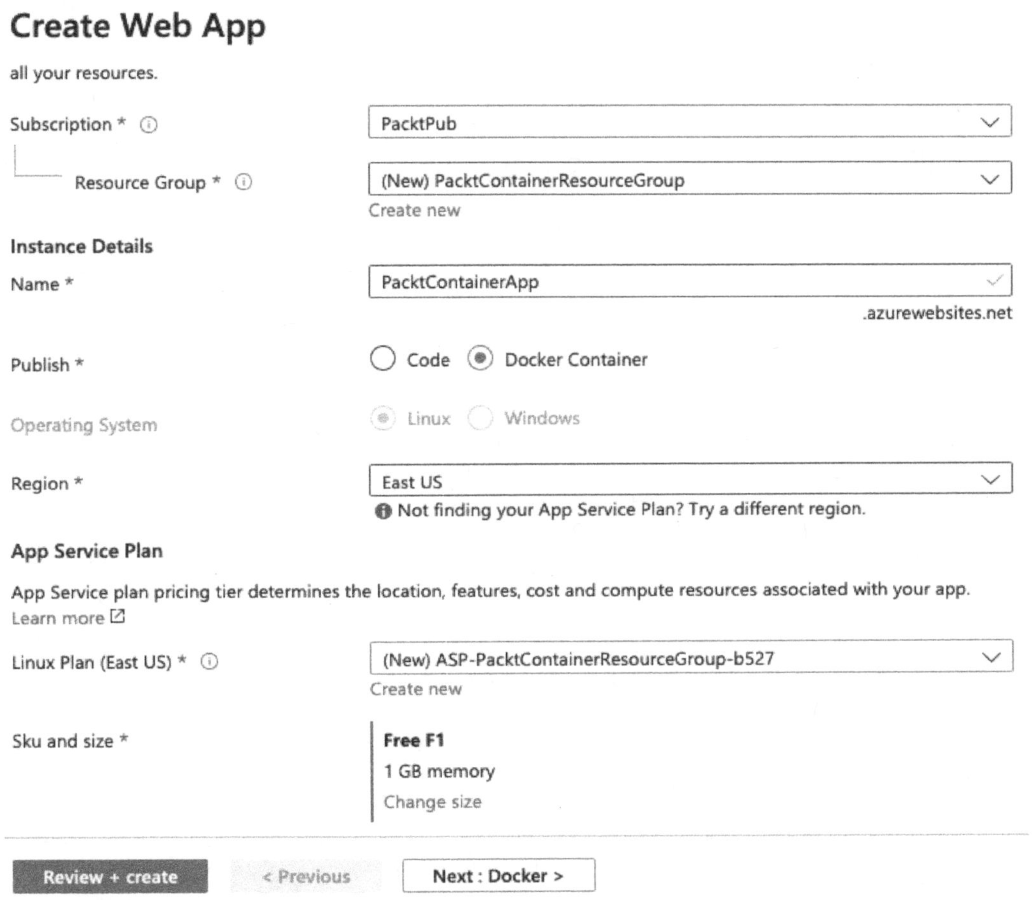

Create Web App

all your resources.

| Subscription * ⓘ | PacktPub ⌄ |

| Resource Group * ⓘ | (New) PacktContainerResourceGroup ⌄ |

Create new

Instance Details

| Name * | PacktContainerApp ✓ |

.azurewebsites.net

| Publish * | ◯ Code ⦿ Docker Container |

| Operating System | ⦿ Linux ◯ Windows |

| Region * | East US ⌄ |

ⓘ Not finding your App Service Plan? Try a different region.

App Service Plan

App Service plan pricing tier determines the location, features, cost and compute resources associated with your app.
Learn more ☑

| Linux Plan (East US) * ⓘ | (New) ASP-PacktContainerResourceGroup-b527 ⌄ |

Create new

| Sku and size * | **Free F1**
1 GB memory
Change size |

| Review + create | < Previous | Next : Docker > |

Figure 13.11 – Web App basic settings

4.   Next, click the **Next: Docker** button and select the following:

   a) **Options**: **Single Container**

   b) **Image Source**: **Azure Container Registry**

   c) **Registry**: Your registry

d) **Image: packtaciapp**

e) **Tag: latest:**

Create Web App

Basics   **Docker**   Monitoring   Tags   Review + create

Pull container images from Azure Container Registry, Docker Hub or a private Docker repository. App Service will deploy the containerized app with your preferred dependencies to production in seconds.

Options	Single Container ⌄
Image Source	Azure Container Registry ⌄

Azure container registry options

Registry *	PacktACIApp20201103125700 ⌄
Image *	packtaciapp ⌄
Tag *	latest ⌄
Startup Command ⓘ	

[Review + create]   [< Previous]   [Next : Monitoring >]

Figure 13.12 – Docker settings

5.   Click **Review + create** and then **Create**.

6.   After it has been created, go to the overview page of `PacktContainerApp`. From the top menu, on the right-hand side, copy the URL:

Figure 13.13 – PacktContainerApp

7.   Open a browser and navigate to the URL. You will see a static page similar to the one shown in the following screenshot:

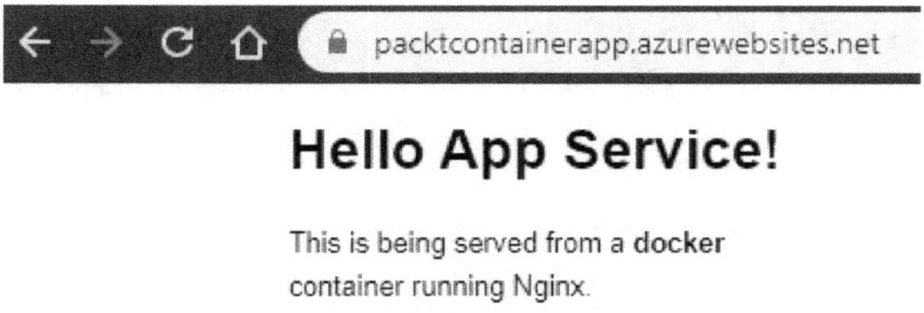

Hello App Service!

This is being served from a **docker**
container running Nginx.

Figure 13.14 – Sample app

In this demonstration, we deployed a container app. In the next section, we are going to look at another way of using container images – AKS.

Understanding AKS

AKS can be used to deploy, scale, and manage Docker containers and container-based applications across a cluster of container hosts. It is based on Kubernetes, which is an open source system that's used for automating the deployment, orchestration, management, and scaling of Linux-based containerized applications. Kubernetes was originally developed by Google and eliminates many of the manual processes involved in deploying and scaling containerized applications. Groups of hosts that run Linux containers can be clustered together, and Kubernetes helps you easily and efficiently manage those clusters.

Apart from this, AKS provides huge benefits. For one, it offers automation, flexibility, and reduced management overhead for administrators and developers and automatically configures all Kubernetes masters and nodes during the development process. It also configures Azure Active Directory integration (it supports Kubernetes **role-based access control** (**RBAC**)), configures advanced networking features, such as HTTP application routing, and handles connections to the monitoring services. Microsoft also handles all the Kubernetes upgrades when new versions become available. However, users can decide when to upgrade to the new versions of Kubernetes in their own AKS cluster to reduce the possibility of accidental workload disruption.

AKS nodes can scale up or down on demand – manually and automatically. For additional processing power, AKS also supports node pools with a **graphics processing unit** (**GPU**). This can be vital for compute-intensive workloads.

AKS can be accessed using the AKS management portal, the CLI, or ARM templates. It also integrates with ACR for Docker image storage and supports the use of persistent data with Azure disks.

In the next section, we are going to deploy an AKS cluster in the Azure portal.

Creating an AKS cluster

In this section, we are going to create an AKS cluster from the Azure portal. We will also deploy a multi-container application that includes a web frontend and a Redis instance in the cluster. To do so, follow these steps:

1. Navigate to the Azure portal by opening `https://portal.azure.com`.

2. Select **Create a resource | Kubernetes Service**.

3. On the **Basics** page, add the following values:

   a) **Subscription**: Select a subscription.

   b) **Resource group**: Create a new one and call it `PacktAKSResourceGroup`.

   c) **Kubernetes cluster name**: `PacktAKSCluster`.

   d) **Region: (US) East US**.

   e) **Availability zones: Zones 1,2**.

   f) **Kubernetes version**: Leave it as the default.

   g) **Node size**: Leave it as the default.

   h) **Node count**: Leave it as the default (**3**):

Figure 13.15 – AKS basic settings

4. Then, select **Authentication** from the top menu. Here, we need to create a new service principal. Create a new one by leaving the default settings as they are, that is, the ones for **(new) default service principal**. You can also choose **Configure service principal** to use an existing one. If you use an existing one, you will need to provide the SPN client ID and secret.

5. Leave **Role-based access control (RBAC)** as **Enabled** and **AKS-managed Azure Active Directory** as **Disabled**:

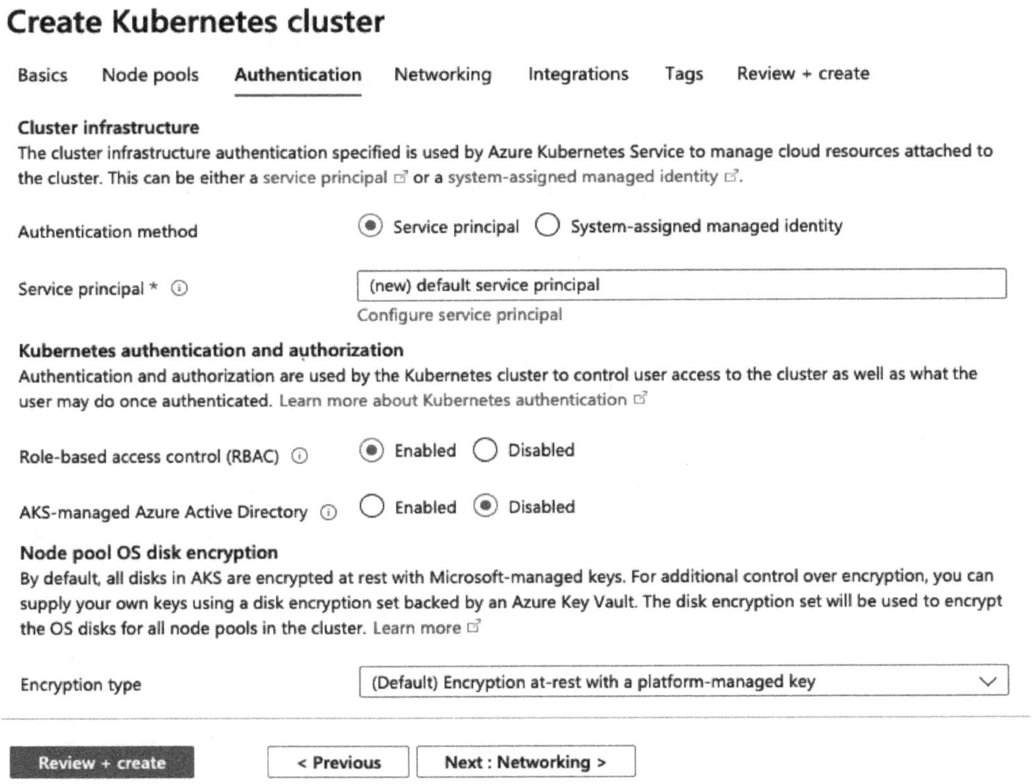

Create Kubernetes cluster

Basics   Node pools   **Authentication**   Networking   Integrations   Tags   Review + create

Cluster infrastructure

The cluster infrastructure authentication specified is used by Azure Kubernetes Service to manage cloud resources attached to the cluster. This can be either a service principal ⊡ or a system-assigned managed identity ⊡.

Authentication method   ● Service principal   ○ System-assigned managed identity

Service principal * ⓘ   (new) default service principal
                               Configure service principal

Kubernetes authentication and authorization

Authentication and authorization are used by the Kubernetes cluster to control user access to the cluster as well as what the user may do once authenticated. Learn more about Kubernetes authentication ⊡

Role-based access control (RBAC) ⓘ   ● Enabled   ○ Disabled

AKS-managed Azure Active Directory ⓘ   ○ Enabled   ● Disabled

Node pool OS disk encryption

By default, all disks in AKS are encrypted at rest with Microsoft-managed keys. For additional control over encryption, you can supply your own keys using a disk encryption set backed by an Azure Key Vault. The disk encryption set will be used to encrypt the OS disks for all node pools in the cluster. Learn more ⊡

Encryption type   (Default) Encryption at-rest with a platform-managed key   ⌄

Review + create   < Previous   Next : Networking >

Figure 13.16 – AKS Authentication settings

6. By default, basic networking is used and Azure Monitor for containers is enabled. Click **Review + create** and then **Create**.

The AKS cluster will now be created. It will take some time before the creation process finishes.

> **Tip**
> Creating new Azure Active Directory service principals may take a few minutes. If you receive an error regarding the creation of the service principal during deployment, this means the service principal hasn't been fully propagated yet. For more information, refer to the following website: `https://docs.microsoft.com/en-us/azure/aks/troubleshooting#im-receiving-errors-that-my-service-principal-was-not-found-when-i-try-to-create-a-new-cluster-without-passing-in-an-existing-one`.

Now that we have created the AKS cluster, we are going to connect to the cluster using a tool called `kubectl`. This must, therefore, be installed on your computer, and the installation depends on the OS of your computer. You can find the installation instructions and links for your specific OS here: `https://kubernetes.io/docs/tasks/tools/install-kubectl/`.

Connecting to the cluster

In this section, we are going to connect to the cluster using `kubectl` from Azure Cloud Shell. The `kubectl` client is preinstalled in Azure Cloud Shell. Let's get started:

1. Navigate to the Azure portal by opening `https://portal.azure.com`.

2. In the top-right menu of the Azure portal, select **Azure Cloud Shell**.

3. First, we need to connect to our Kubernetes cluster. To do so, add the following line of code (the subscription method is optional):

```
az aks get-credentials \
--resource-group PacktAKSResourceGroup \
--name PacktAKSCluster \
--subscription "<your-subscription-id>"
```

4. To verify the connection to your cluster, return a list of the cluster nodes:

```
kubectl get nodes
```

The following output displays the cluster of nodes:

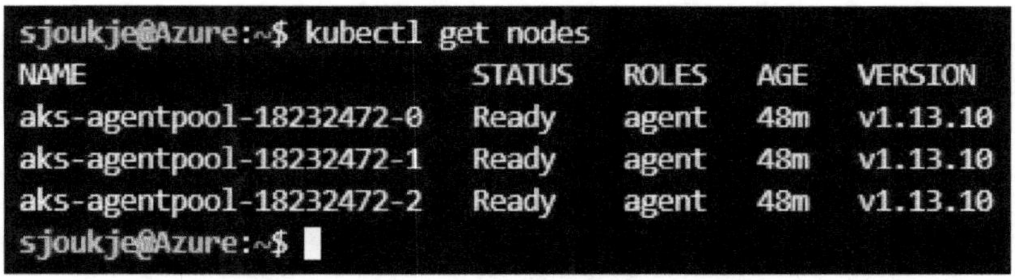

```
sjoukje@Azure:~$ kubectl get nodes
NAME                         STATUS    ROLES    AGE    VERSION
aks-agentpool-18232472-0     Ready     agent    48m    v1.13.10
aks-agentpool-18232472-1     Ready     agent    48m    v1.13.10
aks-agentpool-18232472-2     Ready     agent    48m    v1.13.10
sjoukje@Azure:~$ ▮
```

Figure 13.17 – Cluster of nodes

In this section, we connected to the AKS cluster using the CLI. In the next section, we are going to deploy the application.

Deploying the application

In this section, we are going to deploy the application to the cluster. For this, we need to define a Kubernetes manifest file. This defines the desired state for the cluster, such as what container images to run. We are going to use a manifest to create all the objects that are needed to deploy our sample application. This sample application is the Azure vote application and consists of a Python application and a Redis instance. The manifest will include two Kubernetes deployments: one for the frontend Python application and one for the backend Redis cache. Two Kubernetes services will be created as well, that is, an internal service for the Redis instance and an external service so that we can access the vote application from the internet:

> **Tip**
>
> In this demonstration, we are going to create the manifest file manually and deploy it manually to the cluster. However, in more real-world scenarios, you can use Azure Dev Spaces (https://docs.microsoft.com/en-us/azure/dev-spaces/) to debug your code directly in the AKS cluster. You can use Dev Spaces to work together with others on your team and across OS platforms and development environments.

1. In Cloud Shell, use `nano` or `vi` to create a file named `azure-vote.yaml`.

2. Copy in the following YAML definition. Here, we have the backend application, beginning with the deployment:

```
apiVersion: apps/v1
kind: Deployment
metadata:
name: azure-vote-back
spec:
```

```
replicas: 1
selector:
matchLabels:
app: azure-vote-back
template:
metadata:
labels:
app: azure-vote-back
spec:
nodeSelector: "beta.kubernetes.io/os": linux
containers:
- name: azure-vote-back
image: redis resources:
requests:
cpu: 100m
memory: 128Mi
limits:
cpu: 250m
memory: 256Mi
ports:
- containerPort: 6379
name: redis
---
```

3. Then, we have the service for the backend application:

```
apiVersion: v1
kind: Service
metadata:
name: azure-vote-back
spec:
ports:
- port: 6379
selector:
app: azure-vote-back
---
```

4. Next, we have the deployment for the frontend application:

```
apiVersion: apps/v1
kind: Deployment
metadata:
    name: azure-vote-front
spec:
    replicas: 1
selector:
matchLabels:
app: azure-vote-front template:
metadata:
labels:
app: azure-vote-front
spec:
nodeSelector:
"beta.kubernetes.io/os": linux
containers:
- name: azure-vote-front
image: microsoft/azure-vote-front:v1
resources:
requests:
cpu: 100m
memory: 128Mi
limits:
cpu: 250m
memory: 256Mi
ports:
-containerPort: 80
env:
name: REDIS
value: "azure-vote-back"
---
```

5. After this, we have the service:

```
apiVersion: v1
kind: Service
metadata:
name: azure-vote-front
spec:
type: LoadBalancer
ports:
- port: 80
selector:
app: azure-vote-front
```

6. Now, deploy the application using the `azure-vote` manifest file:

```
kubectl apply -f azure-vote.yaml
```

The application will be deployed. This will result in the following output:

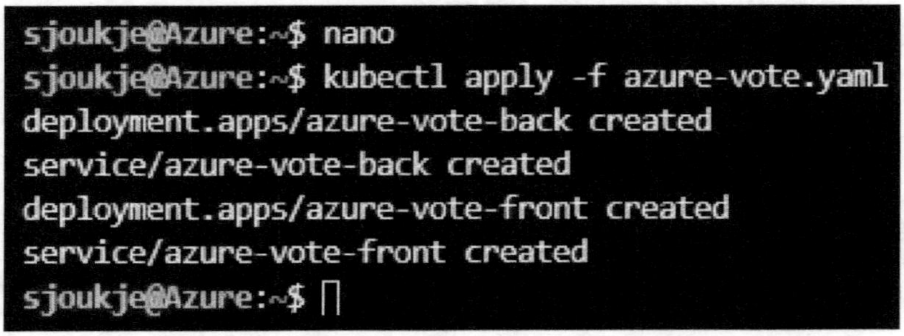

```
sjoukje@Azure:~$ nano
sjoukje@Azure:~$ kubectl apply -f azure-vote.yaml
deployment.apps/azure-vote-back created
service/azure-vote-back created
deployment.apps/azure-vote-front created
service/azure-vote-front created
sjoukje@Azure:~$ []
```

Figure 13.18 – The output of deploying the application

Now that we've deployed the application, we will test it.

Testing the application

When the application runs, the application frontend is exposed to the internet by a Kubernetes service. This process can take a few minutes to complete.

We can monitor its progress by using the following command in the CLI:

```
kubectl get service azure-vote-front --watch
```

When the frontend is exposed to the internet, the following output will be displayed:

```
sjoukje@Azure:~$ kubectl get service azure-vote-front --watch
NAME                 TYPE          CLUSTER-IP    EXTERNAL-IP     PORT(S)        AGE
azure-vote-front     LoadBalancer  10.0.66.19    40.121.48.242   80:31847/TCP   4m
```

Figure 13.19 – Exposed service

Now, you can copy the external IP address and paste it into a browser. This will result in the following output:

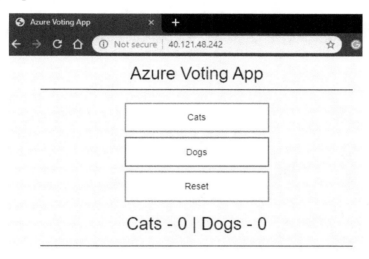

Figure 13.20 – The app running in AKS

Now that we have a working app in AKS, we are going to monitor the health and logs of this application in AKS.

Monitoring the health and logs of the application

When we created the cluster, Azure Monitor for containers was enabled. This monitoring feature provides health metrics for both the AKS cluster and the pods running on the cluster. This is displayed in the Azure portal and consists of information regarding the uptime, current status, and resource usage for the Azure vote pods. Let's take a look:

1. Navigate to the Azure portal by opening https://portal.azure.com.

2. Go to the overview page of PacktAKSCluster.

3. Under **Monitoring**, select **Insights**. Then, in the top menu, choose **Add Filter**:

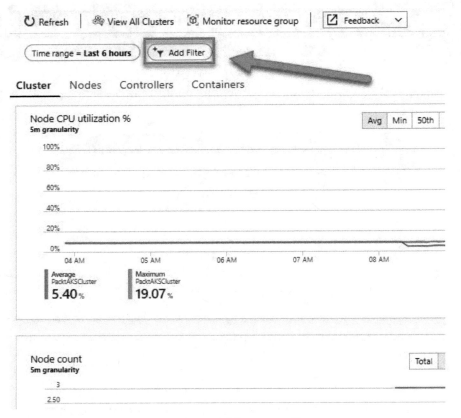

Figure 13.21 – Adding a new filter

4. Select **Namespace** as the property. Then, choose **<All but kube-system>**.

5. Then, from the top menu (under the filters), select **Containers**:

Figure 13.22 – Containers view

6. To see the logs for the `azure-vote-front` pod, select **View container logs** from the dropdown of the containers list. These logs include the `stdout` and `stderr` streams from the container:

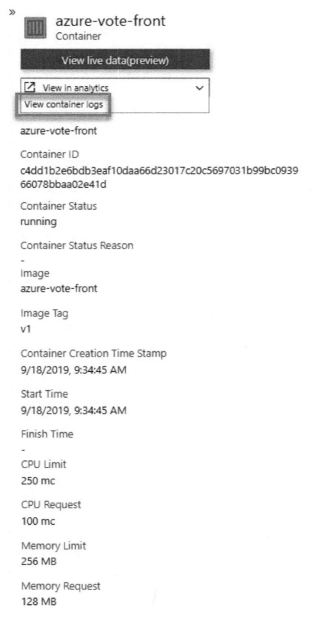

Figure 13.23 – Opening the container logs

7.   This will result in the following output:

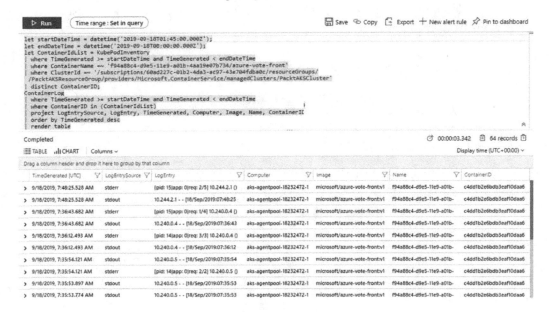

Figure 13.24 – Container logs

These logs are what get the output from the container and include information logs and errors. This concludes this section and this chapter.

Summary

In this chapter, we learned about Docker containers and how to implement an application that runs on an Azure container instance.

We created a container image using a Dockerfile and published it to ACR. Next, we used that image to deploy a web service running as an Azure container instance and an Azure web app.

Finally, we created an AKS cluster and deployed an application to the cluster, before looking at the monitoring capabilities.

In the next chapter, we will continue the *Implement Solutions for Apps* objective by learning how to implement authentication.

Questions

Answer the following questions to test your knowledge of the information contained in this chapter. You can find the answers in the *Assessments* section at the end of this book:

1. You are planning to deploy an application to a container solution in Azure. You want full control over how to manage the containers. Is ACI the right solution for this?

    a) Yes

    b) No

2. Can logs in AKS only be viewed from the CLI?

    a) Yes

    b) No

3. Can you create a Dockerfile using Visual Studio?

    a) Yes

    b) No

4. With Web App for Containers, can you only pull containers from Docker?

    a) Yes

    b) No

Further reading

Check out the following links to find out more about the topics that were covered in this chapter:

*   ACR documentation: `https://docs.microsoft.com/en-us/azure/container-registry/`
*   ACI documentation: `https://docs.microsoft.com/en-us/azure/container-instances/`
*   App Service on Linux documentation: `https://docs.microsoft.com/en-us/azure/app-service/containers/`
*   AKS documentation: `https://docs.microsoft.com/en-us/azure/aks/`

14
Implementing Authentication

In the previous chapter, we covered the last part of the *Implementing Solutions for Apps* objective by covering how to design and develop apps that run in containers. Here, we learned how to implement an application that runs on Azure Container Instances before creating an Azure Kubernetes Service and deploying an application to the cluster.

In this chapter, we are going to cover how to implement authentication for your Web Apps, APIs, functions, and logic apps using certificates, authentication, and more. We are going to implement managed identities for Azure resources service principal authentication and implement OAuth2 authentication.

The following topics will be covered in this chapter:

- Understanding Azure App Service authentication
- Implementing Active Directory authentication
- Implementing authentication by using certificates
- Understanding and implementing OAuth2 authentication and tokens
- Implementing managed identities for Azure resources service principal authentication

Let's get started!

Technical requirements

This chapter's examples will use Azure PowerShell (`https://docs.microsoft.com/en-us/powershell/azure`), Visual Studio 2019 (`https://visualstudio.microsoft.com/vs/`), and Postman (`https://www.getpostman.com/`).

The source code for our sample application can be downloaded from `https://github.com/PacktPublishing/Microsoft-Azure-Architect-Technologies-Exam-Guide-AZ-303/tree/master/Ch14`.

Understanding Azure App Service authentication

Azure App Service provides built-in authentication and authorization support. This makes it easy to sign in users and access data, with minimal to no code changes needing to be made in your Web Apps, APIs, Azure Functions, and mobile backends. It also removes the need for you to have deeper knowledge about security, including encryption, JSON web tokens, federation, and more. This is all handled for you by Azure. However, you are not required to use App Service for authentication and authorization. You can also use security features that come with other web frameworks or even write your own utilities.

The App Service authentication and authorization module runs in the same sandbox as your application code. When it is enabled, it will handle all incoming HTTP requests before they are handled by the application code.

This module handles the following things for your app:

- Authenticates users with the specified provider
- Validates, stores, and refreshes tokens
- Manages the authenticated session
- Injects identity information into request headers

This module is configured using app settings and runs completely separately from the application code. No changes to the application code, SDKs, or specific languages are required.

App Service authentication and authorization use the following security features:

- **User claims**: App Service makes the user claims available for all language frameworks. These claims are injected into the request headers. ASP.NET 4.7 apps use `ClaimsPrincipal.Current` for this. For Azure Functions, the claims are also injected into the headers.

- **Token store**: App Service provides a built-in token store, where tokens are stored for the users of the different apps. When you enable authentication with any provider, this token store is immediately available to your app. In most cases, you must write code to store, collect, and refresh tokens in the application. With the token store, you can just retrieve them when you need them. When they become invalid, you can tell App Service to refresh them.

- **Identity providers**: App Service can also use **federated identity**. A third-party identity provider will then manage the user identities and authentication flow for you. Five identity providers are available by default: Azure Active Directory, Microsoft Accounts, Facebook, Google, and Twitter. Besides these built-in identity providers, you can also integrate another identity provider or your own custom identity solution.

- **Logging and tracing**: When application logging is enabled, the authentication and authorization traces will be directly displayed in the log files. When an unexpected authentication error occurs, all of the details of this error can be found in the existing application logs. When failed request tracing is enabled, you can see exactly what role the authentication and authorization module may have played in a failed request.

In the next section, we are going to enable this feature in the Azure portal.

Implementing Active Directory authentication

By configuring the authentication and authorization feature of the Web App, Windows-integrated authentication is enabled. In this demonstration, we are going to enable this for a Web App. First, we will deploy a Web App to Azure App Service.

Deploying the Web App

In the first part of this demonstration, we are going to deploy a sample Web App. We are going to be using a sample application and will deploy this to Azure using a PowerShell script.

To do this, perform the following steps:

1. First, we need to log into our Azure account, as follows:

```
Connect-AzAccount
```

2. If necessary, select the right subscription, as follows:

```
Select-AzSubscription -SubscriptionId "********-****-
****-****-***********"
```

3. Specify the URL for the GitHub sample application and specify the Web App's name:

```
$gitrepo="https://github.com/Azure-Samples/app-service-
web-dotnet-get-started.git"
$webappname="PacktWebApp"
```

4. Create a Web App:

```
New-AzWebApp `
-Name $webappname `
-Location "East US" `
-AppServicePlan PacktAppServicePlan `
-ResourceGroupName PacktAppResourceGroup
```

5. Then, configure the GitHub deployment from the GitHub repository:

```
$PropertiesObject = @{ repoUrl = "$gitrepo"; branch =
"master";
isManualIntegration = "true";
}
```

6. Deploy the GitHub Web App over the newly created Web App:

```
Set-AzResource `
-PropertyObject $PropertiesObject `
-ResourceGroupName PacktAppResourceGroup `
-ResourceType Microsoft.Web/sites/sourcecontrols `
-ResourceName $webappname/web `
-ApiVersion 2015-08-01 `
-Force
```

7. After deployment, you can obtain the Web App's URL from the **Overview** blade in the Azure portal and paste it into your browser. The output will look as follows:

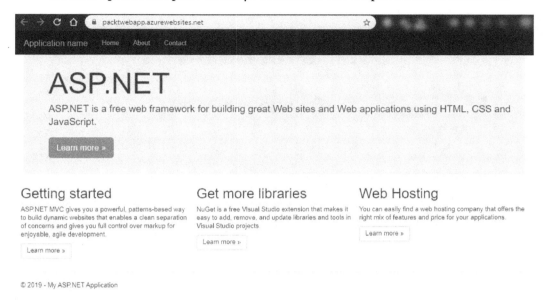

Figure 14.1 – Sample application

Now that we've deployed the Web App, we can enable authentication and authorization for it.

Enabling authentication and authorization

To enable authentication and authorization for your apps, perform the following steps:

1. Navigate to the Azure portal by going to `https://portal.azure.com`.

2. Go to `PacktWebApp`, which we created in the previous section.

3. From the left menu, under **Settings**, select **App Service Authentication** and then turn it **On**, as follows:

Figure 14.2 – Enabling authentication/authorization

4. Under **Action to take when request is not authenticated**, you can select the following options (you can add multiple identity providers to your app; however, you need to do some coding inside your app to make these multiple providers available):

    - **Allow anonymous requests (no action)**

    - **Log in with Azure Active Directory**

- **Log in with Facebook**

- **Log in with Google**

- **Log in with a Microsoft account**

- **Log in with Twitter**

5. In this demonstration, we are going to use **Log in with Azure Active Directory**, so select this one and click the **Save** button at the top of the menu. Our environment can now log into our app using our Microsoft credentials.

6. Now, we need to configure the app to let users log in using their Azure AD credentials. Under **Authentication Providers**, select **Azure Active Directory**:

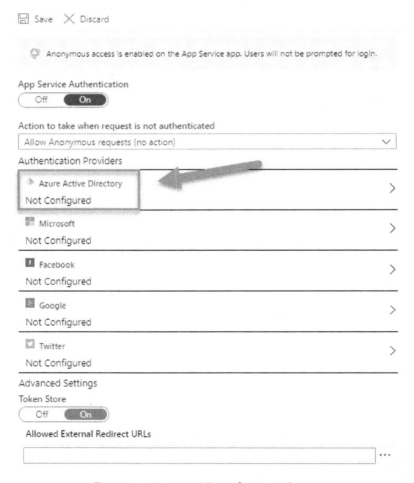

Figure 14.3 – Azure AD configuration button

7. In the next blade, you can select a couple of management mode settings:

    - **Off**: An Azure AD application won't be registered for you.

    - **Express**: An Azure AD application is registered for you automatically using the express settings.

    - **Custom**: You can register an Azure AD application using custom settings that you will provide manually.

8. For this demonstration, select **Express**. Next, you can create a new Azure AD registration (service principal) or select an existing one. We will leave the default settings as-is. Click **OK** and then **Save**:

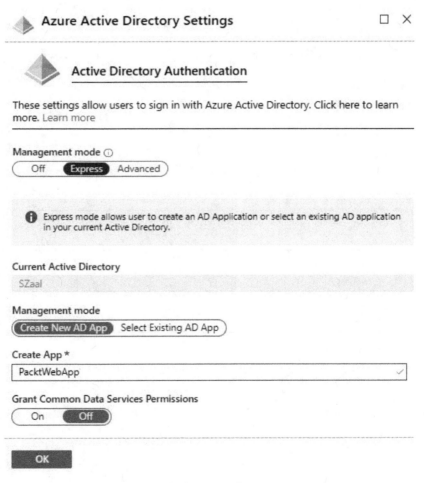

Figure 14.4 – Azure AD configuration settings

9. We are now ready to use Azure AD authentication in `PacktWebApp`.

10. Navigate to the **Overview** blade and click the link for the URL. The website will open. Here, you will have to specify a Microsoft account in order to log in (if you don't get asked to do this, this is because you are already signed into your account). You may also be prompted to allow permissions for the app to your signed in account – this shows that AD authentication is now being used. Click **Accept**:

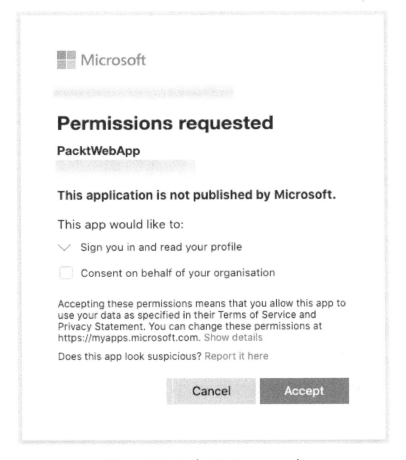

Figure 14.5 – Authentication approval

11. Next, we will amend the code of the Web App to show you how to obtain the details of your logged-in account. In the Azure portal, still on the **WebApp** blade, under **Development Tools**, click **App Service Editor**.

12. **App Service Editor** allows you to directly modify code. First, you will need to click **Go ->**, which will open a new window with your website code inside it.

13. Navigate to `Views\Home\Index.cshtml`, as shown in the following screenshot:

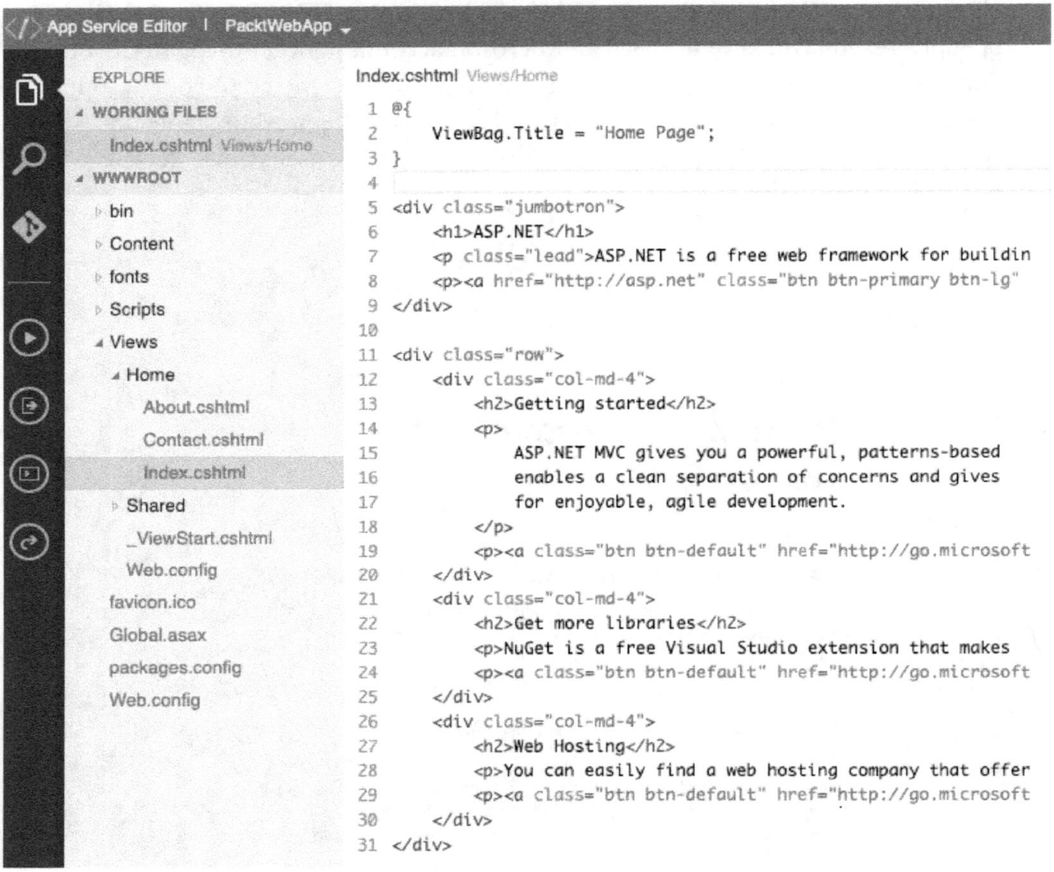

Figure 14.6 – Azure App Service Editor

14. Add the following code before the first *Jumbotron*:

```
@using System.Security.Claims
@using System.Threading

<div class="jumbotron">
    @{
        var claimsPrincipal = Thread.CurrentPrincipal
as ClaimsPrincipal;
        if (claimsPrincipal != null &&
claimsPrincipal.Identity.IsAuthenticated)
```

```
        {
            <h2 style="color:red">Authentication
Succeeded!</h2>
            <div><span><strong>Principal Name:</
strong><br />@claimsPrincipal.Identity.Name</span></div>
            <br />
        }
        else
        {
            <div style="color:blue">Current request
is unauthenticated!</div>
        }
    }
    </div>
```

15. Your page should now look like this:

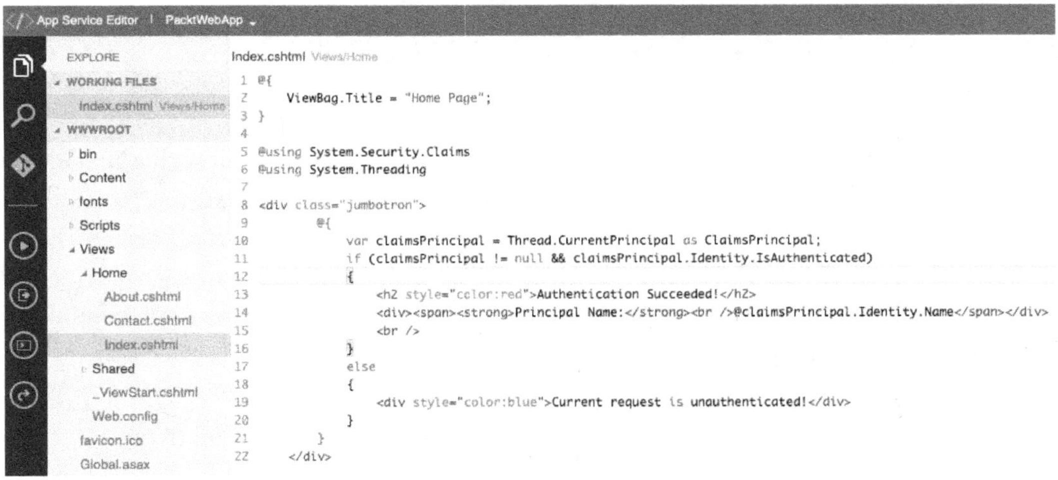

```
Index.cshtml Views/Home
1  @{
2      ViewBag.Title = "Home Page";
3  }
4
5  @using System.Security.Claims
6  @using System.Threading
7
8  <div class="jumbotron">
9      @{
10         var claimsPrincipal = Thread.CurrentPrincipal as ClaimsPrincipal;
11         if (claimsPrincipal != null && claimsPrincipal.Identity.IsAuthenticated)
12         {
13             <h2 style="color:red">Authentication Succeeded!</h2>
14             <div><span><strong>Principal Name:</strong><br />@claimsPrincipal.Identity.Name</span></div>
15             <br />
16         }
17         else
18         {
19             <div style="color:blue">Current request is unauthenticated!</div>
20         }
21     }
22 </div>
```

Figure 14.7 – Updated HTML

16. Navigate to the **Overview** blade and click the link for the URL. The website will open and you will have to specify a Microsoft account in order to log in. Once you've successfully authenticated, the website will display your **Principal Name** and other information, such as claims, as shown in the following screenshot:

Figure 14.8 – Successfully authenticated

With that, we have enabled authentication and authorization. In the next section, we are going to implement authentication using certificates.

Implementing authentication using certificates

By default, Azure secures the `*.azurewebsites.net` wildcard domain with a single SSL certificate. So, when you use the default domain that is generated for your app when you deploy it to Azure App Service, your users will access the app over a secure connection.

When you use a custom domain for your app – for instance, `az-303.com` – you should assign an SSL certificate to it yourself.

You can assign SSL certificates to your app from the Azure portal. To assign a certificate, your app must run in the **Standard**, **Premium**, or **Isolated** App Service plan tiers.

> **Tip**
>
> You can order your SSL certificate from the App Service Certificate Create page directly as well. To order a certificate, please refer to `https://portal.azure.com/#create/Microsoft.SSL`.
>
> You can also create a free certificate and use it at `https://www.sslforfree.com` and convert the certificate into an SSL certificate by going to `https://decoder.link/converter`.

> **Important Note**
>
> For this demonstration, I've added a custom domain to my Web App, obtained a free certificate, and converted it into an SSL certificate using the preceding websites.

To bind an SSL certificate to `PacktWebApp`, perform the following steps:

1. Navigate to the Azure portal by going to `https://portal.azure.com`.

2. Again, go to `PacktWebApp`, which we created in the previous section.

3. From the left menu, under **Settings**, select **TLS/SSL Settings**.

4. In the **TLS/SSL settings** blade, click on **Private Key Certificates (.pfx)**, as follows:

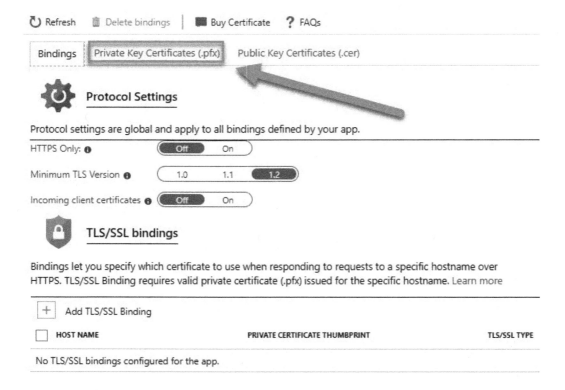

Figure 14.9 – SSL settings

5.   Next, click the **Upload Certificate** button, as follows:

Figure 14.10 – Private certificates

6.   Upload the `.pfx` file from your computer, provide the required password, and click **Upload**, as follows:

Figure 14.11 – Uploading the .pfx file

7. With that, the certificate has been uploaded to Azure. Now, we must set the SSL binding to bind it to the domain. Click **Bindings** in the top menu and click **Add TLS/SSL Binding**, as follows:

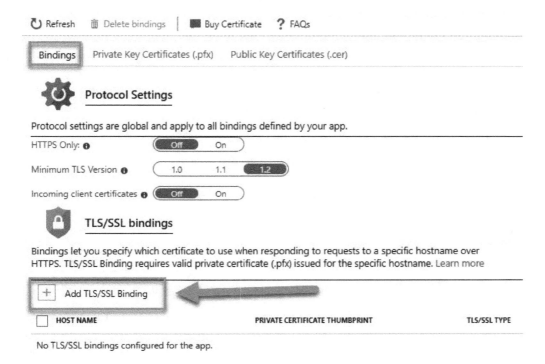

Figure 14.12 – Adding an SSL binding – I

8. To bind the certificate to the domain, you need to specify a few values:

    - **Hostname**: Select the hostname from the drop-down list.

    - **Private Certificate Thumbprint**: Select the uploaded certificate here.

    - **SSL Type**: Select **SNI SSL**.

9. Then, click **Add Binding**, as follows:

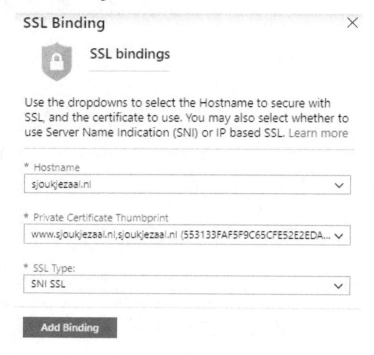

Figure 14.13 – Adding an SSL binding – II

10. If you have a www hostname as well, you should repeat the previous step to bind the same certificate to this.

11. Lastly, we will set one of the protocol settings and switch the website to **HTTPS Only**. This way, the website can only be accessed using HTTPS, which means it is not accessible via HTTP, as shown in the following screenshot:

Figure 14.14 – Protocol Settings

With that, we have covered how to assign an SSL certificate to our app. In the next section, we are going to cover OAuth2.

Understanding and implementing OAuth2 authentication in Azure AD

The Microsoft identity platform supports industry-standard protocols such as OAuth 2.0 and OpenID Connect, as well as open source libraries for different platforms. **Azure Active Directory** (**Azure AD**) uses OAuth 2.0 to allow you to authorize access to resources in your Azure AD tenant. OpenID Connect is then used in your custom applications as middleware to communicate with the OAuth 2.0 framework.

There are two primary use cases in the Microsoft identity platform programming model:

- **During an OAuth 2.0 authorization grant flow**: During this flow, a resource owner grants authorization to the client application. This will allow the client to access the resources from the resource owner.

- **During resource access by the client**: During this flow, the claim values present in the access token are used as a basis for making access control decisions. This is implemented by the resource server.

The OAuth 2.0 authorization code flow is used to perform authentication and authorization in most application types, including Web Apps and natively installed apps. At a high level, the entire authorization flow for an application looks like this:

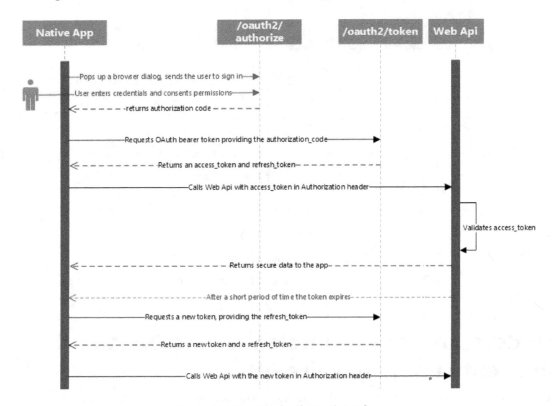

Figure 14.15 – High-level OAuth 2.0 flow

In the next section, we are going to implement OAuth 2.0 authentication.

Implementing OAuth2 authentication

The first step when it comes to implementing OAuth 2.0 authentication in Azure AD is registering an application (service principal) in Azure AD. After registration, permissions can be set to the application, which gives us access to various resources in Azure, such as Microsoft Graph, and more.

Registering the application in Azure AD

In this demonstration, we are going to register an application in Azure AD (also called a service principal). We are going to give this service principal permission to access Azure resources. In this example, we are going to create a user in Azure AD using Microsoft Graph.

We are going to use Postman as an API client to create the requests to Graph. In the *Technical requirements* section at the beginning of this chapter, you can click the respective link to install Postman.

> **Tip**
> For those who are unfamiliar with Postman, please refer to the following website for more information: `https://www.getpostman.com/product/api-client`.

To register the application in the Azure portal, we need to perform the following steps:

1. Navigate to the Azure portal by going to `https://portal.azure.com`.

2. From the left menu, select **Azure Active Directory**.

3. In the **Azure Active Directory Overview** blade, click **App Registrations** and then, in the top menu, **+ New registration**:

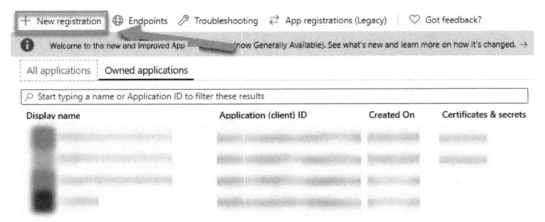

Figure 14.16 – New app registration

4. In the **App registration** blade, add the following values:

a) **Name**: `PacktADApp`.

b) **Supported account types**: Accounts in this organizational directory only. There are three options here: the first will create a single-tenant app. This only has access to the Azure AD tenant where it is created. The second one, **Accounts in any organizational directory**, creates a multitenant app. You can access other Azure AD tenants with this registration as well. The last option creates a multi-tenant app as well, and besides work and school accounts, you can also log in with personal Microsoft accounts, such as outlook.com accounts.

c) **Redirect URI**: Here, you can fill in `https://localhost` because we don't actually have a real application here. This is fine for dev/test scenarios when you're running from your local computer, but in a production environment, you would, of course, use your actual URL. Once the authentication process is complete, the user will be redirected to the specified URL. The following screenshot shows an example of this page:

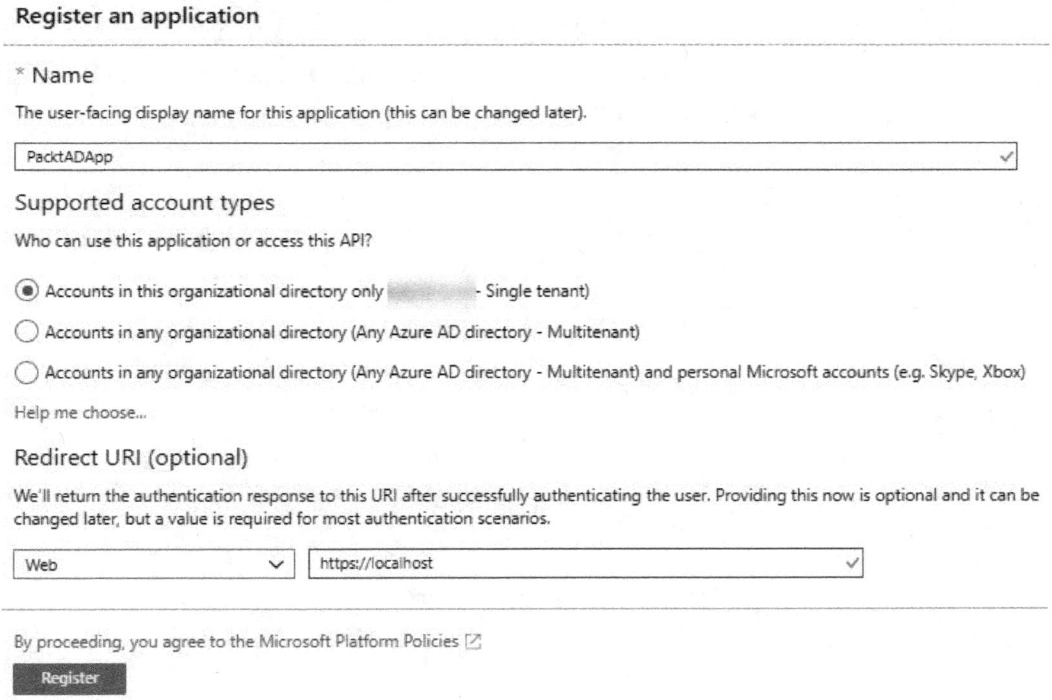

Register an application

* Name

The user-facing display name for this application (this can be changed later).

| PacktADApp | ✓ |

Supported account types

Who can use this application or access this API?

◉ Accounts in this organizational directory only ▇▇▇▇ - Single tenant)

◯ Accounts in any organizational directory (Any Azure AD directory - Multitenant)

◯ Accounts in any organizational directory (Any Azure AD directory - Multitenant) and personal Microsoft accounts (e.g. Skype, Xbox)

Help me choose...

Redirect URI (optional)

We'll return the authentication response to this URI after successfully authenticating the user. Providing this now is optional and it can be changed later, but a value is required for most authentication scenarios.

| Web | ∨ | | https://localhost | ✓ |

By proceeding, you agree to the Microsoft Platform Policies ☑

Register

Figure 14.17 – Registering the application

5. Click **Register**.

6. During registration, Azure AD will assign your application a unique client identifier (**Application ID**). You need this value for the next few sections, so copy it from the **Application** page.

7. Find your registered application in the Azure portal, click **App registrations** again, and then click **View all applications**.

8. The next step is to create an app secret. Therefore, from the **Application** blade, from the left menu, click **Certificates & secrets**. From there, click **+ New client secret**:

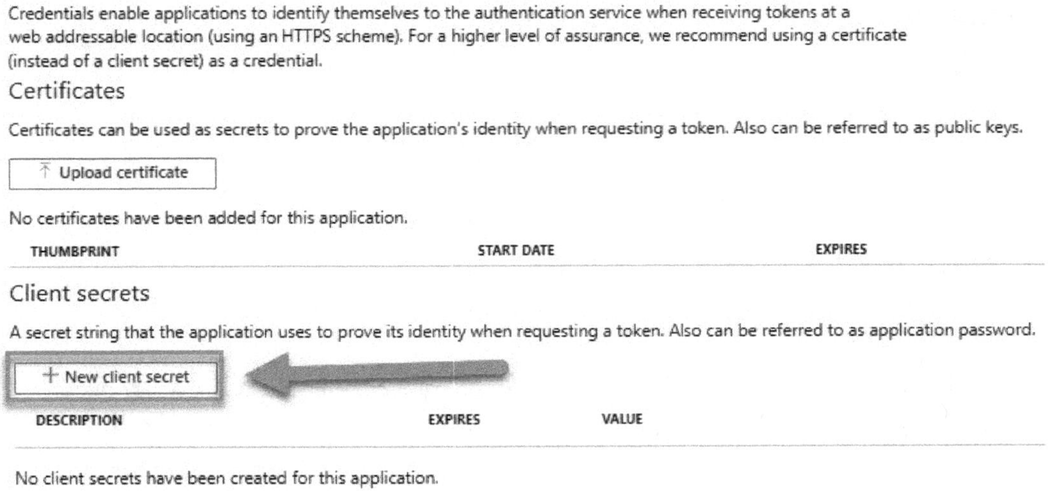

Credentials enable applications to identify themselves to the authentication service when receiving tokens at a web addressable location (using an HTTPS scheme). For a higher level of assurance, we recommend using a certificate (instead of a client secret) as a credential.

Certificates

Certificates can be used as secrets to prove the application's identity when requesting a token. Also can be referred to as public keys.

↑ Upload certificate

No certificates have been added for this application.

THUMBPRINT	START DATE	EXPIRES

Client secrets

A secret string that the application uses to prove its identity when requesting a token. Also can be referred to as application password.

+ New client secret

DESCRIPTION	EXPIRES	VALUE

No client secrets have been created for this application.

Figure 14.18 – Creating a new client secret

9. Add the following values:

   a) **Description**: Key1.

   b) **Expires**: In **1 year**. You can also choose in **2 years** or **never**.

10. Click **Add**.

11. Then, copy the client secret; it's only displayed once:

Credentials enable applications to identify themselves to the authentication service when receiving tokens at a web addressable location (using an HTTPS scheme). For a higher level of assurance, we recommend using a certificate (instead of a client secret) as a credential.

Certificates

Certificates can be used as secrets to prove the application's identity when requesting a token. Also can be referred to as public keys.

↑ Upload certificate

No certificates have been added for this application.

THUMBPRINT	START DATE	EXPIRES

Client secrets

A secret string that the application uses to prove its identity when requesting a token. Also can be referred to as application password.

+ New client secret

DESCRIPTION	EXPIRES	VALUE	
Key1	10/9/2020	qka0NDH6yUBTqay3@9SqkCbBCUBjw:]. 📋	🗑

Figure 14.19 – Client secret

12. Now that we have an application ID and a secret, we can set the appropriate permissions for the application. Therefore, from the left menu, select **API permissions**.

13. In the **API permissions** blade, you will see that one permission has already been added: you are allowed to log in and read your own user profile. Click **+ Add a permission**:

Configured permissions

Applications are authorized to call APIs when they are granted permissions by users/admins as part of the consent process. The list of configured permissions should include all the permissions the application needs. Learn more about permissions and consent

API / Permissions name	Type	Description	Admin Consent Req...	Status	
⋁ Microsoft Graph (1)					⋯
User.Read	Delegated	Sign in and read user profile	-		⋯

Figure 14.20 – Application permissions

14. In the **Request API permissions** blade, select **Microsoft Graph**. Here, you can choose between two different types of permissions: **Delegated permissions** and **Application permissions**. The former gives you access to the API as the signed-in user, which means that all of the permissions that the user has will also apply to the data that can be accessed by the application. The latter basically creates a sort of service account that has access to all the users in a tenant, all security groups, all Office 365 resources, and more. For this example, we want to create a new user in Azure AD. Normal users typically don't have the necessary permissions to access Azure AD, so we have to select **Application permissions** here.

15. After selecting **Application permissions**, all the available permissions will be displayed. We need to unfold the **Directory** item and then choose **Directory. ReadWrite.All**:

Request API permissions

< All APIs

▶ ChannelMessage

▶ Chat

▶ Contacts

▶ DelegatedPermissionGrant

▶ Device

▶ DeviceManagementApps

▶ DeviceManagementConfiguration

▶ DeviceManagementManagedDevices

▶ DeviceManagementRBAC

▶ DeviceManagementServiceConfig

▾ Directory (1)

   ☐ Directory.Read.All
     Read directory data ❶   Allows the app to read and write data in your organization's directory, such as users, and groups, without a signed-in user. Does not allow user or group deletion.

   ☑ Directory.ReadWrite.All
     Read and write directory data 🖑   Yes

▶ Domain

▶ EduAdministration

▶ EduAssignments

▶ EduRoster

▶ ExternalItem

| Add permissions | Discard |

Figure 14.21 – Selecting the appropriate permissions

16. Click **Add permissions**.

17. Because we are using application permissions, an administrator needs to grant admin consent as well. Therefore, you need to click **Grant admin consent for...** and then, in the popup, log in with your administrator credentials and accept the license terms:

Figure 14.22 – Granting admin consent

18. Now, this application has permission to add a new user to the Azure AD tenant:

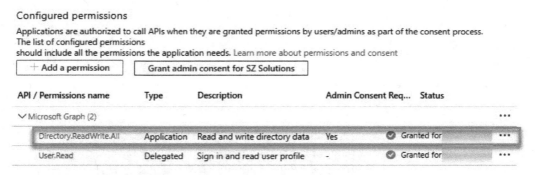

Figure 14.23 – Admin consent granted

19. Finally, select **Azure Active Directory** from the left menu again, and then under **Manage**, click **Properties**. Here, copy the Azure AD tenant ID. We need this to set up the request to Azure AD using Microsoft Graph in the next section.

This concludes the first part of this demonstration. In the next section, we are going to implement tokens.

Implementing tokens

In this part of the demonstration, we are going to make some requests to Microsoft Graph to create a new user in Azure AD. First, we will be creating a request to the login page of Microsoft and providing the Azure AD tenant ID, the application ID, and the secret. This will result in `access_token`, which we can use in the second request to create the actual user.

Therefore, we need to perform the following steps:

1. Open Postman and log in or create an account. For the request URL, add the following (make sure that you replace `{tenant_id}` in the request URL with the correct Azure AD tenant ID, which we copied in the previous section):

    ```
    POST
    https://login.microsoftonline.com/{tenant_id}/oauth2/
    token?api-version=1.0
    ```

2. Click **Body** in the top menu and add the following values:

    ```
    Key: grant_type, Value: client_credentials
    Key: resource, Value: https://graph.microsoft.com
    Key: client_id, Value: <replace-this-with-the-
    application- id>
    Key: client_id, Value: <replace-this-with-the-
    application- secret>
    ```

3. Then, click the **Send** button. In response, you will receive `access_token`. This token will be used to make the actual request to Graph to create the user:

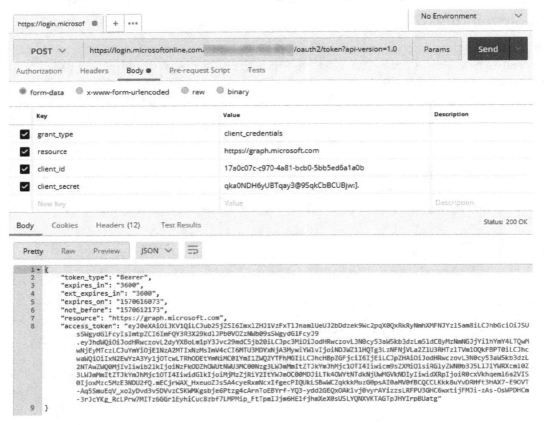

Figure 14.24 – Example Postman screen

4. Open another tab in Postman and add the following to the request URL containing the correct Azure AD tenant ID (which we copied in the previous section):

```
POST https://graph.microsoft.com/v1.0/users
```

5. Click **Headers** in the top menu and add **Key**: **Authorization**, **Value**: `Bearer <access_token>` as values, as shown in the following screenshot:

Figure 14.25 – Adding access_token

6. Click **Body** in the top menu and select **raw** and **JSON**. Then, add the following value (replace the Azure AD tenant name with the name of your Azure AD tenant):

```
{
    "accountEnabled": true,
    "displayName": "PacktUser",
    "mailNickname": "PacktUser",
    "userPrincipalName": "PacktUser@<azure-ad-tenant-
    name>.onmicrosoft.com",
    "passwordProfile" : {
    "forceChangePasswordNextSignIn": true,
    "password": "P@ss@Word"
    }
}
```

The output will be as follows. This shows the full record for the user you have just created:

```
{
    "@odata.context":
    "https://graph.microsoft.com/
    v1.0/$metadata#users/$entity",
    "id": "82decd46-8f51-4b01-8824-ac24e0cf9fa4",
    "businessPhones": [],
    "displayName": "PacktUser",
    "givenName": null,
    "jobTitle": null,
    "mail": null,
    "mobilePhone": null,
```

```
"officeLocation": null,
"preferredLanguage": null,
"surname": null,
"userPrincipalName": "PacktUser@<azure-ad-tenant- name>.
onmicrosoft.com"
}
```

In Postman, this will look as follows – this is the same as the output shown in the preceding code:

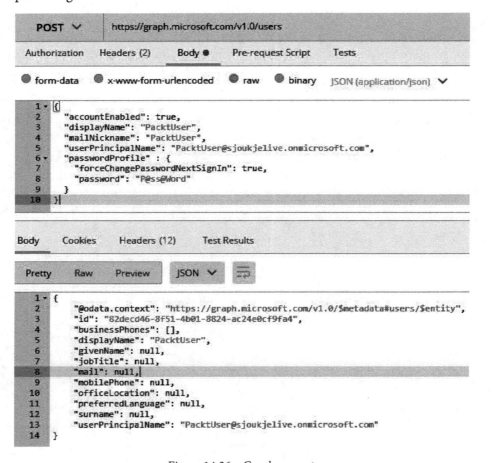

Figure 14.26 – Graph request

Next, we will look at refreshing tokens.

Refreshing tokens

Access tokens are only valid for a short period of time. They must be refreshed to continue having access to resources. You can refresh an access token by submitting another post to the token endpoint, as we did when we requested `access_token`. Refresh tokens are valid for all the resources that we gave consent to in Azure AD via the Azure portal.

Typically, the lifetimes of refresh tokens are relatively long. But in some cases, the tokens are revoked, have expired, or lack sufficient privileges for the desired action. You must ensure that your application handles these errors that are returned by the token endpoint correctly. When you receive a response with a refresh token error, discard the current refresh token and request a new authorization code or access token.

OAuth 2 is a commonly used method for authentication. Next, we will look at using an Azure-specific mechanism for authenticating between components without needing a username and password.

Understanding and implementing managed identities

One of the challenges you'll face when you build applications for the cloud is how to manage your credentials in code for authentication. Keeping those credentials secure is key, so ideally, these credentials will never appear on developer workstations and aren't checked into source control either. You can use Azure Key Vault for securely storing credentials, keys, and secrets, but the application still needs to authenticate to Key Vault to retrieve them.

Managed identities solves this problem. This is a feature of Azure AD that provides an Azure service with an automatically managed identity in Azure AD. You can then use this identity to authenticate to every server that supports Azure AD authentication, including Key Vault, without any credentials in your code.

When you enable managed identities on your Azure resource, such as a virtual machine, Azure Function, or app, Azure will create a service principal and store the credentials of that service principal on the Azure resource itself. When it is time to authenticate, a **Managed Service Identity** (**MSI**) endpoint is called, passing your current Azure AD credentials and a reference to the specific resource.

Managed identities then retrieves the stored credentials from the Azure resource, passes it to Azure AD, and retrieves an access token that can be used to authenticate to the Azure resource or service.

You should note that the service principal is only known inside the boundaries of the specific Azure resource where it is stored. If it needs permissions for other resources as well, you should assign the appropriate role using **role-based access control** (**RBAC**) in Azure AD.

There are two types of managed identities:

- **System-assigned managed identity**: This identity is enabled directly on an Azure service instance. It is directly tied to the Azure service where it is created. It cannot be reused for other services. When the Azure service is deleted, the managed identity is deleted as well.

- **User-assigned managed identity**: This identity is created as a standalone Azure resource. Once the identity has been created, it can be assigned to one or more Azure service instances. Deleting the Azure service will not delete the managed identity.

In the next section, we are going to learn how to enable managed identities for a Web App.

Implementing managed identities for Azure resources service principal authentication

You can enable MSI for your Azure resources in the Azure portal, PowerShell, or the CLI by using ARM templates. In this demonstration, we are going to enable this in the Azure portal for the Azure Web App that we created earlier in this chapter:

1. Navigate to the Azure portal by going to `https://portal.azure.com`.

2. Go to `PacktWebApp`, which we created earlier.

3. From the **Overview** blade of the Web App, from the left menu, under **Settings**, click **Identity**.

4. In the next blade, you can create a system-assigned or user-assigned managed identity. We are going to create a **System assigned** identity for this demo. Change the status to **On** and click **Save**:

Figure 14.27 – Managed identity settings

5. Now that the managed identity has been created, we can assign permissions to it to access the Key Vault. First, we need to create the Key Vault. We will add the Key Vault to the same resource group that the Web App is in. We will do this using the Azure CLI. Open Cloud Shell and enter the following line of code:

```
az keyvault create --name WebAppEncryptionVault -g
"PacktAppServicePlan"
```

6. Once the Key Vault has been created, navigate to it in the Azure portal. From the left menu, click **Access control (IAM)**. Then, click under **Add a role assignment**:

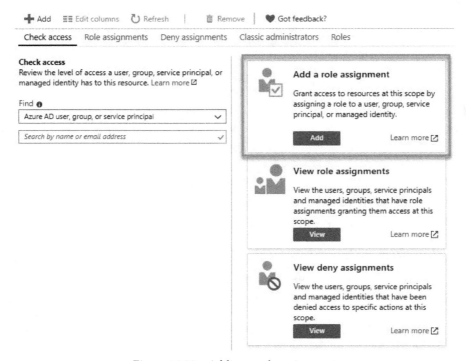

Figure 14.28 – Adding a role assignment

7. Then, add the following values:

    a) **Role: Key Vault Contributor**.

    b) **Assign access to**: Under **System assigned managed identity**, select **App Service**.

    c) **Subscription**: Pick the subscription that the Web App was created for.

8. Then, you can select the managed identity that we created for the Web App:

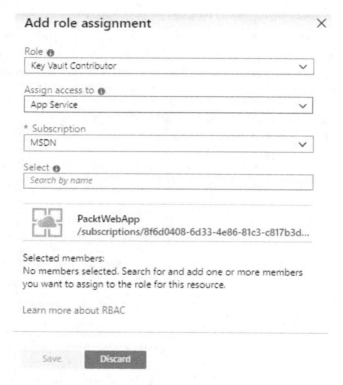

Figure 14.29 – Adding a role to the managed identity

9. Select the managed identity and click **Save**.

10. Now, the managed identity has access to the Key Vault.

From your custom code, you can call the MSI endpoint to get an access token to authenticate the Azure resource as well. For .NET applications, you can use the `Microsoft.Azure.Services.AppAuthentication` library to accomplish this. You can do this by calling the RESTful API as well, but then you have to create the request manually.

> **Tip**
>
> For a sample application, please refer to `https://github.com/ Azure-Samples/app-service-msi-keyvault-dotnet`.

Summary

In this chapter, we covered how to implement the different types of authentication, including Active Directory, certificates, and OAuth. We also looked at token authentication and managed identities for enabling access between Azure components without the need to share secrets and keys.

In the next chapter, we'll begin the *Implementing and Managing Data Platforms* objective. We will start this by looking at Azure Cosmos DB, which is a fully managed NoSQL database.

Questions

Answer the following questions to test your knowledge of the information in this chapter. You can find the answers in the *Assessments* section at the end of this book:

1. Can authentication for App Service only be implemented from Azure AD?

    a) Yes

    b) No

2. If you want to secure your application using SSL certificates, you are required to buy that certificate from the Azure portal.

    a) True

    b) False

3. System-assigned managed identities can be used for multiple Azure resources.

    a) True

    b) False

Further reading

Check out the following links for more information about the topics that were covered in this chapter:

- Authentication and authorization in Azure App Service: `https://docs.microsoft.com/en-us/azure/app-service/overview-authentication-authorization`

- Evolution of the Microsoft identity platform: `https://docs.microsoft.com/en-us/azure/active-directory/develop/about-microsoft-identity-platform`

- Configurable token lifetimes in Azure Active Directory: `https://docs.microsoft.com/en-us/azure/active-directory/develop/active-directory-configurable-token-lifetimes`

- What is managed identities for Azure resources?: `https://docs.microsoft.com/en-us/azure/active-directory/managed-identities-azure-resources/overview`

Section 4: Implement and Manage Data Platforms

In this section, we will look at how to build data solutions using traditional SQL and newer NoSQL technologies. This is followed by mock exam questions and answers.

This section contains the following chapters:

15
Developing Solutions that Use Cosmos DB Storage

In the previous chapter, we covered the last part of the *Implementing Solutions for Apps* objective by learning how to implement authentication for web apps, APIs, and more.

In this chapter, we will be introducing the final objective: *Implementing and Managing Data Platforms*. We are going to cover what a NoSQL database is, learn about the features of Azure's NoSQL offering known as Cosmos DB, and then develop a solution that uses it.

We are also going to learn how to create, read, update, and delete data by using the appropriate APIs, how to set the appropriate consistency level for operations, understand the different partition schemes, and implement a multi-region replica of our database.

The following topics will be covered in this chapter:

- Understanding the differences between NoSQL and SQL
- Understanding Cosmos DB
- Creating, reading, updating, and deleting data by using the appropriate APIs
- Understanding partitioning schemes

- Setting the appropriate consistency level for operations
- Creating replicas

Let's get started!

Technical requirements

This chapter's examples will use Visual Studio 2019 (`https://visualstudio.microsoft.com/vs/`).

The source code for our sample application can be downloaded from `https://github.com/PacktPublishing/Microsoft-Azure-Architect-Technologies-Exam-Guide-AZ-303/tree/master/Ch15`.

Understanding the differences between NoSQL and SQL

Azure Cosmos DB is a **NoSQL** database, so to get the best value from Cosmos DB, we need to understand what NoSQL is and how it differs from traditional SQL databases.

SQL databases are built around the concept of tables, and within each table, you have rows of data split into cells. Each cell contains an individual piece of data within a record, and each row is an entire record.

A key difference from SQL databases is that the table and cell definition, known as the **schema**, must be defined *before* you can enter data into it. So, if you want to store a customer record containing a name and address, first, you need to create a table with each of the columns defined for each data type, such as the first name, address line, city, country, and so on.

Traditionally, SQL tables are built so that they're **relational**. This means that rather than having large records containing all the required information, you split the records into multiple tables and then join them as part of a query. For example, you could store core personal information such as a user's first and last name in one table, but their address details in another table, with a link in each record to join the related tables together.

In a similar fashion, data elements that need to be consistent across records, such as a **Category**, would be defined in a separate table, with a **CategoryId** link in the person record rather than the category itself.

The following screenshot shows an example of how this might look:

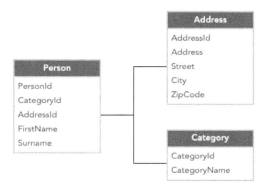

Example Combined Query

Person.PersonId	Person.FirstName	Person.Surname	Category.Name	Address.Street	Address.City
100	Brett	Hargreaves	Active	999 Letsbe Avenue	London

Figure 15.1 – Example relational table

Relational tables can also have **one-to-many relationships** – this means that rather than linking a single address record to a person, you can link multiple address records. They can be used to store separate addresses for billing and shipping, for example.

One of the benefits of a relational table is that any change that's made to a category name, for example, would automatically be reflected across all person records as they only store a link to the category ID, not the data itself.

NoSQL databases, as the name suggests, do not follow these patterns. NoSQL databases store data as **JSON** documents, which are *hierarchical* in structure. This means that related information, such as an address, can be stored alongside the main person record but is *embedded* as a child object, as opposed to being part of a flat structure. Embedding data can also be used for multiple related records; for example, multiple addresses can be stored within the main person record too.

With this in mind, a typical NoSQL JSON document might look like this:

```
{
    "id": "001",
    "firstName": "Brett",
    "surname": "Hargreaves",
    "category": "Active",
    "addresses": [
```

```
        {
            "type": "billing",
            "street": "999 Letsbe Avenue",
            "city": "London"
        },
        {
            "type": "shipping",
            "street": "20 Docklands Road",
            "city": "London"
        }
    ]
}
```

For certain scenarios, NoSQL databases can perform much faster than their SQL counterparts. However, one of the biggest benefits is that you can easily mix the records of completely different schemas.

For example, a container, which is synonymous to a table, might contain the following records – and this would be perfectly valid:

```
[
    {
        "id": "001",
        "firstName": "Brett",
        "Surname": "Hargreaves",
        "category": "Active",
    },
    {
        "id": "002",
        "firstName": "Sjoukje",
        "Surname": "Zaal",
        "profession": "CTO",
    },
    {
        "id": "003",
        "title": "Microsoft Azure Architect Technologies: Exam
Guide AZ303",
        "publisher": "Packt Publishing",
```

```
      }
]
```

Now that you have an understanding of NoSQL, we will investigate Azure's implementation of a NoSQL database – Cosmos DB.

Understanding Cosmos DB

Cosmos DB storage is the premium offering for Azure Table storage. It's a multi-model and globally distributed database service that is designed to horizontally scale and replicate your data to any number of Azure regions. By replicating and scaling the data, Cosmos DB can guarantee low latency, high availability, and high performance anywhere in the world. You can replicate or scale data easily inside the Azure portal by selecting from the available regions on the map.

This high availability and low latency makes Cosmos DB most suitable for mobile applications, games, and applications that need to be globally distributed. The Azure portal also uses Cosmos DB for data storage. Cosmos DB is completely schemaless, and you can use a number of existing APIs with available SDKs to communicate with it. So, if you are using a specific API for your data and you want to move your data to Cosmos DB, all you need to do is change the connection string inside your application and the data will be stored in Cosmos DB automatically.

Cosmos DB offers the following key benefits:

- **Turnkey global distribution**: With Cosmos DB, you can build highly responsive and highly available applications that scale globally. All data is transparently replicated to the regions where your users are, which means they can interact with a replica of the data that is closest to them. Azure regions can easily be added and removed. Data will then seamlessly replicate to the selected regions without any downtime of the application.

- **Always on**: Cosmos DB provides a SLA with 99.99% high availability for both reads and writes. To ensure that your application is designed to fail over in case of regional disasters, Cosmos DB allows you to invoke regional failover using the Azure portal, or programmatically.

- **Worldwide elastic scalability of throughput and storage**: Elastic scalability to both reads and writes is offered by Cosmos DB. By using a single API call, you can elastically scale up from thousands to hundreds of millions of requests per second around the globe, and you only pay for the throughput and storage you need.

- **No schema or index management**: With Cosmos DB, you don't have to deal with schema or index management, which is a painful process for globally distributed apps. Without schema and index management, there is also no downtime for applications while migrating schemas. Cosmos DB automatically indexes all data and serves queries fast.

- **Secured by default**: All the data in Cosmos DB is encrypted at rest and in motion. Cosmos DB also provides row-level authorization.

Cosmos DB supports the following APIs for storing and interacting with your data:

- **SQL API**: With the SQL API, you can use SQL queries as a JSON query language against the dataset inside Cosmos DB. Because Cosmos DB is schemaless, it provides autoindexing of the JSON documents. Data is stored on SSD drives for low latency and is lock-free, so you can create real-time queries for your data. Cosmos DB also supports writing stored procedures, triggers, and **user-defined functions (UDFs)** in JavaScript, and it supports **Atomicity, Consistency, Isolation, Durability (ACID)** transactions inside a collection.

- **MongoDB API**: MongoDB is an open source document database that provides high performance, high availability, and automatic scaling by default. Using it inside Cosmos DB provides automatic sharding, indexing, replication, and encryption of your data on top of this. MongoDB also provides an aggregation pipeline that can be used to filter and transform the data in multiple stages. It also supports creating a full-text index, and you can integrate it easily with Azure Search and other Azure services as well.

- **Gremlin (Graph) API**: The Gremlin API is part of the Apache TinkerPop project, which is an open source graph computing framework. A graph is a way of storing objects (nodes) based on relationships. Each object can have multiple relationships with other objects. You can interact with the data using JavaScript.

- **Table API**: The Azure Table API can be used for applications that have been written for using Azure Table storage but need the additional Cosmos DB features, such as global distribution, automatic indexing, low latency, and high throughput.

- **Cassandra API**: The Cassandra API can be used for applications that have been written for Apache Cassandra. Apache Cassandra is an open source distributed NoSQL database that offers scalability and high availability. Cosmos DB offers no operations management, SLA, and automatic indexing on top of this.

In the future, new APIs will be added to Cosmos DB as well.

In the next section, we are going to create, read, update, and delete data using the appropriate APIs.

Creating, reading, updating, and deleting data using the appropriate APIs

Before we can create, read, update, and delete data using the APIs, first, we need to create a Cosmos DB in Azure. We will do this in the following subsection.

Creating a Cosmos DB

You can create a Cosmos DB using the Azure portal, PowerShell, the CLI, or ARM templates. In this demonstration, we are going to create a Cosmos DB server, database, and container from the Azure portal. To do this, perform the following steps:

1. Navigate to the Azure portal by going to `https://portal.azure.com`.

2. Click **Create a resource**, type **Azure Cosmos DB** into the search bar, and create a new database server.

3. Add the following values:

    a) **Subscription**: Pick a subscription.

    b) **Resource group**: Create a new one and call it `PacktCosmosResourceGroup`.

    c) **Account name**: `packtsqlapi`.

    d) **API**: **Core (SQL)**.

    e) **Notebooks (preview)**: **Off**.

    f) **Location**: **East US**.

    g) **Capacity mode**: **Provisioned throughput**.

    h) **Apply Free Tier Discount**: **Apply** (this is only available for one Cosmos DB per subscription).

    i) **Account Type**: **Non-Production**.

    j) **Geo-Redundancy**: **Disable**.

    k) **Multi-region Writes**: **Disable**.

4. Click **Review + create** and then **Create**.

5. Once the deployment has finished, open the resource. Quickstart is automatically opened and is where you can let Azure automatically create a database and a container, and then download a sample app in the language of your choice:

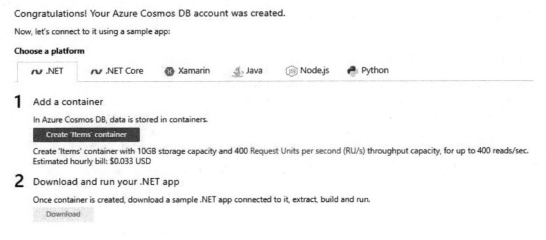

Figure 15.2 – Creating a sample application directly from the Azure portal

6. You can also create your database and container using the Data Explorer. **Click Data Explorer** in the left menu. From the top menu, click **New container**. In the **Add container** blade, add the following values:

    a) **Database id**: `PacktToDoList`

    b) **Throughput**: **400** (provisioning the throughput at the database level means the throughput will be shared across all containers)

    c) **Container id**: `Items`

    d) **Partition Key**: `/category`

7. Click **OK**.

In this demonstration, we created a database server, a database, and a container from the Azure portal. In the next section, we are going to create an application that creates a database and container programmatically. Then, we will create, read, update, and delete data using the Cosmos DB APIs.

Creating the sample application

In this second part of the demonstration, we are going to create the sample application. For this, we are going to create a new console application in Visual Studio 2019. First, we will connect to our Cosmos DB account.

Connecting to our Cosmos DB account

The first step is to connect to the Cosmos DB account that we created in the previous section, as follows:

1. Open Visual Studio and create a new Console App (.NET) project.

2. Name the project PacktCosmosApp.

3. Right-click your project in the **Solution Explorer** window and select **Manage NuGet packages**.

4. Select **Browse**, search for Microsoft.Azure.Cosmos, and install the package.

5. Open Program.cs and replace the references with the following:

```
using System;
using System.Threading.Tasks;
using System.Configuration;
using System.Collections.Generic;
using System.Net;
using Microsoft.Azure.Cosmos;
```

6. Add the following variables to the Program.cs method:

```
// Azure Cosmos DB endpoint.
        private static readonly string EndpointUri =
"<your endpoint here>";
        // Primary key for the Azure Cosmos account.
        private static readonly string PrimaryKey =
"<your primary key>";

        // The Cosmos client instance
        private CosmosClient cosmosClient;
        // The database we will create
        private Database database;
        // The container we will create.
        private Container container;
        // The name of the database and container we will
    create
        private string databaseId = "FamilyDatabase";
        private string containerId = "FamilyContainer";
```

7. Now, go back to the Azure portal. Navigate to the Cosmos DB that we created in the previous step and under **Settings**, select **Keys**. Copy both keys:

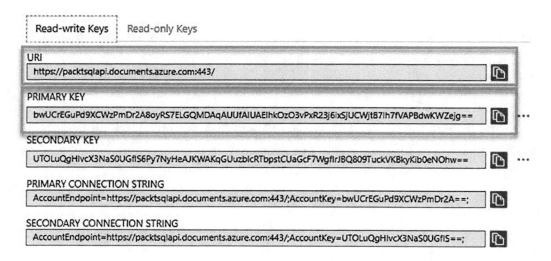

Figure 15.3 – Cosmos DB keys

8. In `Program.cs`, replace `<your endpoint here>` with the URI that you copied. Then, replace `<your primary key>` with the primary key you just copied.

Replace the `Main()` method with the following:

```
public static async Task Main(string[] args)
{

}
```

9. Below the `Main` method, add a new asynchronous task called `GetStartedDemoAsync`, which instantiates our new `CosmosClient`. We use `GetStartedDemoAsync` as the entry point that calls methods that operate on Azure Cosmos DB resources:

```
public async Task GetStartedDemoAsync()
    {
        // Create a new instance of the Cosmos Client
        this.cosmosClient = new
CosmosClient(EndpointUri, PrimaryKey);
    }
```

10. Add the following code to run the `GetStartedDemoAsync` synchronous task from your `Main` method:

```
public static async Task Main(string[] args)
    {
        try
        {
            Console.WriteLine("Beginning
operations...\n");
            Program p = new Program();
            await p.GetStartedDemoAsync();


        }
        catch (CosmosException de)
        {
            Exception baseException =
de.GetBaseException();
            Console.WriteLine("{0} error occurred:
{1}", de.StatusCode, de);
        }
        catch (Exception e)
        {
            Console.WriteLine("Error: {0}", e);
        }
        finally
        {
            Console.WriteLine("End of demo, press any
key to exit.");
            Console.ReadKey();
        }
    }
```

Now, if you run the application, the console will display a message stating **End of demo, press any key to exit**. This means that the application has successfully connected to the Cosmos DB account.

Now that we've successfully connected to the Cosmos DB account, we can create a new database.

Creating a new database

To create a new database, perform the following steps:

1. Now, we can create a database. Copy and paste the CreateDatabaseAsync method below your GetStartedDemoAsync method:

    ```
    private async Task CreateDatabaseAsync()
    {
            // Create a new database
            this.database = await this.cosmosClient.
    CreateDatabaseIfNotExistsAsync(databaseId);
            Console.WriteLine("Created Database: {0}\n",
    this.database.Id);
    }
    ```

2. In the GetStartedDemoAsync method, add the line of code for calling the CreateDatabaseAsync method. The method will now look as follows:

    ```
    public async Task GetStartedDemoAsync()
    {
            // Create a new instance of the Cosmos Client
            this.cosmosClient = new
    CosmosClient(EndpointUri, PrimaryKey);

            //ADD THIS PART TO YOUR CODE
            await this.CreateDatabaseAsync();
    }
    ```

3. Run the application. You will see that the database has been created.

In the next section, we will create the container.

Creating a container

To create the container, do the following:

1. Copy and paste the `CreateContainerAsync` method below your `CreateDatabaseAsync` method. `CreateContainerAsync` will create a new container with the `FamilyContainer` ID, if it doesn't already exist, by using the ID specified in the `containerId` field that's been partitioned by the `LastName` property:

```
private async Task CreateContainerAsync()
    {
        // Create a new container
        this.container = await this.database.
CreateContainerIfNotExistsAsync(containerId, "/
LastName");
        Console.WriteLine("Created Container: {0}\n",
this.container.Id);
    }
```

2. In the `GetStartedDemoAsync` method, add the line of code for calling the `CreateContainerAsync` method. The method will now look as follows:

```
public async Task GetStartedDemoAsync()
    {
        // Create a new instance of the Cosmos Client
        this.cosmosClient = new
CosmosClient(EndpointUri, PrimaryKey);
        await this.CreateDatabaseAsync();
        //ADD THIS PART TO YOUR CODE
        await this.CreateContainerAsync();
    }
```

3. You can now run the application again. You will see that the container has been created.

In the next section, we are going to add items to the container.

Adding items to the container

In this section, we are going to add items to the container using the API. First, we need to create a `Family` class that represents the objects that we are going to store in the container. You can create an item using the `CreateItemAsync` method. When you use the SQL API, items are created and stored as documents. These documents are user-defined arbitrary JSON content. You can then insert an item into your Azure Cosmos container.

Besides the `Family` class, we will also add some subclasses, such as `Parent`, `Child`, `Pet`, and `Address`, which are used in the `Family` class. To do this, perform the following steps:

1. Add a new class to the project and call it `Family.cs`:

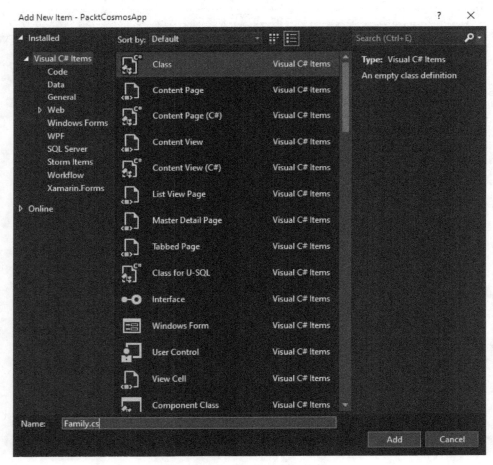

Figure 15.4 – Creating a new class

2. Click **Add**.

3. Replace the references with the following (you should be prompted to install the NuGet package for this; if not, open the NuGet package manager and add it manually):

```
using Newtonsoft.Json;
```

4. Then, add the following code to create the `Family` class:

```
public class Family
{
    [JsonProperty(PropertyName = "id")]
    public string Id { get; set; }
    public string LastName { get; set; }
    public Parent[] Parents { get; set; }
    public Child[] Children { get; set; }
    public Address Address { get; set; }
    public bool IsRegistered { get; set; }
    public override string ToString()
    {
        return JsonConvert.SerializeObject(this);
    }
}
```

5. Now, we can create the `Parent` class below the `Family` class:

```
public class Parent
{
    public string FamilyName { get; set; }
    public string FirstName { get; set; }
}
```

6. Next, create the `Child` class below the `Parent` class:

```
public class Child
{
    public string FamilyName { get; set; }
    public string FirstName { get; set; }
    public string Gender { get; set; }
```

```
        public int Grade { get; set; }
        public Pet[] Pets { get; set; }
    }
```

7. Next, create the Pet class below the Child class:

```
public class Pet
    {
        public string GivenName { get; set; }
    }
```

8. Lastly, create the Address class below the Pet class:

```
public class Address
    {
        public string State { get; set; }
        public string County { get; set; }
        public string City { get; set; }
    }
```

9. In Program.cs, add the AddItemsToContainerAsync method after your CreateContainerAsync method and create the first Family item:

```
private async Task AddItemsToContainerAsync()
        {
            // Create a family object for the Zaal family
            Family ZaalFamily = new Family
            {
                Id = "Zaal.1",
                LastName = "Zaal",
                Parents = new Parent[]
                {
            new Parent { FirstName = "Thomas" },
            new Parent { FirstName = "Sjoukje" }
                },
                Children = new Child[]
                {
            new Child
                {
```

```
                        FirstName = "Molly",
                        Gender = "female",
                        Grade = 5,
                        Pets = new Pet[]
                        {
                            new Pet { GivenName = "Fluffy" }
                        }
                    }
                },
                Address = new Address { State = "WA",
    County = "King", City = "Seattle" },
                IsRegistered = false
            };
```

10. Then, below `item`, add the code for adding `item` to `container`:

```
    try
            {
                // Read the item to see if it exists.
                ItemResponse<Family> zaalFamilyResponse =
    await this.container.ReadItemAsync<Family>(ZaalFamily.Id,
    new PartitionKey(ZaalFamily.LastName));
                Console.WriteLine("Item in database with
    id: {0} already exists\n", zaalFamilyResponse.Resource.
    Id);
            }
            catch (CosmosException ex) when (ex.
    StatusCode == HttpStatusCode.NotFound)
            {
                // Create an item in the container
    representing the Zaal family. Note we provide the value
    of the partition key for this item, which is "Zaal"
                ItemResponse<Family> zaalFamilyResponse =
    await this.container.CreateItemAsync<Family>(ZaalFamily,
    new PartitionKey(ZaalFamily.LastName));

                // Note that after creating the item,
    we can access the body of the item with the Resource
```

property off the ItemResponse. We can also access the RequestCharge property to see the amount of RUs consumed on this request.

```
            Console.WriteLine("Created item in
database with id: {0} Operation consumed {1} RUs.\n",
zaalFamilyResponse.Resource.Id, zaalFamilyResponse.
RequestCharge);
        }
```

11. For the second `Family` item, underneath the previous code (in the same method), add the following:

```
// Create a family object for the PacktPub family
        Family PacktPubFamily = new Family
        {
            Id = "PacktPub.1",
            LastName = "PacktPub",
            Parents = new Parent[]
            {
        new Parent { FamilyName = "PacktPub",
FirstName = "Robin" },
        new Parent { FamilyName = "Zaal", FirstName =
"Sjoukje" }
            },
```

12. Add `children` to `Family`:

```
Children = new Child[]
                {
            new Child
                {
                    FamilyName = "Merriam",
                    FirstName = "Jesse",
                    Gender = "female",
                    Grade = 8,
                    Pets = new Pet[]
                    {
                        new Pet { GivenName = "Goofy" },
                        new Pet { GivenName = "Shadow" }
```

```
                        }
                    },
                    new Child
                    {
                        FamilyName = "Miller",
                        FirstName = "Lisa",
                        Gender = "female",
                        Grade = 1
                    }
                },
                Address = new Address { State = "NY",
    County = "Manhattan", City = "NY" },
                IsRegistered = true
            };
```

13. Then, below item, add the code for adding item to container again:

```
    try
            {
                // Read the item to see if it exists
                ItemResponse<Family>
    packtPubFamilyResponse = await this.container.
    ReadItemAsync<Family>(PacktPubFamily.Id, new
    PartitionKey(PacktPubFamily.LastName));
                Console.WriteLine("Item in database
    with id: {0} already exists\n", packtPubFamilyResponse.
    Resource.Id);
            }
            catch (CosmosException ex) when (ex.
    StatusCode == HttpStatusCode.NotFound)
            {
                // Create an item in the container
    representing the Wakefield family. Note we provide the
    value of the partition key for this item, which is
    "PacktPub"
                ItemResponse<Family>
    packtPubFamilyResponse = await this.container.
    CreateItemAsync<Family>(PacktPubFamily, new
    PartitionKey(PacktPubFamily.LastName));
```

```
                // Note that after creating the item,
we can access the body of the item with the Resource
property off the ItemResponse. We can also access the
RequestCharge property to see the amount of RUs consumed
on this request.
                Console.WriteLine("Created item
in database with id: {0} Operation consumed {1}
RUs.\n", packtPubFamilyResponse.Resource.Id,
packtPubFamilyResponse.RequestCharge);
        }
```

14. Finally, we need to call `AddItemsToContainerAsync` in the `GetStartedDemoAsync` method again:

```
public async Task GetStartedDemoAsync()
    {
        // Create a new instance of the Cosmos Client
        this.cosmosClient = new
CosmosClient(EndpointUri, PrimaryKey);
        await this.CreateDatabaseAsync();
        await this.CreateContainerAsync();
        //ADD THIS PART TO YOUR CODE
        await this.AddItemsToContainerAsync();
    }
```

15. Now, run the application again. The items will be added to the container.

In this section, we added some items to the container. In the next section, we will query the resources.

Querying Azure Cosmos DB resources

In this part of the demonstration, we are going to query the resources. Azure Cosmos DB supports rich queries against JSON documents stored in each container. The following code will show you how to run a query against the items that we stored in the container in the previous section.

To do this, perform the following steps:

1. Copy and paste the `QueryItemsAsync` method after your `AddItemsToContainerAsync` method:

```
private async Task QueryItemsAsync()
    {
            var sqlQueryText = "SELECT * FROM c WHERE
c.LastName = 'Zaal'";

            Console.WriteLine("Running query: {0}\n",
sqlQueryText);

            QueryDefinition queryDefinition = new
QueryDefinition(sqlQueryText);
            FeedIterator<Family>
queryResultSetIterator = this.container.
GetItemQueryIterator<Family>(queryDefinition);

            List<Family> families = new List<Family>();

            while (queryResultSetIterator.HasMoreResults)
            {
                FeedResponse<Family> currentResultSet =
await queryResultSetIterator.ReadNextAsync();
                foreach (Family family in
currentResultSet)
                {
                    families.Add(family);
                    Console.WriteLine("\tRead {0}\n",
family);
                }
            }
    }
```

2. Add a call to `QueryItemsAsync` in the `GetStartedDemoAsync` method:

```
public async Task GetStartedDemoAsync()
    {
            // Create a new instance of the Cosmos Client
```

```
            this.cosmosClient = new
CosmosClient(EndpointUri, PrimaryKey);
            await this.CreateDatabaseAsync();
            await this.CreateContainerAsync();
            await this.AddItemsToContainerAsync();

            //ADD THIS PART TO YOUR CODE
            await this.QueryItemsAsync();
        }
```

3. Run the application; you will see the query results displayed in the console.

With that, we have created a query to retrieve the data from the container. In the next section, we are going to update a `Family` item.

Updating a JSON item

In this part of the demonstration, we are going to update a `Family` item. To do this, add the following code:

1. Copy and paste the `ReplaceFamilyItemAsync` method after your `QueryItemsAsync` method:

```
private async Task ReplaceFamilyItemAsync()
        {
            ItemResponse<Family> PacktPubFamilyResponse =
await this.container.ReadItemAsync<Family>("PacktPub.1",
new PartitionKey("PacktPub"));
            var itemBody = PacktPubFamilyResponse.
Resource;

            // update registration status from false to
true
            itemBody.IsRegistered = true;
            // update grade of child
            itemBody.Children[0].Grade = 6;

            // replace the item with the updated content
            PacktPubFamilyResponse = await this.
container.ReplaceItemAsync<Family>(itemBody, itemBody.Id,
```

```
new PartitionKey(itemBody.LastName));
            Console.WriteLine("Updated Family
[{0},{1}].\n \tBody is now: {2}\n", itemBody.LastName,
itemBody.Id, PacktPubFamilyResponse.Resource);
        }
```

2. Then, add a call to ReplaceFamilyItemAsync in the
    GetStartedDemoAsync method:

```
public async Task GetStartedDemoAsync()
        {
            // Create a new instance of the Cosmos Client
            this.cosmosClient = new
CosmosClient(EndpointUri, PrimaryKey);
            await this.CreateDatabaseAsync();
            await this.CreateContainerAsync();
            await this.AddItemsToContainerAsync();
            await this.QueryItemsAsync();

            //ADD THIS PART TO YOUR CODE
            await this.ReplaceFamilyItemAsync();
        }
```

3. Run the application; you will see that the Family item is being updated.

In this part of the demonstration, we updated an item in the container. In the next section,
we will delete an item from the container.

Deleting an item

To delete a Family item, perform the following steps:

1. Add the DeleteFamilyItemAsync method after the
    ReplaceFamilyItemAsync method:

```
private async Task DeleteFamilyItemAsync()
        {
            var partitionKeyValue = "Zaal";
            var familyId = "Zaal.1";

            // Delete an item. Note we must provide the
```

```
            partition key value and id of the item to delete
                    _ = await this.container.
        DeleteItemAsync<Family>(familyId, new
        PartitionKey(partitionKeyValue));
                Console.WriteLine("Deleted Family
        [{0},{1}]\n", partitionKeyValue, familyId);
                }
```

2. Then, add a call to DeleteFamilyItemAsync in the GetStartedDemoAsync method:

```
        public async Task GetStartedDemoAsync()
            {
                // Create a new instance of the Cosmos Client
                this.cosmosClient = new
        CosmosClient(EndpointUri, PrimaryKey);
                await this.CreateDatabaseAsync();
                await this.CreateContainerAsync();
                await this.AddItemsToContainerAsync();
                await this.QueryItemsAsync();
                await this.ReplaceFamilyItemAsync();

                //ADD THIS PART TO YOUR CODE
                await this.DeleteFamilyItemAsync();
            }
```

3. Run the application; you will see that the item has been deleted.

In this demonstration, we created a Cosmos DB server, database, and container. We also added data to the container, ran a query on the data, updated the data, and then deleted the data.

In the next section, we are going to cover partitioning schemes.

Understanding partitioning schemes

To meet the performance needs of your application, Azure Cosmos DB uses partitioning to scale individual containers in a database. Cosmos DB partitions in a way that the items are divided into distinct subsets called **logical partitions**. These are formed based on the value of the partition key that is added to each item in the container. All of the items that are in a logical partition have the same partition key. Each item in a container has an **item ID** (which is unique within the logical partition). To create the item's index, the partition key and the item ID are combined. This uniquely identifies the item.

> Tip
> If you look at our sample application from the previous section, you will see that the partition key and item ID have been combined.

Besides logical partitions, Azure Cosmos DB also has physical partitions:

- **Logical partitions**: A set of items that have the same partition key is called a logical partition. For instance, if you have a container that stores items that all contain a `Product` property, you can use this as the partition key. Logical partitions are then formed by groups that have the same values, such as `Books`, `Videos`, `Movies`, and `Music`. Containers are the fundamental units of scalability in Azure Cosmos DB. Data that is added to the container, together with the throughput (see the next section) that is provisioned on the container, is automatically (horizontally) partitioned across a set of logical partitions based on the partition key.

- **Physical partitions**: Internally, one or more logical partitions are mapped to a physical partition. This partition consists of a set of replicas, also called a replica set. Each replica set is hosting an instance of the Azure Cosmos DB engine. It makes the data that is stored inside the physical partition highly available, consistent, and durable. The maximum amount of **request units** (**RUs**) and storage is supported by the physical partition.

The following chart shows how logical partitions are mapped to physical partitions that are distributed globally over multiple regions:

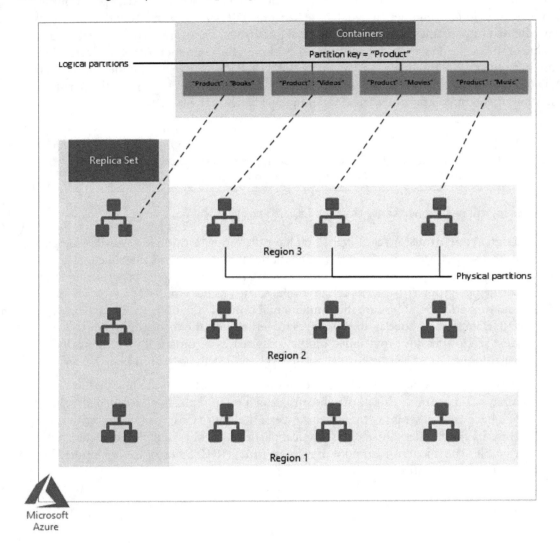

Figure 15.5 – Partitioning in Azure Cosmos DB

In the next section, we are going to cover how to set the appropriate consistency level for operations.

Setting the appropriate consistency level for operations

When you use distributed databases that rely on high availability and low latency, you can choose between two different extreme consistency models: **strong** consistency and **eventual** consistency. The former is the standard for data programmability, but this will result in reduced availability during failures and higher latency. The latter offers higher availability and better performance, but makes it much harder to program applications.

Azure Cosmos DB offers more choices between the two extreme consistency models. Strong consistency and eventual consistency are at two different ends of the spectrum. This can help developers make a more precise choice with respect to high availability and performance. Azure Cosmos DB offers five consistency models, known as **strong**, **bounded staleness**, **session**, **consistent prefix**, and **eventual consistency**:

- **Strong**: Strong consistency is the easiest model to understand. The most recent committed version of an item is returned when the data is read. A partial or uncommitted write is never returned to the client. This will give a lot of overhead because every instance needs to be updated with the latest version of the data, which has a huge price in terms of the latency and throughput of the data.

- **Bounded staleness**: This is a compromise that trades delays for strong consistency. This model doesn't guarantee that all the observers have the same data at the same time. Instead, it allows a lag of 5 seconds (100 operations).

- **Session**: This is the default level for newly created databases and collections. It is half as expensive as strong consistency and it offers good performance and availability. It ensures that everything that is written in the current session is also available for reading. Anything else will be accurate, but delayed.

- **Consistent prefix**: With this level, updates that are returned contain accurate data for a certain point in time, but won't necessarily be current.

- **Eventual**: There's no ordering guarantee for how long it will take to synchronise (like the consistent prefix) and the updates aren't guaranteed to come in order. When there are no further writes, the replicas will eventually be updated.

> Tip
>
> For more information on how to choose the right consistency level for the different APIs that Azure Cosmos DB has to offer, please refer to the article at `https://docs.microsoft.com/en-us/azure/cosmos-db/consistency-levels-choosing`.

The key lesson from this section is that you can control how fast data is replicated around all regions to best support your application requirements. In the next and final section, we will learn how to implement and place our database replicas.

Creating replicas

Azure Cosmos DB is a globally distributed database, and it works by placing replicas of your data in the regions where you need them. In the previous section, we learned how to control the consistency of those replicas. In this section, we will create a replica of our database and set that consistency level. Follow these steps:

1. Navigate to the Azure portal by going to `https://portal.azure.com`.

2. In the search bar, search for and then select **Cosmos DB**.

3. Select the Cosmos DB instance you created.

4. On the left-hand menu, under **Settings**, click **Replicate Globally**.

5. On the right, under **Configure regions**, set **Multi-region writes** to **Enable**.

6. On the right, under **Configure regions**, click **+ Add Region**, as shown in the following screenshot:

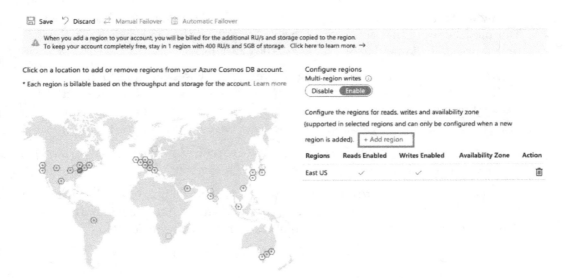

Figure 15.6 – Adding Cosmos DB replication regions

7. A drop-down list will appear under the regions. Here, select **West US**.

8. Click **OK**.

9. **West UK** will now appear under the regions list. Click **Save** at the top of the screen.

The replication process may take some time to complete. When it does, you will be able to see your new replica on the map in a darker colored pin. By default, the replica will use **Session Consistency**. To change this, perform the following steps:

1. Still in the **Cosmos DB** blade, from the left-hand menu, click **Default Consistency**.

2. Along the top, you will see the five different consistency options. Select a new option; for example, **Strong**.

3. For each option, you will see an animation showing the differences between them. With **Strong**, all regions are consistent at the same time.

4. Click **Eventual**. The animation will change to show that some regions take longer to become consistent. Once you have selected your preferred option, click **Save**.

    The following screenshot shows an example of this:

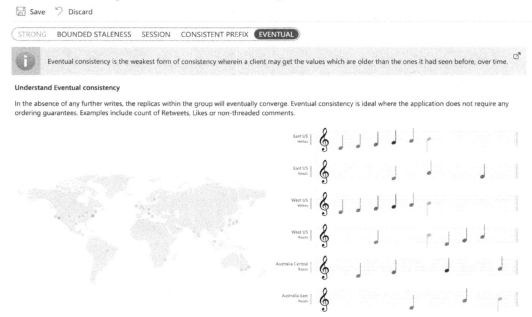

Figure 15.7 – Replica consistency

As we have seen, Cosmos DB is a powerful, fully managed, and fully resilient NoSQL option that provides many options that can be uniquely tailored to your requirements.

Summary

In this chapter, we've learned about Azure's NoSQL offering, known as Cosmos DB, what the differences are between SQL and NoSQL, and how to build a simple solution.

We then covered how to use the different APIs from Visual Studio to create databases, containers, data, and more. We also covered the partition schemes and the different consistency levels that are provided for Azure Cosmos DB.

Finally, we walked through enabling a multi-region replication and configuring its consistency.

In the next chapter, we will continue with this objective by learning how to develop solutions that use a relational database.

Questions

Answer the following questions to test your knowledge of the information in this chapter. You can find the answers in the *Assessments* section at the end of this book:

1. You can only create a Cosmos DB using the Azure portal, PowerShell, or the CLI.

    a) True

    b) False

2. Azure Cosmos DB offers two different consistency models.

    a) True

    b) False

3. A horizontal partition is used to distribute the data over different Azure Cosmos DB instances.

    a) True

    b) False

Further reading

Check out the following links for more information about the topics that were covered in this chapter:

- Azure Cosmos DB documentation: `https://docs.microsoft.com/en-us/azure/cosmos-db/`

- Getting started with SQL queries: `https://docs.microsoft.com/en-us/azure/cosmos-db/sql-query-getting-started`

- Request Units in Azure Cosmos DB: `https://docs.microsoft.com/en-us/azure/cosmos-db/request-units`

- *Getting Behind the 9-Ball: Cosmos DB Consistency Levels Explained*: `https://blog.jeremylikness.com/blog/2018-03-23_getting-behind-the-9ball-cosmosdb-consistency-levels/`

16

Developing Solutions that Use a Relational Database

In the previous chapter, we covered the first part of the *Implement and Manage Data Platforms* objective, by covering how to develop solutions that use Cosmos DB storage. We've covered how to create, read, update, and delete data by using appropriate APIs, and more.

In this chapter, we are continuing with this objective. We are going to cover how to develop solutions that use a relational database. We are going to cover how to provision and configure relational databases, how to configure elastic pools for Azure SQL Database, and how to create, read, update, and delete data tables using code.

We will then set up SQL in a highly available failover group with geo-replication, and then finish off with a look at how to publish on-premises databases to Azure.

The following topics will be covered in this chapter:

- Understanding Azure SQL Database
- Provisioning and configuring an Azure SQL database
- Creating, reading, updating, and deleting data tables using code

- Configuring elastic pools for Azure SQL Database
- Configuring high availability
- Implementing Azure SQL Database managed instances
- Publishing a SQL database

Technical requirements

This chapter will use Visual Studio 2019 (`https://visualstudio.microsoft.com/vs/`) for its examples.

We will also use SQL Server Management Studio and SQL Express, which can be downloaded from here: `https://www.microsoft.com/en-gb/sql-server/sql-server-downloads`.

The source code for our sample application can be downloaded from `https://github.com/PacktPublishing/Microsoft-Azure-Architect-Technologies-Exam-Guide-AZ-303/tree/master/Ch16`.

Understanding Azure SQL Database

Azure SQL Database offers a relational database in the cloud. It uses the last stable SQL Server on-premises database engine. Azure SQL Database can be used for a variety of applications, because it enables you to process both relational data and non-relational structures, such as graphs, JSON, spatial, and XML.

Azure SQL Database offers scalability without causing any downtime for your databases. It offers column-based indexes, which make your queries perform much more quickly.

There is built-in monitoring for your databases and built-in intelligence for increasing the performance of your database automatically, and it provides high availability by providing automatic backups and point-in-time restores. You can also use active geo-replication for global applications.

Azure SQL Database offers the following options for your databases:

- **Elastic database pool**: Elastic pools are a feature that helps in managing and scaling databases that have unpredictable usage demands. All databases in an elastic pool are deployed on the same database server and share the same resources. By managing a pool of databases rather than individual databases, they can share performance and scaling.

- **Single database**: This is a good fit if you have a database with predictable performance. Scaling is done for each database separately.

- **Managed instance**: This is a fully managed instance of the last stable SQL Server on-premises database engine. This option is most suitable for migrating on-premises SQL databases as is into the cloud. It offers 100% compatibility with the latest SQL Server on-premises (Enterprise edition) database engine.

Database elastic pools and single databases have up to three purchasing models when deploying them:

- **Database Transaction Units (DTUs)**: DTUs are a blend of compute, memory, and **input/output (I/O)**. DTU-based purchasing has a wide range of options, from a low-cost development tier to a premium scale for high throughput.

- **vCore**: Enables you to define the number of virtual CPU cores for your database or elastic pool. The vCore purchasing model also allows you the option to use Hybrid Benefit, whereby you can receive discounts for existing SQL licensing that you might already own.

- **Serverless**: With serverless, you can set a minimum and maximum number of vCores and the service will automatically scale up and down in response to demand. There is also the option to automatically pause the service during inactive periods. This can be a cost-effective option when your databases are accessed sporadically.

SQL Server Stretch Database

SQL Server Stretch Database was introduced in SQL Server 2016 and is a feature that can move or archive your cold data from your on-premises SQL Server to Azure SQL Database. This results in better performance for your on-premises server, and the stretched data resides in the cloud, where it is easily accessible for other applications. Within SQL Server, you can mark a table as a Stretch candidate, and SQL Server will move the data to Azure SQL Database transparently. Large transactional tables with lots of historical data can benefit from enabling for Stretch. These are mostly massive tables with hundreds, or millions, of rows in them, which don't have to be queried frequently.

In the next section, we are going to cover how to provision and configure an Azure SQL database.

Provisioning and configuring an Azure SQL database

In this demonstration, we are going to create and configure an Azure SQL database. You can create the database from the Azure portal, PowerShell, the CLI, and ARM templates, as well as by using the .NET libraries. In this demonstration, we are going to create a single database in the Azure portal. Therefore, we need to take the following steps:

1. Navigate to the Azure portal by opening `https://portal.azure.com`.

2. Click **Create a resource**, type `SQL Database` in the search bar, and create a new database server.

3. Add the following values:

   a) **Subscription**: Pick a subscription.

   b) **Resource group**: Create a new one and call it `PacktSQLResourceGroup`.

   c) **Database name**: `PacktSql`.

   d) **Server**: Create a new one and add the following values:

   — **Server name**: `packtsqlserver`

   — **Server admin login**: `PacktAdmin`

   — **Password**: `P@ss@word123`

   — **Location**: East US

   e) **Want to use SQL elastic pool**: No, we are going to use a single database.

   f) **Compute + storage**: Keep the default setting here.

4. Click **Review + create**, and then **Create**.

   The database is now being created.

Now that we have created the database, we can configure a server-level firewall rule.

Creating a server-level firewall rule

When you want to connect to a database from on-premises or remote tools, you need to configure a server-level firewall rule for the database. The firewall is created at the database server level for single and pooled databases. This prevents a client application from connecting to the server or any of its single or pooled databases.

To connect to the server from an IP address outside Azure, you need to create a firewall rule for a specific IP address or range of addresses.

To create a firewall rule, we have to take the following steps:

1. When the deployment is finished, open the resource. The overview blade is automatically displayed. In the top menu, select **Set server firewall**, as shown in the following screenshot:

Figure 16.1 – Overview blade of the database

2. On the **Firewall settings** page, you can add your current client IP address by clicking the **+ Add client IP** button in the top menu. This IP address will then automatically be added to the rules, as can be seen in the following screenshot:

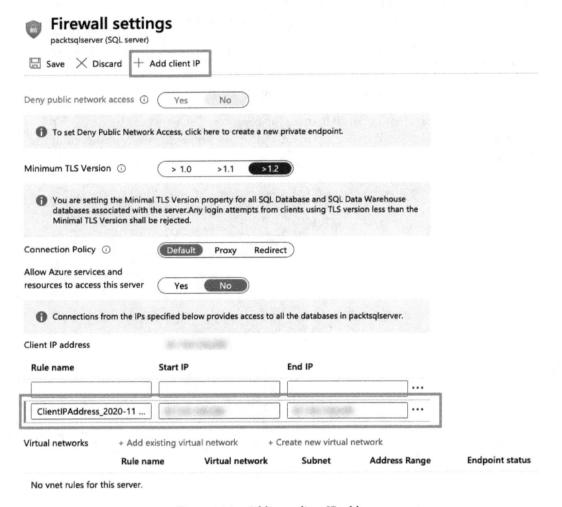

Figure 16.2 – Adding a client IP address

3. Click **Save**.

We can now connect from outside of Azure to the database.

In the next section, we are going to create a new table in the database, using Query Editor.

Creating a table in the database

In this part of the demonstration, we are going to create a new table in the database to store employees. We are going to use Query Editor for this, which is available in the Azure portal.

Therefore, we need to take the following steps:

1. In the settings of Azure SQL Database in the Azure portal (created previously), in the left-hand menu, click **Query editor (preview)**. Log in to the SQL server using the username and password (PacktAdmin and P@ss@word123).

2. On the **Query** screen, add the following line of SQL code to create a new table with the appropriate columns in it:

```
CREATE TABLE Employee (
    Employeeid int IDENTITY(1,1) PRIMARY KEY,
    FirstName varchar(255) NOT NULL,
    LastName varchar(255) NOT NULL,
    Title varchar(255) NOT NULL,
    BirthDate date,
    HireDate date
);
```

3. In Query Editor, this will look as in the following screenshot:

Figure 16.3 – Query Editor in the Azure portal

4. Run the query and the table will be created.

We have created and configured our database and have created a table with columns. In the next section, we are going to create a console application that creates, reads, updates, and deletes data tables programmatically.

Creating, reading, updating, and deleting data tables using code

In this demonstration, we are going to create the sample application. For this, we are going to create a new console application in Visual Studio 2019. We will first connect to our Azure SQL Database account.

Connecting to the Azure SQL database

Follow these steps to connect to establish the connection:

1. Open Visual Studio and create a new console app (.NET) project.

2. Name the project `PacktSQLApp`.

3. Right-click on your project in Solution Explorer and select **Manage NuGet Packages**.

4. Select **Browse**, search for `System.Data.SqlClient`, and install the package.

5. Open `Program.cs` and replace the references with the following:

```
using System;
using System.Data.SqlClient;
```

6. Add the following variable to the `Program.cs` method:

```
static string connectionstring;
```

7. Now, go back to the Azure portal. Navigate to the Azure SQL database that we created in the previous section. Under **Settings**, select **Connection strings**. Copy the connection string, as follows:

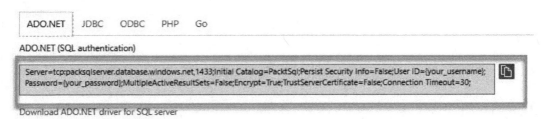

Figure 16.4 – SQL database connection string

8. Below the `Main` method, add a new asynchronous task called `GetStartedDemoAsync`, which instantiates our new `SqlConnection`. We use `GetStartedDemoAsync` as the entry point that calls methods operating on Azure Cosmos DB resources. In the method, replace `<replace-with-your-connectionstring>` with the connection string that you copied. Also, in the connection string, replace `{your-username}` with the username you specified when you created the database, and replace `{your-password}` with the password you provided, as follows:

```
static void GetStartedDemo()
{
    connectionstring = "<replace-with-your-
connectionstring>";
}
```

9. Add the following code to run the `GetStartedDemo` synchronous task from your `Main` method:

```
static void Main(string[] args)
{
    try
    {
        Console.WriteLine("Beginning operations...\n");
        GetStartedDemo();
    }
    catch (SqlException de)
    {
        Exception baseException = de.GetBaseException();
        Console.WriteLine("{0} error occurred: {1}",
de.Message, de);
    }
    catch (Exception e)
    {
        Console.WriteLine("Error: {0}", e);
    }
    finally
    {
        Console.WriteLine("End of demo, press any key to
exit.");
```

```
            Console.ReadKey();
    }
}
```

Now that we have successfully added the connection string to the Azure SQL database, we can create some data in the new database.

Adding items to the database

In this part of the demonstration, we are going to add an employee to the database. First, we need to create an `Employee` class that represents the objects we are going to store in the container. You can create an item using the `CreateItem` method, as follows:

1. Add a new class to the project and call it `Employee.cs`.

2. Replace the references with the following (you should be prompted to install the NuGet package for this; if not, open the NuGet package manager and add it manually):

```
using Newtonsoft.Json;
using System;
```

3. Then, add the following code to create the `Employee` class:

```
public class Employee
{
    public int EmployeeID { get; set; }
    public string FirstName { get; set; }
    public string LastName { get; set; }
    public string Title { get; set; }
    public DateTime? BirthDate { get; set; }
    public DateTime? HireDate { get; set; }
}
```

4. In `Program.cs`, add the following code to create a new employee object before the `Main` method:

```
static Employee NewEmployee = new Employee
{
    FirstName = "Sjoukje", LastName = "Zaal", Title =
"Mrs",
    BirthDate = new DateTime(1979, 7, 7),
```

```
        HireDate = new DateTime(2020, 1, 1)
    };
```

5. In `Program.cs`, add the `AddItemsToDatabase` method after your `GetStartedDemo` method. Then, add the following code to add the new employee to the database. First, create the connection, and then create the query, as follows:

```
static void AddItemsToDatabase()
{
    var connection = new SqlConnection(connectionstring);
    using (connection)
    {
    try
    {
        Console.WriteLine("\nCreate a new employee:");
        Console.
WriteLine("=======================================\n");

        var cmd = new SqlCommand("Insert Employee
(FirstName, LastName, Title, BirthDate, HireDate) values
(@FirstName, @LastName, @Title, @BirthDate, @HireDate)",
connection);
        cmd.Parameters.AddWithValue("@FirstName",
NewEmployee.FirstName);
        cmd.Parameters.AddWithValue("@LastName",
NewEmployee.LastName);
        cmd.Parameters.AddWithValue("@Title",
NewEmployee.Title);
        cmd.Parameters.AddWithValue("@BirthDate",
NewEmployee.BirthDate);
        cmd.Parameters.AddWithValue("@HireDate",
NewEmployee.HireDate);
```

6. Then, open the connection, execute the query, and close the connection, as follows:

```
        connection.Open();
        cmd.ExecuteNonQuery();
        connection.Close();
```

```
            Console.WriteLine("\nFinsihed Creating a new
        employee:");
            Console.
        WriteLine("=======================================\n");
        }
        catch (SqlException e)
        {
            Console.WriteLine(e.ToString());
        }
    }
}
```

7. Finally, we need to call `AddItemsToDatabase` in the `GetStartedDemo` method again, as follows:

```
static void GetStartedDemo()
{
    connectionstring = "<replace-with-your-
    connectionstring>";


    //ADD THIS PART TO YOUR CODE
    AddItemsToDatabase();
}
```

In this part of the demonstration, we have added some items to the database. In the next part, we will query the database.

Querying Azure SQL Database items

In this part of the demonstration, we are going to query the database. The following code will show you how to run a query against the items that we stored in the database in the previous step.

Therefore, we need to take the following steps:

1. Copy and paste the `QueryItems` method after your `AddItemsToDatabase` method. Create the connection again, and then create the query, as follows:

```
static void QueryItems()
{
    var connection = new SqlConnection(connectionstring);
    using (connection)
```

```
        {
            try
            {
                Console.WriteLine("\nQuerying database:");
                Console.
    WriteLine("=========================================\n");


                var cmd = new SqlCommand("SELECT * FROM
    Employee WHERE LastName = @LastName", connection);
                cmd.Parameters.AddWithValue("@LastName",
    NewEmployee.LastName);
```

2. Then, open the connection, execute the query, and close the connection. Write some values to the console app, as follows:

```
                connection.Open();
                cmd.ExecuteNonQuery();


                using (SqlDataReader reader = cmd.
    ExecuteReader())
                {
                    while (reader.Read())
                    {
                        for (int i = 0; i < reader.
    FieldCount; i++)
                        {
                            Console.WriteLine(reader.
    GetValue(i));
                        }
                    }
                    connection.Close();
                }
                Console.WriteLine("\nFinsihed querying
    database:");
                Console.
    WriteLine("=========================================\n");
            }
            catch (SqlException e)
```

```
        {
            Console.WriteLine(e.ToString());
        }
    }
}
```

3. Then, we need to call `QueryItems` in the `GetStartedDemo` method again, as follows:

```
static void GetStartedDemo()
{
    connectionstring = "<replace-with-your-
connectionstring>";

    AddItemsToDatabase();
    //ADD THIS PART TO YOUR CODE
    QueryItems();
}
```

4. Run the application, and you will see the query results displayed in the console.

We have now created a query to retrieve the data from the database. In the next section, we are going to update the `Employee` item.

Updating an Azure SQL Database row

In this part of the demonstration, we are going to update the `Employee` item. Therefore, add the following code:

1. Copy and paste the `UpdateEmployeeItem` method after your `QueryItems` method. Create the connection again, as well as the query, as follows:

```
static void UpdateEmployeeItem()
{
    var connection = new SqlConnection(connectionstring);
    using (connection)
    {
        try
        {
            Console.WriteLine("\nUpdating employee:");
```

```
        Console.
WriteLine("=========================================\n");
```

```
        var cmd = new SqlCommand(" UPDATE Employee SET
FirstName = 'Molly' WHERE Employeeid = 1", connection);
```

2. Execute it as follows:

```
        connection.Open();
        cmd.ExecuteNonQuery();
        connection.Close();

        Console.WriteLine("\nFinished updating
employee");
        Console.
WriteLine("=========================================\n");
        }
        catch (SqlException e)
        {
        Console.WriteLine(e.ToString());
        }
    }
}
```

3. Then, we need to call `UpdateEmployeeItem` in the `GetStartedDemo` method again, as follows:

```
static void GetStartedDemo()
{
    connectionstring = "<replace-with-your-
connectionstring>";

    AddItemsToDatabase();
    QueryItems();

    //ADD THIS PART TO YOUR CODE
    UpdateEmployeeItem();
}
```

In this part of the demonstration, we updated a row in the database. In the next part, we will delete the employee from the database.

Deleting an item

To delete a `Family` item, take the following steps:

1. Add the `DeleteEmployeeItem` method after your `UpdateEmployeeItem` method. Create the connection and a query, as follows:

```
static void DeleteEmployeeItem()
{
    var connection = new SqlConnection(connectionstring);
    using (connection)
    {
        try
        {
            Console.WriteLine("\nDeleting employee:");
            Console.
WriteLine("==========================================\n");

            var cmd = new SqlCommand(" Delete FROM dbo.
Employee where Employeeid = 1", connection);
```

2. Then, open the connection and execute the query. Write some details to the console app, as follows:

```
connection.Open();
            cmd.ExecuteNonQuery();
            connection.Close();

            Console.WriteLine("\nFinished deleting
employee");
            Console.
WriteLine("==========================================\n");
        }
        catch (SqlException e)
        {
            Console.WriteLine(e.ToString());
        }
```

```
        }
    }
```

3. Next, we need to call `UpdateEmployeeItem` in the `GetStartedDemo` method again, as follows:

```
static void GetStartedDemo()
{
    connectionstring = "<replace-with-your-
connectionstring>";

    AddItemsToDatabase();
    QueryItems();
    UpdateEmployeeItem();

    //ADD THIS PART TO YOUR CODE
    DeleteEmployeeItem();
}
```

4. If you now run the application, the console will create a new employee, then query the database. Next, it will update the employee, and finally, it will delete the employee from the database.

We have now created an application that can create, read, update, and delete items from an Azure SQL database. In the next section, we are going to configure elastic pools for the database.

Configuring elastic pools for Azure SQL Database

While the usage demands for an Azure SQL Database server are unpredictable during creation, SQL database elastic pools are a simple, cost-effective solution for managing and scaling multiple databases. The databases in an elastic pool share a set of resources at a set price.

You can configure elastic pools for an Azure SQL database in two different ways. The first is to enable it during the creation of the database server and database. The other option is to enable it on an existing database. In this demonstration, we are going to enable it on the database that we created in the first demonstration of this chapter. Therefore, we need to take the following steps:

1. Navigate to the Azure portal by opening `https://portal.azure.com`.

2. Navigate to the overview blade of the Azure SQL database that we created in the first demonstration of this chapter. In the top menu, click on the link that is displayed next to **Elastic pool**, as shown in the following screenshot:

Figure 16.5 – Elastic pool settings

3. You will now be redirected to the Azure SQL Database server settings. In the overview blade, in the top menu, select **+ New pool**, as shown in the following screenshot:

Figure 16.6 – Adding a new pool

4. In the elastic pool blade, add the following values:

   a) **Elastic Pool Name**: `PacktElasticPool`.

   b) **Compute + storage**: Here, you can select the pool's service tier. This determines the features available to the elastics in the pool. It also determines the maximum amount of resources available for each database, as shown in the following screenshot:

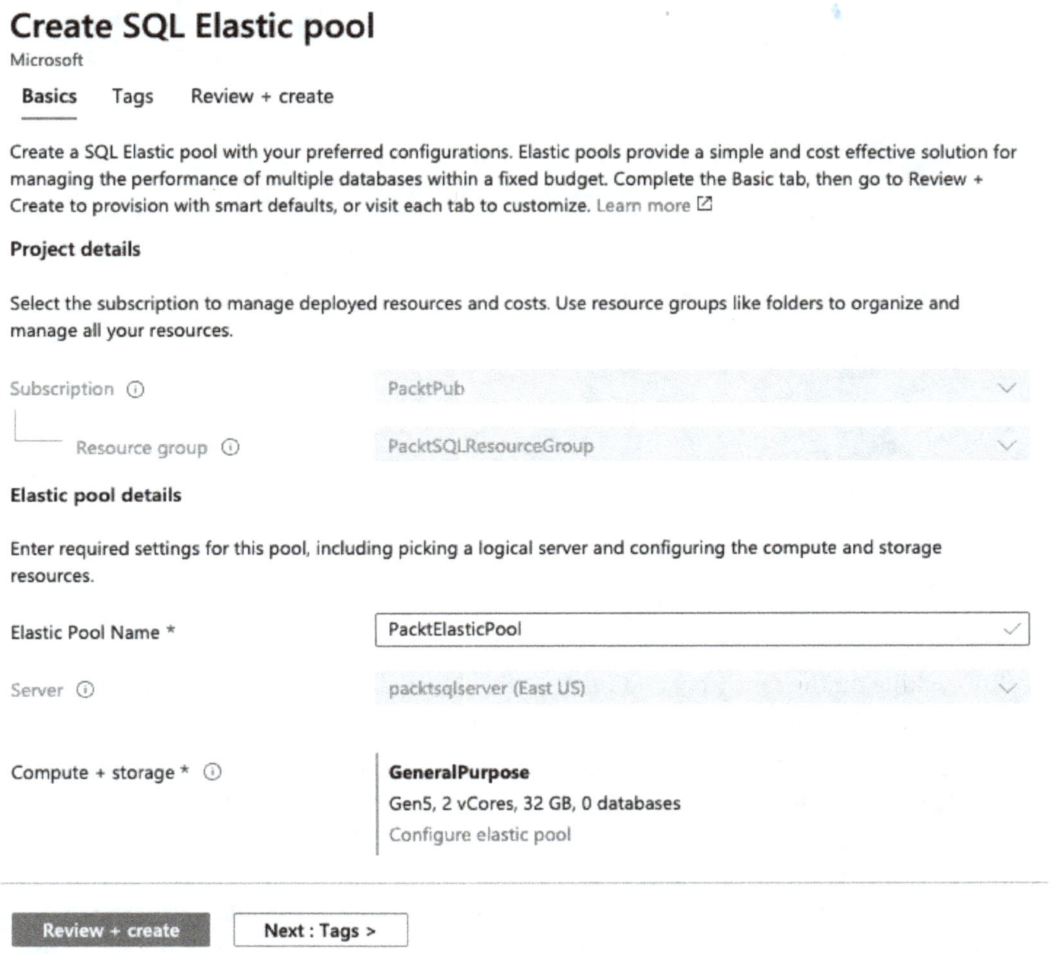

Create SQL Elastic pool

Microsoft

Basics   Tags   Review + create

Create a SQL Elastic pool with your preferred configurations. Elastic pools provide a simple and cost effective solution for managing the performance of multiple databases within a fixed budget. Complete the Basic tab, then go to Review + Create to provision with smart defaults, or visit each tab to customize. Learn more ☑

Project details

Select the subscription to manage deployed resources and costs. Use resource groups like folders to organize and manage all your resources.

Subscription ⓘ   PacktPub   ∨

      Resource group ⓘ   PacktSQLResourceGroup   ∨

Elastic pool details

Enter required settings for this pool, including picking a logical server and configuring the compute and storage resources.

Elastic Pool Name *   PacktElasticPool   ✓

Server ⓘ   packtsqlserver (East US)   ∨

Compute + storage * ⓘ   **GeneralPurpose**
                                         Gen5, 2 vCores, 32 GB, 0 databases
                                         Configure elastic pool

[Review + create]   [Next : Tags >]

Figure 16.7 – Configuring the elastic pool

5. Click **Review + create** and then **Create** to create the pool.

6. When the pool is created, on the SQL Server blade, in the left-hand menu, under **Settings**, select **SQL elastic pools**. There, you can see the newly created pool, as follows:

Figure 16.8 – Overview of elastic pools

7. By clicking on the newly created pool, you will be redirected to the settings of the pool. There, you can monitor and configure it.

8. On the overview blade of the elastic pool, you can monitor the utilization of the pool and its databases. This monitoring includes recent alerts and recommendations (if available) for the elastic pool, and monitoring charts showing the resource usage of the elastic pool. If you want more information about the pool, you can click on the available information in this overview. Clicking on the available notifications will take you to a blade that shows the full details of the notification. Clicking on the **Resource utilization** chart will take you to the Azure Monitor view, where you can customize the metrics and time window shown in the chart.

9. If you would like to monitor the databases inside your pool, you can click on **Database Resource Utilization** in the **Monitoring** section on the left-hand side of the resource menu.

10. Under **Settings**, select **Configure**. Here, you can make any combination of the following changes to the pool: add or remove databases to and from the pool, change the service tier of the pool, scale the performance (DTU or vCore) and storage up or down, review the cost summary to view any changes to your bill as a result of your new selections, and set a min (guaranteed) and max performance limit for the databases in the pools, as shown in the following screenshot:

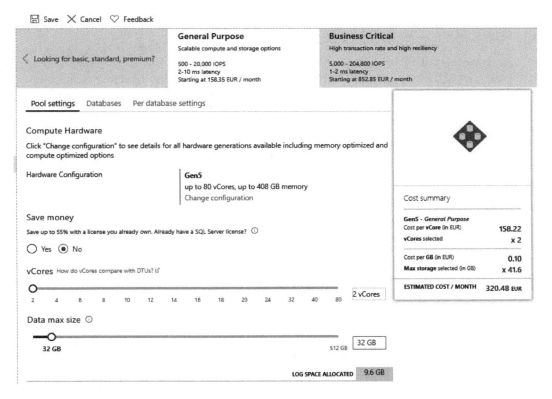

Figure 16.9 – Configuring the elastic pools

In this section, we have covered how to create elastic pools. In the next section, we are going to cover how to create highly available databases.

Configuring high availability

Even when your databases are hosted in Azure, there is still a chance that failures and outages will occur. In the case of an outage (such as a total regional failure, which can be caused by a natural disaster, an act of terrorism, war, a government action, or a network or device failure external to the data centers of Microsoft), your data still needs to be accessible.

To create highly available SQL Server databases in Azure, you can use failover groups and active geo-replication, described in the following list:

- **Active geo-replication**: Geo-replication is a business continuity feature that allows you to replicate the primary database and up to four read-only secondary databases in the same or different Azure regions. You can use the secondary databases to query data or for failover scenarios when there is a data center outage. Active geo-replication has to be set up by the user or the application manually.

- **Failover groups**: Failover groups are a feature that automatically manages the failover for the database. They automatically manage the geo-replication relationship between the databases, the failover at scale, and the connectivity. To use failover groups, the primary and secondary databases need to be created within the same Azure subscription.

Before you can create a failover group, you need to have a replica.

Creating a SQL replica

In the next demonstration, we will set up geo-replication:

1. Navigate to the Azure portal by opening `https://portal.azure.com`.

2. Navigate to the overview blade of the Azure SQL server that we created in the first demonstration of this chapter by searching for `Azure SQL Server` and selecting your SQL server.

3. In the SQL Server blade, on the left-hand menu, click **Geo-Replication**.

4. You will see the current region that your SQL server is in, and a list of all other regions you can replicate it to. Scroll down the list until you find **West US** and click it.

5. You must now configure a new server in the new region by clicking **Target server** and entering the following details:

    a) **Server Name**: `packtsqlserver-westus`

    b) **Server admin login**: `PacktAdmin`

    c) **Password**: `P@ss@word123`

6. Click **Select**.

7. Your screen should look like the following. Click **OK**:

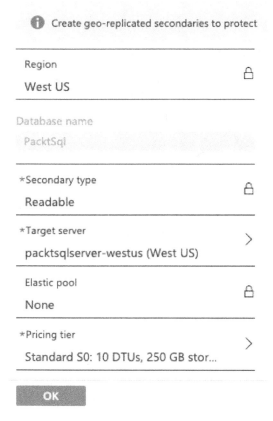

Figure 16.10 – Creating a SQL replica

8. Your SQL server replica will be created. Once completed, you will be taken to the main geo-replication screen, showing your primary and secondary servers.

Now that we have a server replica, we can create a failover group.

Creating a SQL database failover

While replicas are set in the SQL database blade, failover groups are set at the server level:

1. Navigate to the Azure portal by opening `https://portal.azure.com`.

2. Navigate to the overview blade of the Azure SQL server that we created in the first demonstration of this chapter by searching for `SQL Server` and selecting your primary SQL server.

3. On the left-hand menu, click **Failover groups**.

4. Click **+ Add group**.

5. Create your failover group with the following details:

    a) **Failover group name**: `packtsqlfailover`.

    b) **Secondary Server**: The secondary server that you created in the previous demonstration (`packtsqlserver-westus`).

    c) **Read/Write failover policy**: **Automatic**.

    d) **Read/Write grace period**: **1 hours**.

    e) **Databases with the group**: Click and then select your database.

6. Click **Create**.

7. Wait a few minutes for the failover group to be created. Once completed, it will appear on the screen in a list. Click on it to see the details – you will see a screen similar to the following:

packtsqlfailover
packtsqlserver

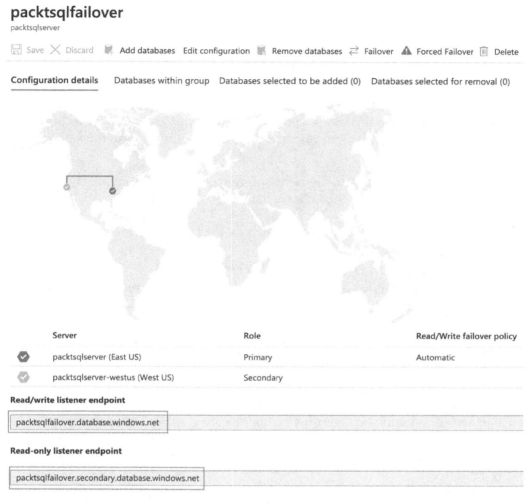

🖫 Save   ✕ Discard   📝 Add databases   Edit configuration   🗑 Remove databases   ⇄ Failover   ⚠ Forced Failover   🗑 Delete

Configuration details   Databases within group   Databases selected to be added (0)   Databases selected for removal (0)

	Server	Role	Read/Write failover policy
✅	packtsqlserver (East US)	Primary	Automatic
✅	packtsqlserver-westus (West US)	Secondary	

Read/write listener endpoint

packtsqlfailover.database.windows.net

Read-only listener endpoint

packtsqlfailover.secondary.database.windows.net

Figure 16.11 – SQL failover groups

With your SQL failover group created, you have a number of options for accessing it.

The first option is to use the existing connection strings you used when creating the sample application.

However, because the replica database is on a different server, your application would need to update its connection string to the new location in the event of a failover.

If you look at the preceding screenshot, again you will notice two areas highlighted – **Read/write listener endpoint** and **Read-only listener endpoint**. By changing your connection string to the read/write listener endpoint, such as `packtsqlfailover.database.windows.net` in this example, in the event of a failover, your application will automatically be directed to the active server.

The read-only listener is useful for spreading a load across servers when you only need to read from your database. In this way, when combined with frontend user interfaces spread across regions, you can achieve faster response times by directing users to their closest replica.

Failover will occur automatically in the event of failure of the primary server, but you can also trigger a failover by clicking the **Failover** button:

Figure 16.12 – SQL failover

In this section, we have covered how to use geo-replication and failover groups. In the next section, we are going to cover an Azure SQL Database managed instance.

Implementing Azure SQL Database managed instances

Azure SQL Database managed instances provide a near-100% compatibility with the latest SQL Server on-premises (Enterprise edition) database engine. Azure will then handle all the updates and patches for your databases, such as when using PaaS services. It also provides a native **virtual network** (**VNet**) implementation that addresses common security concerns.

This allows customers to easily lift and shift their on-premises SQL databases to Azure, with minimal database and application changes. Databases can be migrated using the **Database Migration Service** (**DMS**) in Azure to a completely isolated instance, with native VNet support. Because of the isolation, this instance offers high security for your databases.

A managed instance is available in two service tiers:

- **General purpose**: This tier offers normal performance and I/O latency for applications. It offers high-performance Azure Blob storage (up to 8 TB). The high availability is based on Azure Service Fabric and Azure Blob storage.

- **Business critical**: This tier is designed for applications with low I/O latency requirements. It also promises a minimal impact of any underlying maintenance operations on the workload. It comes with SSD local storage (up to 1 TB on Gen4, and up to 4 TB on Gen5). High availability is based on Azure Service Fabric and Always On availability groups. For read-only workloads and reporting, it also offers built-in additional read-only database replicas. For workloads with high-performance requirements, in-memory **Online Transactional Processing (OLTP)** is used.

To create a managed instance SQL server, perform the following steps:

1. Navigate to the Azure portal by opening `https://portal.azure.com`.
2. Click + **Create Resource**.
3. Search for and select **Azure SQL Managed Instance**.
4. Click **Create**.
5. Complete the following details:

    a) **Subscription**: Select your subscription.

    b) **Resource Group**: Create a new resource group called `PacktSQLMIResourceGroup`.

    c) **Managed Instance name**: `packtsqlmanagedinstance`.

    d) **Region: East US**.

    e) **Compute + Storage**: Leave as default.

    f) **Managed Instance admin login**: `PacktAdmin`.

    g) **Password**: `P@ss@word1234567`.

6. Click **Next: Review + Create**.
7. Click **Create**.

> **Tip**
> SQL Managed Instance is relatively expensive. If you follow through these examples, make sure you delete the instance if it's no longer needed!

A SQL managed instance will now be created for you; however, it can take a couple of hours to create.

In the final section of this chapter, we will look at how to deploy databases into Azure.

Publishing a SQL database

One of the requirements for the AZ-303 exam is knowing how to migrate a database from an existing on-premises database into Azure.

There are actually a number of different ways we can achieve this, the most obvious being to perform a backup and restore operation.

However, SQL Server Management Studio contains a built-in wizard that makes the process of moving databases easy.

In this example, we are going to *publish* a database that resides on an on-premises database server to Azure:

> **Important note**
> For the following example, I have set up a Windows server running SQL Server Express and SQL Server Management Studio. I have also opened my Azure SQL Server firewall to allow my Windows server to connect. A T-SQL script to create and populate a simple database can be found at the GitHub link in the *Technical requirements* section.

1. On a computer, run SQL Server Management Studio.

2. Connect to the local server.

3. Right-click **Databases** and choose **New Database**.

4. In **Database name**, enter SimpleDb and click **OK**.

5. Right-click the new **SimpleDb** database and choose **New Query**.

6. In the new query window, enter the following (the code is in the SimpleDb.sql file in CH16 in the AZ-303 GitHub repository):

```
CREATE TABLE [dbo].[Items](
[Id] [int] NOT NULL,
[Category] [nvarchar](max) NULL,
[Name] [nvarchar](max) NULL,
[Description] [nvarchar](max) NULL,
[Completed] [bit] NULL
) ON [PRIMARY] TEXTIMAGE_ON [PRIMARY]
GO
INSERT [dbo].[Items] ([Id], [Category], [Name],
[Description], [Completed]) VALUES (1, N'personal',
N'groceries', N'Get Milk and Eggs', 0)
```

```
GO
INSERT [dbo].[Items] ([Id], [Category], [Name],
[Description], [Completed]) VALUES (2, N'work', N'SQL
Lecture', N'Create SQL Lecture', 0)
GO
```

7. Above the query window, click **Execute**.

8. Right-click the database and choose **Tasks | Deploy database to Microsoft Azure SQL Database…**, as in the following screenshot:

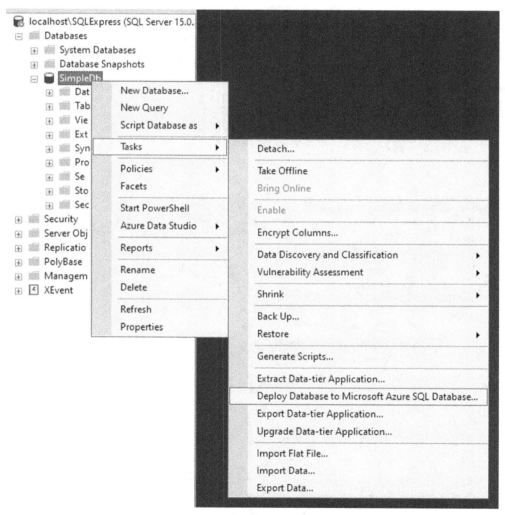

Figure 16.13 – Publishing a database

9. In the deploy database wizard, click **Next**.

10. On the next screen, click **Connect** and enter the details for your Azure SQL server:

    a) **Server Name**: Your SQL server name plus `.database.windows.net` – for example, `packtsqlserver.database.windows.net`

    b) **Login**: `PacktAdmin`

    c) **Password**: `P@ss@word123`

11. Click **Connect**.

12. Now, enter the details for the new SQL database size:

    a) **Edition of Microsoft Azure SQL Database: Basic**

    b) **Maximum database size (GB): 2**

    c) **Service Objective: Basic:**

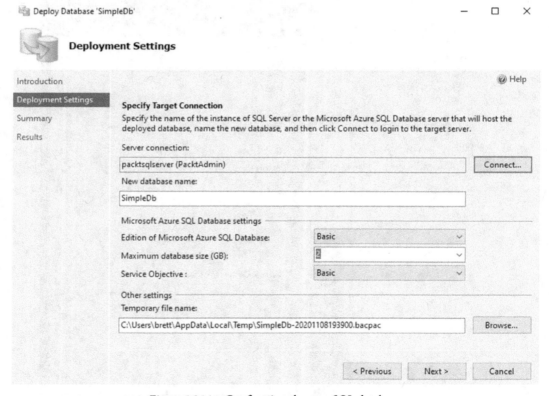

Figure 16.14 – Configuring the new SQL database

13. Click **Next**.

14. Confirm the summary details and click **Next**.

Your database will now be copied into Azure. You can confirm that the database has been created by going to the Azure SQL databases view – your database will now be listed, as in the following example:

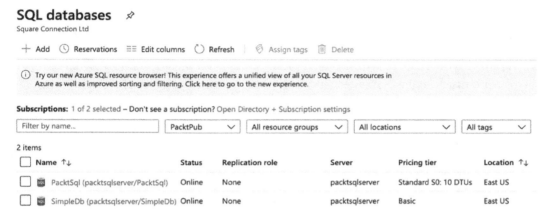

Figure 16.15 – SQL databases list

In this demonstration, we have seen how to migrate a database from an on-premises server into Azure using the SQL Studio Manager deploy option.

Summary

In this chapter, we covered how to develop solutions that use a relational database. We covered how to create an Azure SQL database, as well as how to add, remove, update, and query data from a custom application.

We saw how to enable elastic pools, create geo-redundant replicas, and enable failover groups to protect from region failures.

Finally, we looked at SQL managed instances, and how we can publish databases from on-premises servers to Azure.

With the knowledge gained throughout these chapters, you should now be able to pass the AZ-303 exam. Don't forget to look at the *Further reading* sections at the end of each chapter, because there is a lot of extra information listed there that could be covered in the exam as well.

Questions

Answer the following questions to test your knowledge of the information in this chapter. You can find the answers in the *Assessments* section at the end of this book:

1. Can an elastic pool only be created during the creation of the Azure SQL database?

    a) Yes

    b) No

2. Is an Azure SQL Database managed instance based on the last stable SQL Server on-premises database engine?

    a) Yes

    b) No

3. Do failover groups automatically manage the geo-replication relationship between the databases, the failover at scale, and the connectivity?

    a) Yes

    b) No

Further reading

You can check out the following links for more information about the topics that are covered in this chapter:

*   What is the Azure SQL Database service?: https://docs.microsoft.com/en-us/azure/sql-database/sql-database-technical-overview
*   Azure SQL Database service tiers: https://docs.microsoft.com/en-us/azure/sql-database/sql-database-service-tiers-general-purpose-business-critical
*   Elastic pools help you manage and scale multiple Azure SQL databases: https://docs.microsoft.com/en-us/azure/sql-database/sql-database-elastic-pool
*   What is Azure SQL Database Managed Instance?: https://docs.microsoft.com/en-us/azure/sql-database/sql-database-managed-instance

Mock Exam
Questions

1. You have an Azure subscription that has eight VMs deployed in it. You need to configure monitoring for this and want to receive a notification when the **Central Processing Unit** (**CPU**) or available memory reaches a certain threshold value. The notification needs to be sent using an email and needs to create a new issue in the corporate issue tracker. What is the minimum number of action groups and alerts that you need to create in order to meet these requirements?

    A) Eight alerts and one action group

    B) Two alerts and two action groups

    C) One alert and two action groups

    D) One alert and one action group

2. You have a Windows Server 2016 machine deployed inside an availability set. You need to change the availability set assignment for the VM. What should you do?

    A) Migrate the VM to another Azure region.

    B) Assign the VM to a new availability set.

    C) Redeploy the VM from a recovery point.

    D) Move the VM to a different availability set.

3. You have an Azure Application Gateway deployed that currently load balances all traffic on port 80 to a single backend pool. You now have a requirement to load balance all traffic that includes /Video/* in the path to be forwarded to a different backend pool. What should you do?

    A) Create a new backend pool, and then create a new basic rule and include the /Video/* path and the new backend pool.

    B) Create a new backend pool, and then create a new path-based rule and include the /Video/* path and the new backend pool.

    C) Create a new Application Gateway and traffic manager and load balance all requests that contain the /Video/* path to the correct target.

    D) Add the /Video/* path to the default rule.

4. Your company wants to deploy a storage account. You need to ensure that the data is available in the case of the failure of an entire data center. The solution must be the most cost-effective. What should you do?

    A) Configure geo-redundant storage.

    B) Configure local redundant storage.

    C) Configure read-access geo-redundant storage.

    D) Configure zone-redundant storage.

5. You need to assign a static IPv4 address for a Windows Server VM named PacktVM1 running in a VNet named PacktVNet1. What should you do?

    A) Modify the IP configuration of the VNet interface associated with the PacktVM1 VM.

    B) Edit the address range of the PacktVNet1 VNet.

    C) Connect to the PacktVM1 VM by using WinRM and run the Set-NetIPAddress cmdlet.

    D) Connect to the PacktVM1 VM by using Remote Desktop Protocol and edit the VM's virtual network connection properties.

6. You need to add another administrator who will be responsible for managing all **Infrastructure-as-a-Service (IaaS)** deployments in your Azure subscription. You create a new account in Azure AD for the user. You need to configure the user account to meet the following requirements: read and write access to all Azure IaaS deployments; read-only access to Azure AD; no access to Azure subscription metadata. The solution must also minimize your access maintenance in the future. What should you do?

    A) Assign the owner role at the resource level to the user account.

    B) Assign the global administrator directory role to the user account.

    C) Assign the VM operator role at the subscription level to the user account.

    D) Assign the contributor role at the resource group level to the user account.

7. You have Azure Site Recovery configured for failover protection for 7 on-premises machines to Azure in case of an accident. You want to ensure that only 10 minutes of data is lost when an outage occurs. Which PowerShell cmdlet should you use for this?

    A) `Edit-AzureRmSiteRecoveryRecoveryPlan`

    B) `Get-AzureRmSiteRecoveryPolicy`

    C) `Get-AzureRmSiteRecoveryRecoveryPlan`

    D) `Update-AzureRmSiteRecoveryPolicy`

8. Your organization has Azure resources deployed in the West US, West Europe, and East Australia regions. The company has four offices located in these regions. You need to provide connectivity between all the on-premises networks and all the resources in Azure using a private channel. You configure a VPN gateway for each Azure region and configure a site-to-site VPN for each office and connect to the nearest VPN gateway. You then configure virtual network peering. You need to ensure that users have the lowest traffic latency. Does this solution meet your goal?

    A) Yes

    B) No

9. Your company has an Azure AD tenant and an on-premises AD that are synced
    using Azure AD Connect. The security department notices a high number of logins
    from various public IP addresses. What should you do to reduce these logins?

    A) Enable Azure AD smart lockout.

    B) Add all the public IP addresses to conditional access and use location blocking to
    deny all login attempts.

    C) Create a conditional access rule to require MFA for all risky logins labeled
    medium risk and above.

    D) Turn on Azure MFA fraud alerts.

10. You have an Azure App Service API that allows users to upload documents to the
    cloud with a mobile device. A mobile app connects to the service by using REST
    API calls. When a document is uploaded to the service, the service extracts the
    document metadata. Usage statistics for the app show a significant increase in app
    usage. The extraction process is very CPU-intensive. You plan to modify the API
    to use a queue. You need to ensure that the solution scales, handles request spikes,
    and reduces costs between the spikes. What should you do?

    A) Configure a CPU-optimized VM and install the Web App service on the new
    instance.

    B) Configure a series of CPU-optimized VMs and install the extraction logic for the
    app to process a queue.

    C) Move the extraction logic to an Azure function and then create
    a queue-triggered function to process the queue.

    D) Configure Azure Container Instances to retrieve the items from the queue and
    run the extraction logic across a pool of VM nodes.

11. You want to create a group of resource group managers in the Azure portal. Which
    RBAC role do you need to assign to them in order to manage all the resource groups
    in the Azure subscription?

    A) Contributor

    B) Reader

    C) Owner

    D) Monitoring reader

12. Your company has an application that requires data from a blob storage to be moved from the hot access tier to the archive access tier to reduce costs. Which type of storage account do you need to create?

    A) A general-purpose V2 storage account

    B) A general-purpose V1 storage account

    C) Azure File storage

    D) Azure Blob storage

13. You are planning data security for your Azure resources. The confidentially of code on your VMs must be protected while the code is being processed. Which feature should you use for this?

    A) Azure Batch

    B) Azure Confidential Compute

    C) Azure Container Instances

    D) Azure Disk Encryption

14. You have two Azure resource groups, named `ResourceGroup1` and `ResourceGroup2`. The `ResourceGroup1` resource group contains 20 Windows Server VMs and all the VMs are connected to an Azure Log Analytics workspace named `Workspace1`. You need to write a log search query that collects all security events with the following properties: all security levels other than 8 and with the Event ID `4672`. How should you write your query?

    A) `SecurityEvent | where Level == 8 | and EventID == 4672`

    B) `SecurityEvent 4672 | where  Level <> 8 | where EventID ==4672`

    C) `SecurityEvent 4672 | where Level == 8 |summarize EventID==4672`

    D) `SecurityEvent | where Level <> 8 | and EventID == 4672`

15. You are using an Azure Logic App to integrate SharePoint Online, Dynamics, and an on-premises Oracle database. You are informed that the logic app access key has been compromised. What should you do?

    A) Delete the logic app and redeploy it.

    B) Only allow internal IP addresses to access the logic app.

    C) Add a resource lock.

    D) Regenerate the access key.

16. You have two subscriptions named `subscription 1` and `subscription 2`. Each subscription is associated with a different Azure AD tenant. `subscription 1` contains a virtual network named `VNet 1`. `VNet 1` contains an Azure VM named `VM1` and has an IP address space of `10.0.0.0/16`. `subscription 2` contains a virtual network named `VNet 2`. `VNet 2` contains an Azure VM named `VM2` and has an IP address space of `10.0.0.0/24`. You need to connect `VNet 1` to `VNet 2`. What should you do first?

    A) Move `VM2` to `subscription 1`.

    B) Provision virtual network gateways.

    C) Move `VNet 1` to `subscription 2`.

    D) Modify the IP address range of `VNet 2`.

17. Your company has a VM that is stored inside a resource group. You need to deploy additional VMs in the same resource group. You are planning to deploy them using an ARM template. You need to create a template from the original VM using PowerShell. Which cmdlet should you use?

    A) `Export-AzResourceGroup`

    B) `Get-AzResourceGroupDeployment`

    C) `Get-AzResourceGroupDeploymentOperation`

    D) `Get-AzResourceGroupDeploymentTemplate`

18. You are developing an app that references data that is shared across multiple Azure SQL databases. The app must guarantee transactional consistency for changes across several sharding key values. You need to manage the transactions. What should you implement?

A) Elastic database transactions with horizontal partitioning

B) Distributed transactions coordinated by **Microsoft Distributed Transaction Coordinator (MSDTC)**

C) Server-coordinated transactions from a .NET application

D) Elastic database transactions with vertical partitioning

19. You create a VM called VM1 with a Premium SSD operating system disk. You enable Azure Disk Encryption for the VM and then you add a Premium SSD data disk. Is the data disk automatically encrypted?

A) Yes

B) No

20. Your company has an application that uses an Azure SQL database to store information. The company has also deployed System Center Service Manager. You need to configure an alert when the database reaches 80% of CPU usage. When this alert rises, you want your administrator to be notified using email and SMS. You also need to create a ticket in the corporate issue tracker automatically when the alert arises. Which two actions should you perform?

A) Configure System Center Service Manager with Azure Automation.

B) Configure one action group with three actions, one for email, one for SMS, and one for creating the ticket.

C) Configure an IT Service Management Connector.

D) Configure two actions groups, one for email and SMS, and one for creating the ticket.

21. A VM named `PacktVM1` is deployed in a resource group named `PacktResourceGroup1`. The VM is connected to a VNet named `PacktVNet1`. You plan to connect the `PacktVM1` VM to an additional VNet named `PacktVNet2`. You need to create an additional network interface on the `PacktVM1` VM and connect it to the `PacktVNet2` VNet. Which two Azure **Command-line Interface (CLI)** commands should you use?

    A) `az vm nic add`

    B) `az vm nic create`

    C) `az network update`

    D) `az network nic create`

22. You need to grant access to an external consultant to some resources inside your Azure subscription. You plan to add this external user using PowerShell. Which cmdlet should you use?

    A) `New-AzADUser`

    B) `New-AzureADMSInvitation`

    C) `Get-AzADUser`

    D) `Get-AzureADMSInvitation`

23. You are planning to migrate your on-premises environment to Azure using Azure Site Recovery. You have already created a storage account, a virtual network, a Recovery Services vault, and a resource group in the Azure portal. You now need to grant the cloud engineer the requisite privileges to perform the migration. Which two built-in roles should you use, using the principle of least privilege?

    A) Site Recovery Contributor

    B) Network Contributor

    C) Reader

    D) Virtual Machine Contributor

24. You use Azure AD Connect to synchronize all AD domain users and groups with Azure AD. As a result, all users can use **Single Sign-on (SSO)** to access applications. You should reconfigure the directory synchronization to exclude domain services accounts and user accounts that shouldn't have access to the application. What should you do?

A) Rerun Azure AD Connect and configure OU filtering.

B) Stop the synchronization service.

C) Remove the domain services and user accounts manually.

D) Configure conditional access rules in Azure AD.

25. You configure Azure Application Gateway to host multiple websites on a single instance of the Application Gateway. You create two backend server pools, named `PacktPool1` and `PacktPool2`. Requests for `http://Packt1.info` should be routed to `PacktPool1`, and requests for `http://Packt2.info` should be routed to `PacktPool2`. Users only see the content of `PacktPool2`, regardless of the URL they use. You need to identify which component is configured incorrectly. What should you check?

A) The CName resource record

B) The backend port settings

C) The routing rule

D) The SSL certificate

26. Your company is developing a .NET application that stores information in an Azure Storage account. You need to ensure that the information is stored in a secure way. You ask the developers to use a **Shared Access Signature (SAS)** when accessing the information. You need to ensure that the required configurations on the storage account comply with security best practices. Which statement is false?

A) You need to configure a stored access policy.

B) To revoke an SAS, you can delete the stored access policy.

C) You should set the SAS start time to now.

27. You need to use an Azure logic app to receive a notification when an administrator modifies the settings of a VM in a resource group, `ResourceGroup1`. Which three components should you create next in the Logic Apps Designer? Pick the three components and set them in the correct order.

    A) An action

    B) An Azure Event Grid trigger

    C) A condition control

    D) A variable

28. Your company has an Azure AD tenant and an on-premises AD that are synced using Azure AD Connect. Your on-premises environment is running a mix of Windows Server 2012 and Windows Server 2016 servers. You use Azure MFA for multi-factor authentication. Users report that they are required to use MFA while using company devices. You need to turn MFA off for domain-joined devices. What should you do?

    A) Enable SSO on Azure AD Connect.

    B) Create a conditional access rule to allow users to use either MFA or a domain-joined device when accessing applications.

    C) Configure Windows Hello for Business on all domain-joined devices.

    D) Add the company external IP address to the Azure MFA Trusted IPs list.

29. You maintain an existing Azure SQL Database instance. Management of the database is performed by an external party. All cryptographic keys are stored in Azure Key Vault. You must ensure that the external party cannot access the data in the SSN column of the **Person** table. What should you do?

    A) Enable **AlwaysOn** encryption.

    B) Set the column encryption setting to **disabled**.

    C) Assign users to the public fixed database role.

    D) Store the column encryption keys in the system catalog view in the database.

30. You have an Azure resource group named `PacktResourceGroup1` that contains a Linux VM named `PacktVM1`. You need to automate the deployment of 30 additional Linux machines. The VMs should be based on the configuration of the `PacktVM1` VM. Which of the following solutions will meet the goal?

    A) From the VM Automation's script blade, click **Deploy**.

    B) From the **Templates** blade, click **Add**.

    C) From the resource group's policy blade, click **Assign**.

31. You have an Azure subscription that contains two different VNets. You want the VNets to communicate through the Azure backbone. Which solution should you choose?

    A) VNet peering

    B) Site-to-site VPN

    C) Point-to-site VPN

    D) Azure Expressroute

32. You are using Azure Application Gateway to manage traffic for your corporate website. The Application Gateway uses the standard tier, with an instance size of **medium**. You are asked to implement WAF to guard the website against SQL injection attacks and other vulnerabilities. To configure WAF, which two actions should you perform?

    A) Enable WAF in detection mode.

    B) Change the Azure Application Gateway to an instance size of **large**.

    C) Enable WAF in prevention mode.

    D) Change the Azure Application Gateway tier.

33. You have VMs deployed inside a Hyper-V infrastructure and you are planning to move those VMs to Azure using Azure Site Recovery. You have the following types of machines. Can all these types of machines be moved using Azure Site Recovery?

    — Windows VMs Generation 2

    — Linux VMs Generation 2

    — Windows VMs with BitLocker installed on them

    A) Yes

    B) No

34. You have a web app named PacktApp. You are developing a triggered app service background task using the WebJobs SDK. This task will automatically invoke a function in code whenever any new data is received in the queue. Which service should you use when you want to manage all code segments from the same Azure DevOps environment?

    A) Logic Apps

    B) A custom web app

    C) Web Jobs

    D) Functions

35. You are developing a workflow solution using Azure technologies. Which solution is the best fit if you want to use a collection of ready-made actions?

    A) Azure Functions

    B) Logic Apps

    C) Web Apps

36. You are creating a new Azure Functions app to run a serverless C# application. This function has an execution duration of 1 second and a memory consumption of 256 MB, and executes up to 1 million times during the month. Which plan should you use?

    A) The Linux App Service plan

    B) The Windows Consumption plan

    C) The Windows App Service plan

    D) The Kubernetes App Service plan

37. You plan to create a Docker image that runs on an ASP.NET Core application named `PacktApp`. You have a setup script named `setupScript.ps1` and a series of application files, including `PacktApp`. You need to create a Dockerfile document that calls the setup script when the container is built and runs the app when the container starts. The Dockerfile document must be created in the same folder where `PacktApp.dll` and `setupScript.ps1` are stored. In which order do the following four commands need to be executed?

   A) `Copy ./.`

   B) `WORKDIR /apps/PacktApp`

   C) `FROM microsoft/aspnetcore:2.0`

   D) `RUN powershell ./setupScript.ps1 CMD ["dotnet", "PacktApp.dll"]`

38. The HR department uses multiple Azure subscriptions within your tenant. The head of HR wants to be able to have read access to all components in anything built by their team. How can you achieve this?

   A) Assign the head of HR read access on every individual subscription.

   B) Create an Active Directory group and add all HR subscriptions to it. Make the head of HR the admin of that group.

   C) Create an Azure management group called HR. Ensure that all HR subscriptions are under that group. Grant read access to the head of HR for that management group.

   D) Create a resource group in each subscription. Add the *Read* RBAC role to the head of HR for each resource group.

39. A VM has been deleted and nobody knows who did it or why. How can you investigate what happened and who did it?

   A) In the **Virtual machine** blade, go to the **delete items** view.

   B) Go to the **subscription activity** log view. Filter events on **Operation – Delete Virtual Machine**.

   C) Go to the resource group the VM was deployed in, and then go to the **Deployments** tab.

   D) On the **subscription** blade, go to the **Security** view. Search for `delete` `events`.

40. You are creating a solution that stores data in a database. The application is dynamic and needs to store data using different schemas, as opposed to using a well-defined, static schema. Which Azure native database technology is the best choice?

    A) Azure SQL

    B) Azure Cosmos DB

    C) Azure Blob storage

41. Your security team wants you to ensure that all subscriptions contain a Key vault and a VNET that routes all traffic to a central hub containing a firewall. You need to prevent users from changing or deleting the VNET. How can this be achieved?

    A) Create an Azure blueprint that contains the desired configuration. Set the Blueprint to **readonly**.

    B) Create an ARM template that contains the desired configuration. Run that template against all new subscriptions.

    C) Manually create the desired configuration. Create an RBAC role to specifically deny access to network components.

    D) Manually create the desired configuration. Create an alert if any network component is deleted or modified.

42. You have an application that uses a global Cosmos DB. Performance is not as important as ensuring that all replicas of the database are always up to date. Which is the best consistency model?

    A) Strong

    B) Bounded staleness

    C) Session

    D) Consistent prefix

    E) Eventual

43. You are building a new solution that uses an Azure SQL backend database. The database itself must be protected from an entire region outage, and any failover must be fully automatic. How do you configure the Azure SQL Server and Database to achieve this?

    A) Build your SQL database in an elastic pool. Use the individual database connection string as this will be the same in the event of a failover.

    B) Set up geo-replication on the server with the replica in another region. Use the normal database connection string.

    C) Set up geo-replication on the server, and then create a database failover group. Use the normal database connection string.

    D) Set up geo-replication on the server, and then create a database failover group. Use the read/write listener endpoint connection string.

44. You are developing a 3-tier application, and you need to ensure that the backend services, middle tier, and frontend UI are as secure as possible from a networking perspective. Which TWO options will achieve this?

    A) Build all three tiers within a single subnet. Set up the apps themselves to only communicate with each upstream or downstream tier.

    B) Build all three tiers within a single VNET. Set up **Network Security Groups (NSGs)** on the VNET to only allow communication between tiers.

    C) Build all three tiers within a single VNET, but separated into subnets. Set up NSGs on each subnet to only allow communication between tiers.

    D) Build all three tiers within a single subnet. Group each tier into an **Application Security Group (ASG)**. Set up NSGs on the subnet to only allow communication between servers defined in the ASGs.

45. You have multiple subscriptions and solutions in your Azure tenant. You need to ensure that all your services within the Azure platform are protected at the network level. Which of the following options would achieve this with minimal setup and administration?

    A) Define a single NSG that defines your firewall rules. Assign this NSG to all subnets.

    B) Create a hub subscription that contains an Azure Firewall with your firewall rules applied. Configure all other subscriptions to route traffic through the hub subscription.

    C) Create an Azure Firewall in each subscription and implement a standard set of rules on each.

    D) Define ASGs. Group the server types into ASGs. On each NSG, use the ASGs to control traffic.

46. What kinds of VMs can you update with Azure Update Management?

    A) Windows

    B) Linux

    C) Both

47. You have been asked to protect a VM workload in Azure. The Recovery Point Objective is 1 hour (the business cannot lose more than an hour's worth of data). Which is the best backup solution for this requirement and why?

    A) Azure Backup, because your VM can be backed up every hour

    B) Azure Site Recovery, because data can be copied to another region

    C) Azure Backup, because restoring is instant

    D) Azure Site Recovery, because snapshots are replicated every 5 minutes

48. You are setting up an Azure Bastion service for the purpose of secure communication with your Azure VMs. Which statement is correct?

    A) Azure Bastion can be used to connect to VMs on another VNET without any additional configuration.

    B) Azure Bastion cannot be used to connect to VMs on another VNET.

    C) Azure Bastion can be used to connect to VMs on another VNET if that VNET is peered to the VNET Azure Bastion.

    D) Azure Bastion can only connect to VMs on the same subnet.

49. You have been asked to build a solution that can load-balance and protect services that span two different regions. Which TWO of the following options can be used?

A) Azure Traffic Manager + Azure Front Door

B) Azure Application Gateway

C) Azure Traffic Manager + Azure Application Gateway

D) Azure Front Door + Azure Load Balancer

50. Which of the following Azure SQL options use native VNET integration?

A) SQL Managed Instance

B) SQL Single Database

C) SQL Elastic Pool

D) SQL Hyperscale

Mock Exam Answers

1. **D** – You should create one alert and one action group for this. One alert can contain multiple metrics-based conditions, and a single action group can contain more than one notification or remediation step, so you can create the metrics for both the CPU and memory in one alert. You can use one action group for sending out the email and creating an issue in the corporate issue tracker.

2. **C** – You should redeploy the VM from a recovery point. VMs can only be assigned to an availability set during initial deployment.

3. **B** – You should create a path-based rule for this.

4. **D** – You should configure a storage account with **zone-redundant storage (ZRS)** replication. This makes a synchronous copy of the data between three different zones in the same region.

5. **A** – You should modify the IP configuration of the virtual network interface associated with `PacktVM1`.

6. **D** – You should assign the Contributor role at the resource group level to the user account. This provides the user with full read/write access at the resource group level, but doesn't grant the user any permissions in the subscription or Azure AD levels.

7. **D** – You should use the `Update-AzureRmSiteRecoveryPolicy` cmdlet. This has the recovery points method in it, which you can set to specify the maximum amount of time that data will be lost for.

8. **A** – Yes: Because you configure a VPN gateway for each region, this solution meets the goals. This will result in the lowest traffic latency for your users.

9. **C** – You should create a conditional access rule to require MFA authentication for all risky logins labeled medium-risk and above. Azure AD can apply risk levels to all sign-in attempts using a selection of parameters. You can use conditional access to enforce sign-in requirements based on those levels.

10. **C** – You should move the extraction logic to an Azure function. This is the most scalable and cost-effective solution.

11. **C** – You should assign the **Owner** role to the group of resource group managers.

12. **A** – You need to configure a general-purpose V2 storage account to move data between different access tiers.

13. **C** – You should use Azure Confidential Compute for this requirement.

14. **B** – The correct query should be `SecurityEvent | where Level <> 8 | where EventID == 4672`.

15. **D** – You should regenerate the access key. This will automatically make the old access key invalid.

16. **B** – You need virtual network gateways to connect VNets that are associated with different Azure AD instances.

17. **A** – You should use the `Export-AzResourceGroup` cmdlet. This captures the specified resource group as a template and saves it to a JSON file.

18. **A** – You should implement Elastic database transactions with horizontal partitioning.

19. **A** – The data disk is automatically encrypted using the premium disks.

20. **B and C** – You need to create one action group and you need to configure the **IT Service Management Connector** (**ITSMC**). This connector connects the System Center Service Manager with Azure.

21. **A and D** – You should use `az network nic create` to create a new NIC. Then you should use `az vm nic add` to attach the NIC to `PacktVM1`.

22. **B** – You should use the `New-AzureADMSInvitation` cmdlet to add an external user to your Azure AD tenant using PowerShell.

23. **A and D** – You need to grant the cloud engineer the Virtual Machine Contributor role to enable the replication of a new VM. You do this by creating a new VM inside the Azure portal. You should also grant the Site Recovery Contributor role to the engineer. This way, the engineer has permission to manage the site recovery vault without permission to create new vaults or assign permissions to other users.

24. **A** – You should rerun Azure AD Connect. This will perform OU filtering and refresh the directory schema.

25. **B** – You should check the routing rule. The backend port settings are configured incorrectly.

26. **A is True** – You need to configure a stored access policy.
    B is True – To revoke an SAS, you can delete the stored access policy.
    C is False – When you set the timer to now, there can be differences in the clock of the servers hosting your storage account. This can lead to access problems for a short period of time.

27. **B, C and A** – You should set them in the following order:

    1. An Azure Event Grid trigger

    2. A condition control

    3. An action

28. **B** – You should create a conditional access rule to allow users to use either MFA or a domain-joined device when accessing applications. The rule will not force MFA when using a domain-joined device.

29. **C** – You should enable **AlwaysOn** encryption.

30. **A and B** – You can deploy the ARM template of the VM from the VM's Automation script blade and you can deploy the template from the **Templates** blade in the Azure portal.

31. **A** – VNet peering is the only solution that makes it possible to communicate directly through the Azure backbone.

32. **C and D** – You should enable WAF in prevention mode and change the application gateway tier. The standard tier doesn't support the ability to configure WAF. Prevention mode actively blocks SQL injection attacks.

33. **B – No** – you can only use Azure Site Recovery for the Windows VMs Generation 2 machines that you have installed inside your Hyper-V environment. The rest are not supported.

34. **C** – You should use Web Jobs to manage all the code segments from the same DevOps environment.

35. **B** – You should use Logic Apps.

36. **B** – You should use the Windows Consumption plan. This plan supports per-second resource consumption and execution.

37. **C, B, A, and D** – The script should look like the following:

```
FROM microsoft/aspnetcore:2.0

WORKDIR /apps/PacktApp

Copy ./.

RUN powershell ./setupScript.ps1 CMD ["dotnet", "PacktApp.
dll"]
```

38. **C** – Management groups are the easiest way to manage access to multiple subscriptions.

39. **B** – The Azure Activity Log will display all actions that have taken place within subscriptions, and filtering options can help to find specific events.

40. **B** – Azure Cosmos DB is a NoSQL database, which means it is schema-less.

41. **A** – Azure Blueprint allows you to define base configurations and lock those configurations as **readonly** or **denydelete**.

42. **A** – Strong provides the best consistency level, but at the expense of performance.

43. **D** – For a fully automated failover, you must set up geo-replication on the server, create a database failover group, and use the read/write listener endpoint to connect to rather than the normal database connection string.

44. **C or D** – You can limit traffic between servers by either separating servers into subnets and using NSGs to control the flow of traffic, or by assigning servers to ASGs and define rules in an NSG to control the traffic.

45. **B** – Azure Firewall is a robust and powerful firewall product. Building it in a hub-spoke model can help minimize administration and costs.

46. **C** – Azure Update Management can update both Linux and Windows VMs.

47. **D** – Azure Site Recovery replication takes a crash-consistent snapshot every 5 minutes, meaning that the most recent data on disk is up to 5 minutes old.

48. **C** – Azure Bastion can connect to VMs on another VNET if that VNET is peered.

49. **C and D**. Azure Traffic Manager and Azure Front Door are both ideal solutions for load balancing across regions. They can both be combined with Azure Load Balancer and Azure Application Gateway, which are both used in local load balancing.

50. **A** – SQL Managed Instances are always built with VNET integration.

Assessments

Chapter 1

1. **Yes**—Log Analytics is integrated in Azure Monitor. However, the data is still stored inside the Log Analytics Workspace.

2. **No**—you can't use SQL. You need to use the Kusto Query Language to query the data.

3. **No**—Action Groups are unique sets of recipients and actions that can be shared across multiple alert rules.

4. **IP Flow verify**—Next Hop only proves connectivity with the next appliance and doesn't allow ports to be defined. Packet capture is for capturing data as it flows in and out of a VM.

5. **Budgets from Subscription**

Chapter 2

1. **No**—You can also download the Azure Storage Explorer for Linux and macOS.

2. **No**—You can also configure storage accounts to be accessed from on-premises networks.

3. **No**—You can change the replication type of your storage account later as well, from the Azure portal, PowerShell, or the CLI.

Chapter 3

1. **Yes**—you use VM scale sets to automate the deployment of multiple VMs.

2. **Yes**—by using Availability Sets, you can spread VMs across different fault and update domains.

3. **Yes**—you use resource providers to deploy different artifacts in Azure using ARM templates.

4. **False**—when running VMs on a dedicated host, the cost of compute and hardware is covered by the dedicated host. However, depending on your licensing agreement, you may be charged for OS licenses.

Chapter 4

1. **No**—there is a maximum of 10 dynamic public IP addresses and 10 static public IP addresses allowed per basic subscription. The default limits vary by subscription offer (Pay-As-You-Go, CSP, or Enterprise Agreement); however, they can be extended by contacting support.

2. **Yes**—by defining user-defined routes, you can adjust the routing between the different resources in your VNet, according to your needs.

3. **Yes**—Azure supports dual-stacked IPv4/IPv6 addressing; however, the subnets for IPV6 must be /64 in size.

Chapter 5

1. **No**—VNet peering uses the backbone infrastructure of Azure; there is no need to create gateways.

2. **No**—VNet-to-VNet uses a virtual network gateway, which is set up with a public IP address.

3. **Yes**—VNet peering doesn't use a virtual network gateway, so it doesn't have any bandwidth limitations. Those limitations typically belong to the gateway

Chapter 6

1. **Yes**—You need to use New-AzureADMSInvitation to add a guest user to your Azure AD tenant.

2. **No**—Azure AD join can be used without connecting an on-premises Active Directory to Azure AD.

3. **Yes**—When you add a custom domain to Azure AD, you need to verify it by adding a **TXT** record to the DNS settings of your domain registrar. After adding this record, you can verify the domain in the Azure portal.

Chapter 7

1. **Yes**—trusted IP addresses are meant for bypassing MFA for certain IPs. This way, you can disable MFA for users who log in from the company intranet, for instance.

2. **No**—you can use Conditional Access policies to enable MFA for users and applications.

3. **Yes**—fraud alerts can only be enabled for MFA server deployments.

Chapter 8

1. **No**—password hash synchronization is enabled by default if you use Express Settings during the Azure AD Connect installation.

2. **No**—password sync needs to be enabled on the on-premises domain controller and in the Azure portal.

3. **No**—you can install Azure AD Connect when the on-premises forest name doesn't match one of the Azure AD custom domain names, but you will receive a warning during installation that SSO is not enabled for your users.

Chapter 9

1. **Yes**—Azure Migrate can be used to migrate web applications using the Web Application Migration Assistant.

2. **Yes**—The Azure Migrate Assessment tools are capable of visualizing server dependencies.

3. **No**—You can also use Azure Migrate to migrate physical machines.

4. **Yes**—Azure Site Recovery can provide protection for Hyper-V, VMWare, and AWS-hosted VMs.

Chapter 10

1. **Yes**—Azure Load Balancer can only balance traffic within a region. To balance traffic between regions, you must use either Azure Traffic Manager or Azure Front Door.

2. **Yes**—Azure Application Gateway includes an optional web application firewall.

3. **No**—NSGs only work with IPs, service tags, or virtual networks. Azure Firewall can be used to block or deny access using an FQDN.

4. **No**—You can access VMs on another VNET, but the VNETs must be peered

Chapter 11

1. **No**—to assign permissions to users, you need to use RBAC.

2. **Yes**—you can use Azure Policy to check whether all of the VMs in your Azure subscription use managed disks.

3. **No**—custom policies are created in JSON.

4. **Yes**—you can create up to 10,000 management group levels, up to six levels deep.

Chapter 12

1. **False**—Developers don't have to add additional code to log information to the different web server logs. This only needs to be done for application logs.

2. **False**—WebJobs are not yet supported on Linux.

3. **No**—For executing small coding tasks, such as image processing, Azure Functions is the most suitable option.

Chapter 13

1. **No**—AKS allows you to have more control over the containers than ACI.

2. **No**—you can view them from the Azure portal.

3. **Yes**—Docker files can be created directly in Visual Studio.

4. **No**—you can also pull containers from a private Azure container registry

Chapter 14

1. **No**—You can set this directly from App Service.

2. **False**—You can attach all kinds of SSL certificates to Azure App Service, regardless of where they come from.

3. **False**—System-assigned managed identities can only be assigned to the scope of the Azure resource where they are created.

Chapter 15

1. **False**—You can also use the APIs to create a database and container in Cosmos DB.

2. **False**—Azure Cosmos DB offers five different consistency models: strong, bounded staleness, session, consistent prefix, and eventual consistency.

3. **True**—Horizontal partitioning is used to distribute the data over different Azure Cosmos DB instances.

Chapter 16

1. **No**—an elastic pool can also be configured for databases that already exist.

2. **Yes**—Azure SQL Database managed instances are based on the last stable SQL Server on-premises database engine.

3. **Yes**—failover groups automatically manage the geo-replication relationship between the databases, the failover at scale, and the connectivity.

Other Books You May Enjoy

If you enjoyed this book, you may be interested in these other books by Packt:

Learn Azure Administration

Kamil Mrzygłód

ISBN: 978-1-83855-145-2

- Explore different Azure services and understand the correlation between them
- Secure and integrate different Azure components
- Work with a variety of identity and access management (IAM) models
- Find out how to set up monitoring and logging solutions
- Build a complete skill set of Azure administration activities with Azure DevOps
- Discover efficient scaling patterns for small and large workloads

Learn Azure Sentinel

Richard Diver, Gary Bushey

ISBN: 978-1-83898-092-4

- Understand how to design and build a security operations center
- Discover the key components of a cloud security architecture
- Manage and investigate Azure Sentinel incidents
- Use playbooks to automate incident responses
- Understand how to set up Azure Monitor Log Analytics and Azure Sentinel
- Ingest data into Azure Sentinel from the cloud and on-premises devices
- Perform threat hunting in Azure Sentinel

Leave a review - let other readers know what you think

Please share your thoughts on this book with others by leaving a review on the site that you bought it from. If you purchased the book from Amazon, please leave us an honest review on this book's Amazon page. This is vital so that other potential readers can see and use your unbiased opinion to make purchasing decisions, we can understand what our customers think about our products, and our authors can see your feedback on the title that they have worked with Packt to create. It will only take a few minutes of your time, but is valuable to other potential customers, our authors, and Packt. Thank you!

Index